Life in Stone

Edited by Rolf Ludvigsen

Life in Stone: A Natural History of British Columbia's Fossils

UBCPress / Vancouver

Printed in Canada on acid-free paper ∞

ISBN 0-7748-0577-3 (hardcover)
ISBN 0-7748-0578-1 (paperback)

Canadian Cataloguing in Publication Data

Main entry under title:
Life in Stone

Includes bibliographical references and index.
ISBN 0-7748-0577-3 (bound)
ISBN 0-7748-0578-1 (pbk)

1. Fossils — British Columbia. 2. Paleontology — British Columbia. I. Ludvigsen, Rolf, 1944-
QE705.C3L83 1996 560′.9711 C96-910492-8

UBC Press gratefully acknowledges the ongoing support to its publishing program from the Canada Council, the Province of British Columbia Cultural Services Branch, and the Department of Communications of the Government of Canada.

UBC Press
University of British Columbia
6344 Memorial Road
Vancouver, BC V6T 1Z2
(604) 822-3259
Fax: 1-800-668-0821
E-mail: orders@ubcpress.ubc.ca
http://www.ubcpress.ubc.ca

Contents

Preface

There is no shortage of information about Cordilleran fossils. Over the past century, paleontologists have published hundreds of papers in scientific journals, as well as many reports and monographs on their research on the paleobiologic and geologic significance of fossils collected from localities throughout British Columbia. Nevertheless, this information is reaching only a limited audience. Written for geologists and paleontologists, these publications reside in the libraries of large universities and are not generally available in regional libraries or bookstores. As a result, the broad audience of curious naturalists does not read them.

The information about Cordilleran fossils contained in these specialized publications should, in fact, be of interest to many residents and visitors to British Columbia. After all, each paper or report is a window through which the reader can view strange inhabitants of ancient ecosystems from the distant geological past. This is local history, to be sure, but it is local history writ large.

Convinced that Cordilleran fossil plants and animals are potentially as interesting to BC naturalists as are living plants and animals, in the fall of 1992 I wrote to about two dozen paleontologists actively working in the province with a proposal to compile a multi-authored book on the fossils of British Columbia. The book was to be written in accessible English to reach a broad audience and, of course, it had to be well illustrated. I thought it was important to cover the full range of BC fossils through the entire Phanerozoic Eon – familiar groups such as dinosaurs, mastodons, trilobites, and ammonoids, to be sure, but also little-appreciated fossil groups such as conodonts, brachiopods, radiolarians, pollen, and insects. I recognized that a single book could not possibly deal with all of the thousands of different fossils known in the province, nor with all paleontological themes and biological aspects presented by these fossils. But I believed that it could explore the nature, significance, and meaning of the fossil groups that are particularly well represented in the rocks of British Columbia. The response I received from the paleontologists was gratifying and overwhelmingly positive. This volume on Cordilleran fossils is the result.

Despite good intentions, the coverage of BC fossil groups of different ages is not complete. I had hoped to include chapters on Vendian multicellular fossils; Paleozoic shelly fossils such as brachiopods, corals, and cephalopods; Permian and Carboniferous fusilinid foraminiferans; Triassic marine reptiles; and Cenozoic (including Pleistocene) mollusks. But for a variety of reasons – not the least being the diminishing numbers of paleontologists in Canadian universities and geological surveys – this goal was not possible.

This book has been prepared under the aegis of the British Columbia Paleontological Alliance, an association of amateur and professional paleontolgists recently formed to increase understanding of fossils of the Cordillera. In assembling and editing the book, I acknowledge, first and foremost, the paleontologists who considered it important to communicate their knowledge and enthusiasm about various fossil groups to a broad audience. Their contributions here illustrate a notable conceptual distinction, namely, that 'popularize' does not mean 'trivialize.'

The project was supported by a grant from the Canadian Geological Foundation, a highly unusual private agency that supports endeavours that raise public interest, understanding, and involvement with the geological sciences. I am grateful to Richard Hebda of the Royal British Columbia Museum and Peter Milroy of UBC Press for supporting and encouraging this project from its inception. And finally, I am greatly indebted to Louise Bell of Denman Island, whose talent and editorial insights have been instrumental in converting the submitted manuscripts into a unified whole.

Life in Stone

1

Introduction: Deep Time in the Cordillera

Rolf Ludvigsen

> Oh, these mountains! Like great sitting hens, squatting immovably, unperturbed, staring, guarding their precious secrets till something happens.
>
> Emily Carr, Victoria, 27 July 1933
> *The Journals of an Artist*

British Columbia is defined by mountains. Essentially it is a great sea of ranges, peaks, ridges, and hills that extends from the Pacific to the crest of the Main Ranges of the Rocky Mountains. Interrupted here and there by high plateaus and scored by narrow lakes and rivers, these mountains make up the western backbone of the Americas – the Cordillera – the broad expanse of ranges that guards the Pacific coast from incursions from the east.

Terror, delight, or awe! Mountains affect us in different ways. Many people see them as a challenge, as a source of creativity, or as the focus of mystical and spiritual forces. Others fear them or see them only as frustrating barriers. But regardless of how they are viewed, British Columbia's mountains can never be ignored. We live among them and they loom as an ever-present backdrop to our lives.

Stripped to its essentials, a mountain is a piece of the earth's crust that sticks out of the ground – and a window to the history of our planet. That's the reason geologists converge on mountain belts. Here, as nowhere else, are unparalleled opportunities to examine and sample great expanses of rock and to study the relationship among rock units. Paleontologists are drawn to mountains because they expose layers of rock, or strata, that are quite literally stony archives safeguarding the biologic heritage of the earth.

Geologists and paleontologists have worked in the mountains of British Columbia for a century and a half. Throughout this period, one question has remained fundamental: How and when did this broad mountain system form? At first, plausible explanations proved elusive, but pieces of the puzzle began to fall into place in the 1970s as the plate tectonic revolution transformed thinking throughout the geologic community. Old familiar concepts of a static earth were swept away and replaced by new models involving continually moving plates, shifting continents, and creation and destruction of oceanic crust. Mountains were now explained as a consequence of subduction: oceanic crust plunging back to the mantle beneath the continental crust. Subduction alone, however, could not satisfactorily explain the variety and sequence of rocks found across the 800-kilometre-broad Cordillera in British Columbia.

Over the past dozen years a new and intriguing explanation has taken form. This model, a younger sibling of plate tectonics known as terrane tectonics, has brought new understanding to the nature and origin of these and other mountain belts. This theory holds that the Cordillera is a crazy quilt of terranes, regions of distinct rocks, each of which originated out in the ancient Pacific Ocean. When the Atlantic Ocean began to open, some 150 million years ago, these terranes collided with North America, which was steadily moving northwestward into the Pacific at the rate of five to ten centimetres a year. Like bugs on the windshield of a speeding car, the terranes accumulated on the edge of the continent to build out the thick belt of mountain ranges known as the Cordillera.

The Fossils: Life in Stone

The Cordillera of British Columbia is a vast storehouse of fossils. Thousands of exposures of sedimentary rocks throughout the province contain fossil shells, scales, bones, teeth, and leaves. Some of these fossils – the bones of mammals and reptiles, entire ammonoids, and complete fishes and fern fronds, for example – are large and truly impressive, but most fossils are small and rather common shells, bits of bone, or blackened fragments of plants. Nevertheless, even a small, drab fossil shell changes into a treasured and unique icon of a former age, once its nature and antiquity are made clear.

Almost all of the fossils dealt with in this book are body fossils of animals or plants (Figure 1.1). Body fossils are the actual remains of part of an animal or plant – shells, bones, teeth, wood, or foliage. Since they are already made of minerals, shells and skeletons are readily preserved in sedimentary rocks. The most common minerals are calcium carbonate (the shells of bivalves, ammonoids, brachiopods, gastropods, corals, and crustaceans), calcium phosphate (bones and teeth of vertebrates), and silicon dioxide (skeletons of radiolarians and some sponges). In general, body fossils of plants are preserved as carbonized impressions of leaves and needles or as petrified wood, in which case the spaces between the cell walls are filled with the mineral silica.

Trace fossils constitute the other major category of fossils. These features in rock (for example, footprints, burrows, borings, and bite marks) give clues about the behaviour or activities of ancient animals (Figure 1.2).

Fossils – whether body or trace – are crucial to deciphering the history of the earth. In the Cordillera and elsewhere, they are used to date rock units, to gauge the passage of geologic time, and to correlate formations.

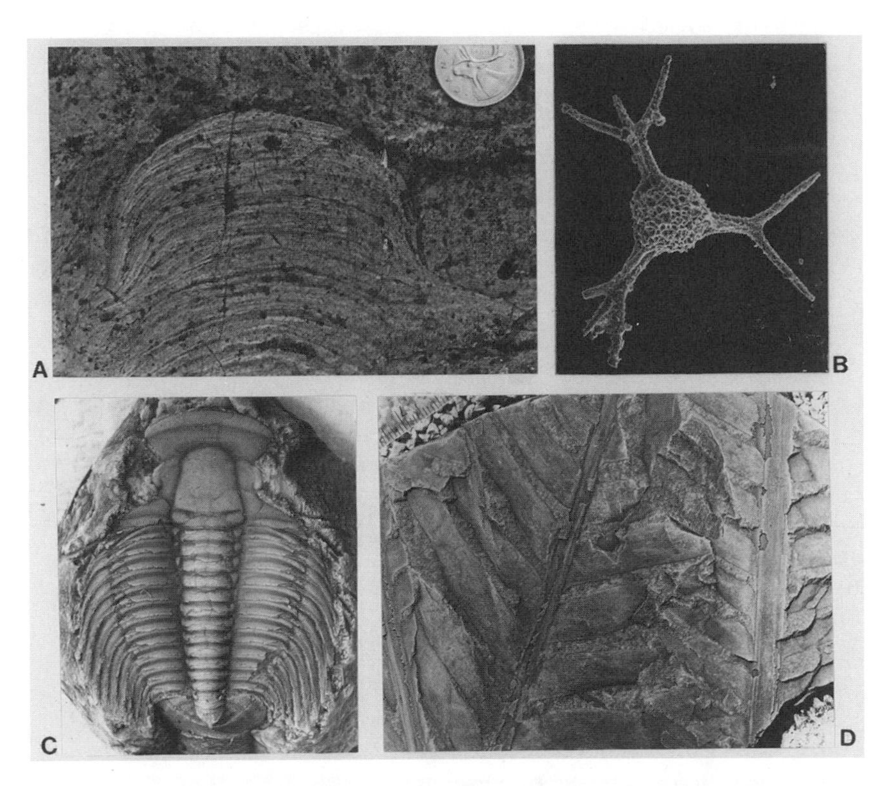

Figure 1.1 **Well-preserved fossils representing four of the five organic kingdoms.** These specimens are representative of fossils contained in rocks of British Columbia; only the kingdom Fungi lacks an abundant fossil record. (A) Kingdom Monera. Bacteria and blue-green algae produce laminated structures called stromatolites. This domal stromatolite is from Vendian strata on Mt. Fitzwilliam, Mount Robson Provincial Park. (B) Kingdom Protoctista. This silicious skeleton of the radiolarian *Kalherosphaera* was etched from Upper Triassic limestone exposed near Campbell River, Vancouver Island. The specimen measures 0.5 mm between the spine tips. (C) Kingdom Metazoa. This moulted exoskeleton of the trilobite *Labiostria* was collected from Upper Cambrian shales exposed near the Bull River, east of Cranbrook. Specimen is 2 cm long. (D) Kingdom Metaphyta. These fronds of the cycad *Pseudoctenis* came from Upper Cretaceous shales at an old mine dump near Cumberland, Vancouver Island. The slab shown is 7 cm high.
Sources: (A) Photograph by Martin Tetz, Amoco Canada; (B) Scanning electron micrograph (SEM) by Fabrice Cordey, Geological Survey of Canada; (C) Collections of the Department of Geology, University of Alberta; (D) Collections of the Royal British Columbia Museum

Fossils furnish direct evidence about the nature of ancient environments and about the composition of past animal and plant communities. And even though the fossil record in rock is far from complete, it provides the best information possible about the history of life and about the

Figure 1.2 **Trace fossils preserve evidence of activities of ancient animals.** These annulated and meandering trails, called *Scolicia,* were probably made by heart urchins burrowing in fine sand. The slab is from Upper Cretaceous rocks on Denman Island. It measures 40 cm across.
Source: Collections of the School of Earth and Ocean Sciences, University of Victoria

course that evolution has followed. But fossils are more than information-laden scientific objects. To many people, fossils are natural images and sculptures of great beauty.

Deep Time

Geologists talk rather glibly of events that happened millions of years ago. But, truth be told, no one really comprehends time of such magnitude. The term 'deep space' conjures up the unimaginably vast distances between stars and galaxies, measurable in millions of light years but way beyond comprehension by the human mind. In his book *Basin and Range,* the American writer John McPhee coined a parallel term, 'deep time,' to capture the full scope of geologic time, sweeping back beyond the paltry crust of recent history that is calibrated by mere human records.

Within the Cordillera of British Columbia, deep time extends back a full billion years. Clocked by radiometric dating of minerals in millions of years, its passage is best shown by the successions of fossils preserved in the shales, sandstones, and limestones exposed in mountainsides, at roadcuts, and along rivers throughout the province. These fossil floras and faunas are representative of intervals of geologic time such as the Mesozoic Era, Devonian System, and Eocene Series. These units are arranged in a hierarchy – eons divided into eras, eras into systems, systems into series, series into stages, and stages into zones. Each of these units can be divided further into early, middle, and late portions (when the reference is to geologic time) and into lower, middle, and upper (when the reference is to rocks).

The classification of all living and fossil organisms is hierarchically arranged in categories, or taxa, starting with the species, then genus, family, order, class, phylum, and finally kingdom. As an example, our species *Homo sapiens* (along with extinct species such as *H. erectus*) belongs to the genus *Homo,* the family Hominidae, the order Primates, the class Mammalia, the phylum Vertebrata, and the kingdom Metazoa. By convention, species and genera are italicized in print, but not higher categories.

The nature of the fossils preserved in rocks is the basis for a division of all earth history into two eons (Figure 1.3). Rocks of the Cryptozoic Eon ('eon of hidden life,' also commonly called Precambrian Eon) contain minute fossils of single-celled organisms without a nucleus – bacteria and algae belonging to the kingdom Monera. Their visible record consists of fossilized mats of algae called stromatolites. It was only late in the Cryptozoic that the kingdom Protoctista appeared. Protoctistans are single-celled organisms such as foraminiferans, radiolarians, and ameboids, all with a nucleus.

The base of the Phanerozoic Eon ('eon of evident life') is defined by the appearance of large, conspicuous fossils of multicellular animals (kingdom Metazoa) and multicellular plants (kingdom Metaphyta). The fifth kingdom, Fungi, is represented by minute fossils of fungal spores. The Phanerozoic Eon is divided into three eras, again based on the fossil content of the rocks: Paleozoic ('ancient life'), Mesozoic ('middle life'), and Cenozoic ('recent life'). The eras, in turn, comprise systems – from Vendian, starting 650 million years ago, to Triassic, starting 250 million years ago, to Quaternary, starting two million years ago. All thirteen Phanerozoic systems are represented by rocks containing fossils in the BC Cordillera.

The names of these time units are part of the basic working vocabulary of paleontologists and stratigraphic geologists for referencing the age of rocks in the Cordillera and elsewhere (Figure 1.3).

A Mari usque ad Mare

Canada became a continental reality on a hot summer day in 1871. On 20 July of that year, the Terms of Union was signed by representatives of the fledgling Canadian government and the Crown colony of British Columbia. A crucial clause of this agreement was a commitment of the federal government to construct a transcontinental railroad from the St. Lawrence River, across the Canadian Shield and the prairies and through the mountain passes, to tidewater on the Pacific Ocean. A decade later this enterprise unfolded as the defining endeavour of Canada as a nation

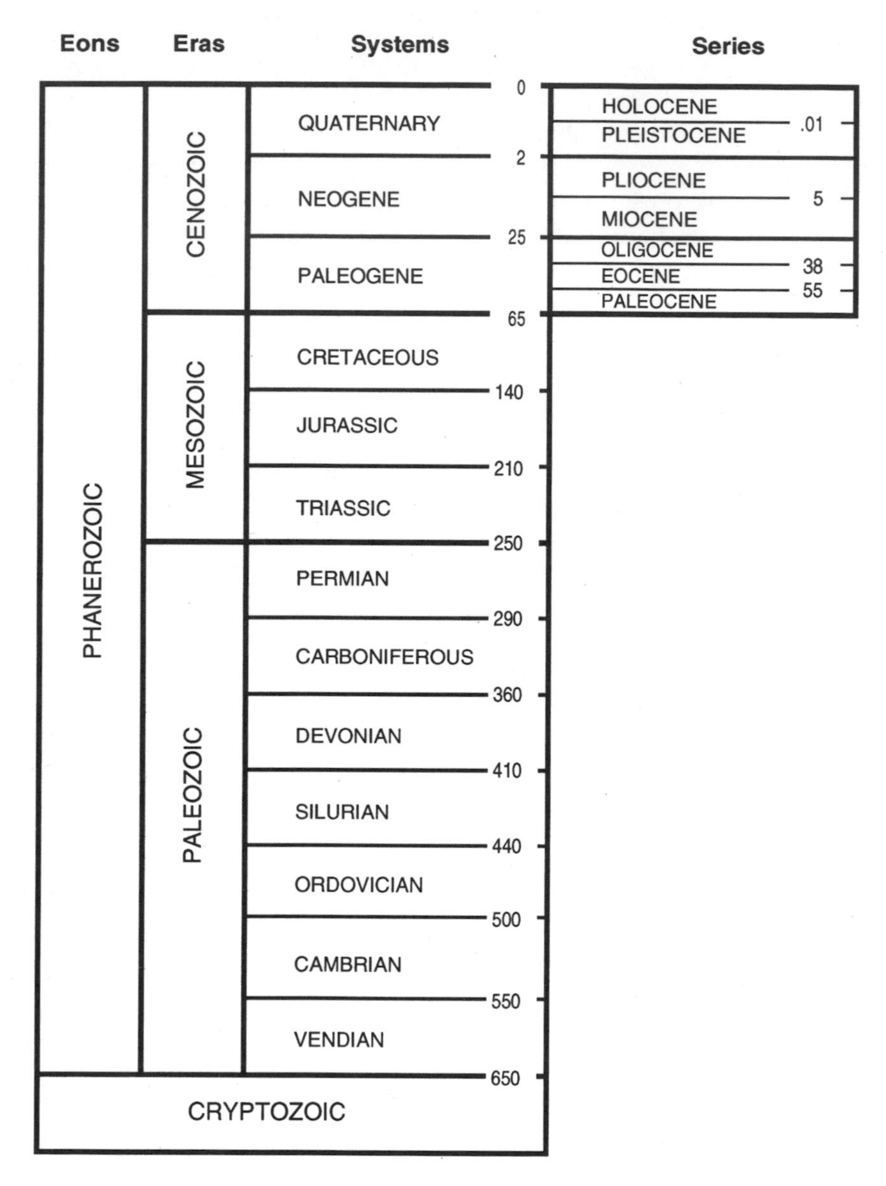

Figure 1.3 **The geologic time scale of eons, eras, systems, and series in millions of years**

– Sir John A's National Dream. (And a century later, Pierre Berton's too.) But that was not all. The pragmatic colonial delegation insisted on yet another clause – the prompt initiation of a geological survey of the new province. That task fell to the woefully understaffed Geological Survey of Canada (GSC), already reeling from recently being assigned responsibility for all of Rupert's Land. With only five field geologists and a single paleontologist at his disposal, Director Alfred Selwyn now had to try to meet the demands of mining men and other entrepreneurs for concise information on the occurrence and geology of mineral deposits from the Atlantic all the way to the Pacific (Zaslow 1975).

On the day that the Terms of Union was signed in Ottawa, Selwyn was already in British Columbia. Accompanied by the geologist James Richardson, he was steaming up the Fraser River en route to Yale, the terminus of steamboat navigation and the staging point for the Interior. That summer Selwyn explored central BC as far east as the Yellowhead Pass through the Rocky Mountains, before descending the North Thompson and Fraser rivers by canoe to New Westminster. Richardson continued up the Cariboo Road to Quesnel and then travelled to Vancouver Island to begin study of the coalfields at Nanaimo and Cumberland, a task that was to keep him occupied for the next four years. That fall, Selwyn returned to Montreal with new appreciation of the magnitude of the task that the GSC faced in that maze of western mountains.

Selwyn thought that the requirements of prospectors and mine developers in British Columbia could best be met by having a professional geologist stationed year-round in the province. In George Mercer Dawson he found an extraordinary man for this task. Superb geologist as well as accomplished paleontologist, geographer, botanist, zoologist, anthropologist, ethnologist, artist, and photographer, Dawson was perhaps the only natural science polymath ever produced in Canada. In spite of a significant physical handicap, Dawson proved himself to be an indefatigable geological explorer whose stamina in the field exceeded that of most men. Beginning work for the GSC in 1875 from his base in Kamloops, he quickly demonstrated a remarkable grasp of the regional geology of vast tracts of the Interior, Vancouver Island, the Queen Charlotte Islands, the Yukon, and the Rocky Mountains. Wherever he went in the Cordillera he collected rock and mineral samples, fossils, and Native artifacts, all the while filling numerous field notebooks with observations and data. Dawson rose through the ranks to succeed Selwyn as director of the Survey in 1895 – at an annual salary of $3,200. He emerged as the most able administrator in the early years of that organization and undoubtedly the best loved. Dawson died at the age of fifty-one in 1901. 'The Little Doctor,' as he was affectionately known to prospectors in the West, lives on in the names of Dawson City in the Yukon and Dawson Creek in British Columbia.

The Paleontologists

If you ask a classroom of grade-five students what they want to be when they grow up, odds are pretty good that at least a few of them will declare confidently that they will become paleontologists. Despite this level of interest, only a few paleontologists have ever been employed as professionals in Canada. In British Columbia in the mid-1990s, only a dozen paleontologists are currently on staff at the three main universities and at the GSC. This number of paleontologists is hardly sufficient to do justice to the astonishing variety of fossils known from the province. Luckily, Canadian paleontologists from outside the province and some foreign paleontologists have decided to work on Cordilleran fossils. And these professionals are now joined by an increasing number of amateur paleontologists.

Geological Survey of Canada

Since the summer of 1871, the GSC has participated in much of the paleontological work accomplished in the Cordillera. Numerous fossils were collected by early GSC field geologists for identification and study by pioneer Survey paleontologists Elkanah Billings and J. Frederick Whiteaves. As paleontologists are fond of pointing out, however, field geologists rarely collect enough top-quality fossil specimens. So, early in the this century when GSC paleontologists were encouraged to venture into the Cordillera to collect their own fossils, the volume and significance of the fossil collections increased markedly.

Frank McLearn began work on ammonoids in Triassic rocks of the Peace River area after the First World War. In these rocks he documented the most complete Triassic marine succession in the world. In the 1930s, Charles Sternberg, scion of a celebrated fossil-collecting family, mapped Cretaceous dinosaur footprints along the Peace River. After the Second World War, Hans Frebold and, later, Howard Tipper started work on Jurassic fossils of the Cordillera. Jack Usher documented the rich Cretaceous ammonoid faunas of Vancouver Island at the same time as George Jeletzky began studying Cretaceous and Cenozoic mollusks throughout the Cordillera, work which was to occupy him for more than thirty years. Walter Bell described Cretaceous and Cenozoic plants of Vancouver Island and the Peace River area. Bruce Cameron studied microfossils – the fossils of micro-organisms – such as foraminiferans and conodonts. Tim Tozer expanded on McLearn's work and in 1967 published a synthesis of Triassic time based on ammonoids and bivalves of the Canadian Cordillera and Arctic. In the 1970s, Bill Fritz began work on Cambrian trilobite faunas of the eastern Cordillera, while Brian Norford concentrated on Ordovician and Silurian fossils. Willy Norris and Alan Pedder studied Devonian fossils and Wayne Bamber and Walter Nassichuk, Late Paleozoic fossils. The current complement of GSC Cordilleran paleontologists continues and expands the work of its predecessors – Terry Poulton works on Jurassic mollusks, Jim Haggart on Cretaceous ammonoids, and Mike Orchard on Late Paleozoic and Triassic conodonts. Their biostratigraphic determinations provide most of the precise dating of rock units that is necessary to outline the complex geologic history of the Cordilleran terranes.

Universities in British Columbia

Academic paleontologists at BC universities are also working on Cordilleran fossils. Paul Smith and his students at the University of British Columbia concentrate on Jurassic ammonoids, especially from the Queen Charlotte Islands. Ted Danner, also of UBC, works on Late Paleozoic fossils, particularly fusilinids. Rolf Mathewes of Simon Fraser University, Glenn Rouse of UBC, and Richard Hebda of the University of Victoria and the Royal British Columbia Museum specialize in Cenozoic and Pleistocene plant fossils. Chris Barnes at the University of Victoria has begun study of the Ordovician conodonts of British Columbia.

Outside British Columbia

Many Canadian paleontologists from outside the province have made important contributions to knowledge of BC fossils. In the 1930s, Loris Russell of the National Museum of Canada and later of the Royal Ontario Museum began work on the sparse Cenozoic mammal faunas of the province. Richard Harington of the Canadian Museum of Nature in Ottawa has been studying Pleistocene mammals throughout the province for many years. Alan McGugan of the University of Calgary has documented the Cretaceous foraminiferans of Vancouver Island. Since 1975 Des Collins of the Royal Ontario Museum has been collecting Burgess Shale fossils from different localities in Yoho National Park. Paleontologists from the University of Alberta have made a special study of the exquisite fossils preserved in shales and chert from Eocene lakes of the BC Interior – Ruth Stockey and her students focus on the diverse flowering plants and Mark Wilson on the fishes and insects. Jim Basinger from the University of Saskatchewan studies the conifers preserved in the same rocks.

In a paleontological rescue mission, dinosaur trackways along the Peace River, slated for flooding behind the Bennett Dam in the late 1970s, were mapped and some footprints collected by Phil Currie and a team from the Royal Tyrrell Museum of Palaeontology in Alberta. Other paleontologists at the Tyrrell – Don Brinkman, Betsy Nicholls, and Andy Neuman – are investigating Triassic marine reptiles and fishes from exposures near Wapiti Lake. Chris McGowan of the Royal Ontario Museum is studying the ichthyosaurs from the same localities. Rare body fossils of some of the earliest animals from Vendian rocks of the Cariboo Mountains are now being studied by Guy Narbonne of Queen's University.

Outside Canada

Foreign paleontologists have worked on BC fossils for many years and continue to do so. Charles D. Walcott, head of the Smithsonian Institution in Washington, began collecting Cambrian trilobites of the Rocky Mountains in 1907. Two years later he discovered superbly preserved soft-bodied fossils in the Middle Cambrian Burgess Shale in Yoho National Park; their study was to occupy him until his death in 1927. Starting in the late 1930s, Teiichi Kobayashi of the University of Tokyo made a special study of Cambrian and Ordovician trilobites of eastern British Columbia. Bobb Schaeffer of the American Museum of Natural History in New York collected important Triassic fishes near Wapiti Lake in the 1960s. The fossils of the Burgess Shale continue to receive a lot of deserved attention. Harry Whittington of Cambridge University and his colleagues Simon Conway Morris and Derek Briggs have nearly finished a massive revision of the paleobiology of these fossils in more than fifty papers and monographs based on Walcott's original specimens and new material collected by the GSC in the 1960s. Harvard University paleontologist Stephen Jay Gould brought the significance of the Burgess Shale fossils to a broad audience in his provocative book *Wonderful Life* published in 1989.

Amateurs

The ranks of past and present Cordilleran paleontologists have now been joined by another group – amateur paleontologists. Although often (and mistakenly) dismissed as dilettantes, amateurs (the root is the Latin word *amator,* or 'lover') are those who pursue an activity out of interest instead of for financial gain.

Some of the most important fossils of British Columbia are now being discovered by amateurs. An example comes from Vancouver Island. Cretaceous strata on the island are renowned among Canadian paleontologists as the source of diverse and important ammonoid faunas, but despite a hundred years of intense collecting, fossil reptile bones were not recognized in these rocks. Then, in a remarkable series of linked discoveries starting in the late 1980s, a group of knowledgeable amateurs uncovered fossil reptile bones at different localities in the Courtenay area. Partial skeletons of half a dozen marine reptiles – plesiosaur, turtle, and mosasaur – have now been either located or collected, including a magnificent ten-metre-long elasmosaur now on display at the Courtenay Museum. These and other discoveries announce that amateurs are now assuming an important role in BC paleontology.

British Columbia Paleontological Alliance

In 1993, professional paleontologists at universities, museums, and the GSC joined with amateur paleontologists and fossil collectors throughout the province to form the British Columbia Paleontological Alliance. Now numbering more than 250 members, the Alliance encourages responsible collecting of fossils and strives to enhance knowledge about fossils and paleontology in the Cordillera.

The mountains are still there, squatting immovably and unperturbed. Emily Carr would still feel at home, but something has happened in the Cordillera since her time. Rocks and fossils throughout the Cordillera have been collected, studied, and analyzed by successions of geologists and paleontologists. The conclusions of these earth scientists, coupled with scientific insight and imagination, have resulted in exciting new explanations of the formation of the Cordillera. Rich fossil faunas and floras are testaments to the passage of familiar and unusual life-forms on the western edge of Canada. Truly, the mountains are beginning to disclose their precious secrets of deep time.

Reference

Zaslow, M. 1975. *Reading the Rocks: The Story of the Geological Survey of Canada 1842-1972*. Toronto: Macmillan

2

Fossils and Museums: Windows into Ancient Worlds

Richard J. Hebda and David A.E. Spalding

It's hard to think of fossils without thinking of museums. We think of great bony skeletons, raised high in an equally great hall, hard proof that fossils are real and once were great beasts. Or paintings stretched across a wall, upon which crawl and float unimaginable creatures in ancient unfathomable seas. And laid out in front of the scene, rocks with strange but regular patterns, tangible evidence that creatures like those fleshed out on the wall once patrolled the seas. Perhaps small dusty cases and, displayed behind the glass, stony mysterious objects, begging to be touched.

The discipline of paleontology, or paleobiology, would not exist without museums, for in those museums are stored the objects of study themselves. Unlike many living plants and animals, fossils are difficult, and often impossible, to study in detail in the field. They can be observed in rocks and collected, but in order to discover their identity and significance, paleontologists usually must take fossils to a laboratory to prepare them for study. Once the diagnostic or essential features are revealed, the fossil is compared to previously described material and identified. Although fossils can be identified through comparison to illustrations, definitive analysis occurs in a museum or institution with a collection. Furthermore, unique or diagnostic specimens must then remain available for future study and comparison; this step can only happen in a properly managed collection, which usually means one held by a museum.

Fossil collections also serve many other purposes, chief among them being education of the public at large about the earth's ancient life and its changes. Museums and similar institutions are the contact point between the amateur and the expert. They also represent a door into the greater world of fossils, placing the fossil specimen or collection in its proper place in the full scope of fossils and their history on the planet. A fossil fish found in the Interior of British Columbia, for example, may be shown to have lived in a lake surrounded by redwoods, ginkgos, and poplars.

Many people consider that fossils constitute part of our shared natural heritage, believing that these wonders of ancient history belong to us all and should be accessible to all. Through its responsibility for the care of collections and public education, the museum serves as custodian of that common heritage and reveals it to society. Few of the important values of fossils are realized unless the specimens reach a museum in the first place.

A History of Museums and Fossils

Fossils and museums have had a long relationship (Rudwick 1972). One of the first documented collections of objects of natural history was made by Johann Kentmann (1518-74). Entitled *Arca rerum fossilium* ('ark of fossil objects'), this sixteenth-century collection contained several fossils, among other objects dug out of the earth. The renowned Swiss naturalist Conrad Gesner (1516-65) used specimens from this ark and his own collections for the first book about fossils, which was published in 1565. This volume, and many that followed, catalogued fossils from private and institutional museums. Paleontological historians believe that these early collections were probably the beginning of modern scientific museum collections. In those days, fossils were primarily of interest as natural curiosities (Rudwick 1972).

During the seventeenth and eighteenth centuries, fossils continued to be collected and described. As more and more exotic living creatures became known to the European world during this era of exploration, scientists realized that fossils were the remains of living beings. Thus these curiosities, housed in museums, gained importance for holding insights into the history of life itself.

Museum collections proved instrumental in the early developments of various theories of evolution. In the late 1700s Georges Cuvier (1769-1832), working with collections at the Musée d'Histoire Naturelle, showed through his work on fossil elephants, including mammoths, that many species had become extinct. In the 1840s, the first dinosaurs were described by William Buckland (1784-1856) and Gideon Mantell (1790-1852) in England. Recognizing their value, the British Museum bought Mantell's dinosaur collections. These great creatures served as the basis for reconstruction of life-size models based on contemporary interpretations, models later displayed at the Crystal Palace in London.

Fossils fascinated Charles Darwin (1809-82); he believed that fossils provided support for his theories on evolution. Indeed, once his ideas were published, there ensued a great hunt for missing links, for which museum collections provided the raw materials. Using one of the most important museum fossil specimens, Thomas Huxley (1825-95) demonstrated that *Archaeopteryx* was a link between birds and reptiles.

From the late 1800s to the present, fossils continued to arrive at museums. Great expeditions were undertaken, and volumes of material

Figure 2.1 **C.F. Newcombe's cabinet of fossils from the early twentieth century**
Source: Collections of the Royal British Columbia Museum

were shipped to the world's major institutions for future study. One such massive collection was made on the ridge between Mount Field and Wapta Mountain in the Rocky Mountains by Charles D. Walcott (1850-1927) beginning in 1909. Sixty thousand fossils found their way to the Smithsonian Institution in Washington, DC. These fossils proved to be one of the most important records of early ancient life; we now know them as the Burgess Shale fauna.

During the exciting paleontological discoveries at the end of the last century, the residents of Vancouver Island decided that they wanted to be part of the museum movement, like their relatives in Europe and eastern North America. The result was the establishment of the British Columbia Provincial Museum (now the Royal British Columbia Museum). Through the efforts of scientific explorers such as G.M. Dawson of the Geological Survey of Canada and C.F. Newcombe of the museum, it accumulated extensive fossil collections from the beginning (Figure 2.1). Several of the early fossils provided original material for the description of previously unknown creatures.

Keeping Fossil Collections

For the visitor to a museum the first contact with fossils may be an exhibit that displays only a small part of the collection. Behind the scenes there is usually much more. Most fossils, no matter what their nature, are part of rocks or stones, which are usually stored in drawers within cabinets or on shelves (Figure 2.2). Fossils can remain in drawers for decades, needing little additional care. If there are no facilities to house them in drawers, or if the museum is without staff to curate the collection, the specimens are simply left in boxes or bags. Provided they are not moved too much, the fossils remain relatively safe.

Figure 2.2 **Cabinet drawer mainly of Jurassic ammonoids from the Queen Charlotte Islands**
Source: Collections of the Royal British Columbia Museum

Each separate fossil piece or group of pieces in a collection is identified in a unique manner. Often a reference number is discreetly but clearly inscribed on the rock containing the specimen. Sometimes a card with the number and other information is attached. The number provides access to crucial data about the fossil, such as its identity and where it was found, which are recorded in hard copy or computer files. Until about five years ago, the information for most fossils at the Royal British Columbia Museum was recorded on cards. Now all collection information is entered into a computer data base, where it is easily accessible and from which it can be exported to a national data base via the Canadian Heritage Information Network.

The Amateur Collector and the Museum

People collect fossils for various reasons: Fossils are curious objects with a mysterious past; they look interesting; they are durable, lasting for a long time; and they can be transported easily. But the amateur collector has a responsibility to ensure that the many potential values of the specimens are considered. Unfortunately, many fossils end up in a drawer or on a window sill. Eventually, the record of when and where they were found, or who collected them, is lost. Without this information, fossils are valuable only as curiosities or artifacts; there is no potential for scientific study, exhibit, or education.

When an amateur collector discovers a fossil or fossil site, it should be reported to the nearest museum that maintains fossil collections. Two immediate benefits ensue. First, staff at the museum are able to direct the collector to sources of information and experts who can identify and explain the fossil. Second, the process of assessing the value and significance of the fossil or fossil site can begin. It is only through this route that a particular fossil increases our knowledge of ancient life.

In British Columbia, amateur collectors perform a vital role. The province is vast, its population is small and concentrated in limited areas, and professional paleontologists are few. Consequently, the well-informed amateur provides the feet and eyes for scientists. To a significant extent, museums in the province must rely on the public to discover the location of sites and the fossils they harbour, and bring this information to the attention of a museum.

Important scientific discoveries are made by amateurs. On 12 November 1988, Vancouver Island amateur paleontologist Mike Trask and his daughter Heather discovered the remains of a large creature near Courtenay. Trask removed some of the fossilized bone to his home and contacted the Royal British Columbia Museum. With help from an Alberta museum, now the Royal Tyrrell Museum of Palaeontology, the bones were discovered to belong to an elasmosaur, a marine reptile in the plesiosaur group. Courtenay is only the second area on the west coast of North America where such creatures have been found.

Sound collecting means collecting more than just the fossils themselves. It requires recording sufficient information so that the fossil site can be located precisely and so that the fossil can be properly catalogued and entered into a museum's collections. Making the necessary observations and recording them effectively adds to the excitement of collecting fossils and to the learning experience.

At minimum, the location where the fossil was found should be noted as precisely as possible. In the ideal case, the amateur collector takes a 1:50,000 National Topographic Series map sheet into the field and marks the fossil locality on the map. Then, the map, or a copy of it, is turned in with the fossil to the museum. Without a map, the collector should write out instructions for reaching the site, clearly indicating the distance and direction from easily recognized landmarks. A car odometer can be used when leaving the site to measure the distance to the nearest reference point. People frequently bring fossils to the Royal British Columbia Museum for identification and as donations. Often, however, it is difficult to establish where the material came from, because inadequate information was recorded.

Other useful data to gather include the nature of the surrounding rock, how the strata are arranged, and the size and appearance of the exposure. A photograph of the site often helps in relocating it. As well, the date should be noted and the names of those present at the time of the collection. For people intending to collect fossils often, a published guide for amateur fossil collectors is useful. An excellent example is *West Coast Fossils* (Ludvigsen and Beard 1994), which describes and illustrates many fossils from Vancouver Island.

When fossils are collected, they are usually wrapped in newsprint or cloth to protect them from physical damage. Specimens should not be changed in any way to improve their appearance. Instead, they should be kept as they are found, because the adhering rock may hold valuable clues to the nature and identity of the fossil. The surrounding matrix may contain microscopic fossils, for example, that could establish the age of the deposit.

If the fossil is large and not easily moved, it is best left in the ground. When a collector tries to extract this kind of specimen, its scientific and exhibit value can be destroyed. Experts know how to do this work. Even for them, it often takes months to extract vertebrate fossils intact. Anyone discovering a fossil like this should simply take a photograph and inform the local museum.

Fossils in BC Museums

Of perhaps 300 museums in the province, about thirty are known to keep fossils, but only a few of these give much priority to paleontology. Data analyzed over the past two years suggest that only 12 per cent of the 140,000 or more BC fossil specimens reside in public collections within the province. Most of these are in collections of the universities and at the Royal British Columbia Museum.

Few museums in the province have exhibits of fossil material and fewer still have the staff or funds to maintain an active program in fossils. Nevertheless, consideration is now being given to having several museums and visitor centres focus their programs on paleontology.

The Royal British Columbia Museum

Founded in 1886, the Royal British Columbia Museum was the first major provincial museum to be established west of Ontario. From the start it was the intention that the museum include 'geological and natural history specimens' (Corley-Smith 1984), and early volunteers such as Rev. G.W. Taylor, Walter Harvey, and Dr. C.F. Newcombe brought in many fossil specimens. The museum has never had a paleontologist on staff, however, and following the days of Newcombe there was no systematic initiative to acquire material. Until recently, in fact, the fossil collection remained largely dormant.

Today the museum houses more than 10,000 fossil specimens, representing sites across the province and throughout the geologic time scale. The holdings contain specimens of scientific importance and of educational and exhibit value. Cushioned in its drawers are numerous mollusks that populated the ancient seas, whose record is now exposed on the Queen Charlotte Islands and Vancouver Island. Many of the specimens were obtained a century and more ago by G.M. Dawson of the Geological Survey of Canada and C.F. Newcombe, from the Jurassic rocks around Skidegate Inlet. Newcombe and others also donated Cretaceous fossil mollusks and plants acquired on the east side of Vancouver Island and the adjacent Gulf and San Juan islands.

Although the Burgess Shale is within British Columbia, the museum holds only a small collection of the fascinating creatures from this locality (Whittington 1985; Gould 1989). Nevertheless, these important specimens are lent out for research and educational purposes, and appear occasionally in the fossil displays of the museum.

The most spectacular collection consists of Triassic fossil fishes from Wapiti Lake in eastern British Columbia near the Alberta border. These hundreds of exquisitely preserved individuals range from small minnow-sized creatures to a coelacanth contained on a single rock slab more than a metre long.

The collections also include dinosaur tracks from the Peace River country and southeastern British Columbia and a four-metre-long ichthyosaur – an extinct fish-like lizard – from Holberg Inlet, north Vancouver Island. There are even teeth of a herbivorous sea-cow that swam off the shores of Sooke twenty-five million years ago.

Some of the most fascinating fossil material is also the youngest. The collection includes several tusks and numerous molars belonging to mammoths and mastodons from ice-age deposits in the province. A superbly preserved skull of an extinct bison (Figure 2.3), affectionately known as 'Bison George,' represents the giant creatures that roamed south Vancouver Island only 11,750 years ago. Between 12,000 and 13,000 years ago, the sea stood much higher than today. In this cool late-glacial water lived a diverse community of clams, snails, and barnacles. Hundreds of their shells are part of the collections.

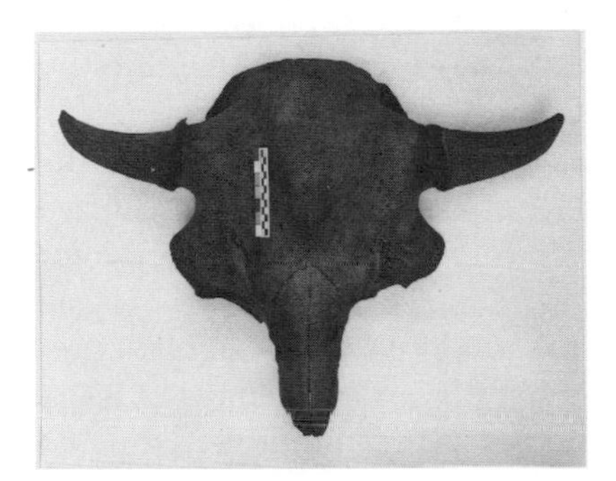

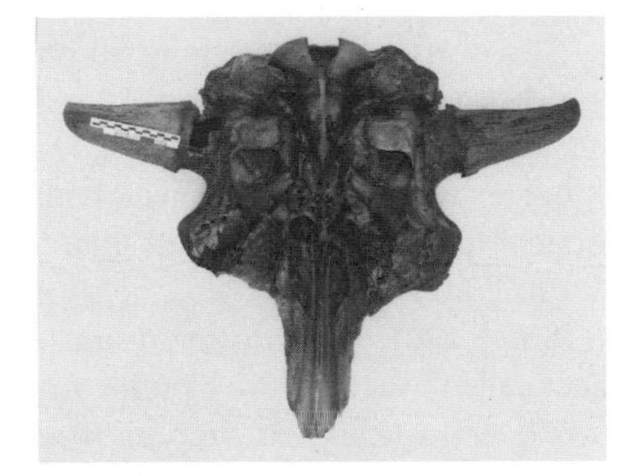

Figure 2.3 **Upper and lower views of skull of extinct giant bison, 11,750 years old, recovered from limy lake deposits, Saanich Peninsula, Vancouver Island**

Source: Collections of the Royal British Columbia Museum

Despite these collections, the Royal British Columbia Museum has no permanent paleontological exhibits, aside from a cast slab of dinosaur footprints in the courtyard and a reconstructed woolly mammoth plus a few associated specimens in the natural history gallery. Several times each year, however, fossils appear as part of special and temporary events held by the museum.

Other Museums in British Columbia

Identified fossils are exhibited in only about ten other museums in the province. Included among these are: the M.Y. Williams Geological Museum in the Department of Geological Sciences of the University of British Columbia, with its display of a full-size hadrosaur and collections of early trilobites, Jurassic ammonoids, and Cambrian archaeocyathids (sponge-like marine organisms); the Hudson's Hope Museum, with fossil dinosaur footprints; the Kelowna Museum, with casts of important hominid and *Archaeopteryx* fossils from other parts of the world, and an Eocene bird from British Columbia; and the Princeton Museum, with numerous local plant fossils of Cenozoic age.

Among other regional museums, the most extensive collections and exhibits can be found at the Courtenay and District Museum. This museum has held fossil collections for about twenty-five years, but only recently have they become a priority. Mike Trask's 1988 elasmosaur discovery, now on display at the Courtenay Museum, precipitated major paleontological activity at the museum and in the region by the local amateur community. Provincial grants at that time supported the preliminary study and excavation of the elasmosaur remains by Rolf Ludvigsen, Mike Trask, and Elizabeth Nicholls. Then, further grants in 1990 sustained the excavation of the remainder of the elasmosaur. The help of over forty local volunteers was notable in this effort.

Discovery of this single fossil has dramatically increased the interest in paleontology on Vancouver Island and attracted coverage in radio, television, and print media. As a result of the enthusiasm for fossils, the Courtenay Museum has developed displays of local ancient life, organized special programs, and made plans to build a new permanent home for its important collections.

Since 1991, the Courtenay museum has produced a multi-faceted project called 'Ancient Life – public involvement in the management of B.C.'s fossil heritage.' With support from various sources, the project fostered provincial awareness of fossils, promoted the establishment of a temporary 'fossil house' next to the museum, and hosted a travelling exhibit from the Royal Tyrrell Museum. As well, the project sponsored school programs, field trips, and lectures, and initiated a provincial inventory and management study. Fossil enthusiasts associated with the project established the very active Vancouver Island Paleontological Society and assisted in establishing the province-wide British Columbia Paleontological Alliance.

Recently an extensive fossil exhibit was constructed. It features a large mural depicting the elasmosaur, a mosasaur, and swimming ammonoids. With the mural are displayed numerous fossil remains from the area. The Courtenay museum also houses a modest but remarkable fossil collection, containing fine vertebrate, invertebrate, and plant specimens.

The Vancouver Island Paleontological Museum has recently been established at Qualicum Beach through the efforts of Graham and Tina Beard. Based on their personal collections, the museum contains 5,000 or more high-quality plant and animal fossils representing ancient life of the Late Cretaceous, seventy to ninety million years ago, found in rocks on the east side of Vancouver Island.

BC Fossils in Museums outside British Columbia

Most BC fossils collected for scientific purposes are housed outside British Columbia; about a third are elsewhere in Canada; and slightly more than half are in the United States. Many of these collections represent the major research of the past; others are still growing, through projects active in the province today. As in the past, most paleontological research in the province is conducted by institutions based outside British Columbia.

Elsewhere in Canada, major BC collections reside at the Royal Tyrrell Museum, the Royal Ontario Museum, and the Geological Survey of Canada. In 1976, staff of the Provincial Museum of Alberta learned that the site of important dinosaur footprints at the Peace River Canyon was threatened by proposed flooding for a new reservoir. In cooperation with the Royal British Columbia Museum, the provincial Archaeological Survey, and BC Hydro, important material was discovered and rescued. The research documented more than 1,700 footprints, many of them parts of extended trackways. ACCESS, Alberta's educational television station, helped make a movie of the site, showing some of the finest trackways from the air. As an unexpected bonus, paleontologists discovered the oldest known bird footprints and what may be marsupial tracks. Some specimens went to the Royal British Columbia Museum and others to the Provincial Museum of Alberta. Later, after the development of the Royal Tyrrell Museum of Palaeontology, the Alberta collections were transferred to Drumheller.

The Museum at Drumheller also holds other important BC collections. Don Brinkman and Andrew Neuman have developed an extensive program of studying Triassic fishes and marine reptiles in the Wapiti Lake area. Some of their 4,000 fish specimens are included in the permanent collection at Drumheller and were part of a substantial temporary exhibit on view in 1993. Paul Johnston has augmented this collection by adding hundreds of ancient invertebrates from Upper Ordovician rocks collected at Top-of-the-World Provincial Park in southeastern British Columbia. Notable fossils from this site are extinct sponge-like colony-forming organisms called aulacerid stromatoporoids.

The Royal Ontario Museum has been active in British Columbia since at least 1935 when Loris Russell undertook several important studies of mollusks and mammals from Cenozoic deposits. Other highlights include an expedition by Gordon Edmond into the Peace River Canyon in 1965, during which he collected dinosaur trackways. Des Collins and associates from this museum have made extensive collections of Burgess Shale fossils since the 1970s. And Chris McGowan discovered and excavated several fossil ichthyosaurs from sites at Williston Lake in northeastern British Columbia.

The largest collections of BC fossils are held by the Geological Survey of Canada at the Cordilleran Section in Vancouver, the Institute of

Sedimentary and Petroleum Geology in Calgary, and the headquarters in Ottawa. The Survey has tens of thousands of specimens generated by the field work of hundreds of its scientists. The section in Vancouver, for example, houses large collections of BC conodonts, ammonoids, and bivalves. In Ottawa, the headquarters has a large collection of Burgess Shale material as well as most of the specimens from early exploration of our province.

BC fossils have found their way to several institutions outside the country. In the 1890s, Henry Woodward (1832-1921) of the British Museum (Natural History) described fossil crustaceans from Vancouver Island and the Queen Charlotte Islands. The Smithsonian Institution, as already mentioned, retained thousands of Walcott's specimens from the Burgess Shale. Last, the Burke Museum at the University of Washington recently acquired large collections of Cenozoic fossil plants from British Columbia.

These collections, both within and outside British Columbia, attest to the abundance and significance of fossils in the province. Many have been collected by amateurs, a reflection of the pleasure associated with collecting and the eagerness of people to know about ancient life. The role of museums is to provide or verify information as it relates to individual discoveries. As well, the museum brings these discoveries to the wider public, inviting them to become involved by attending exhibits, participating in hands-on programs, or planning paleontological projects.

References

Corley-Smith, P. 1984. *The Ring of Time*. British Columbia Provincial Museum. Special Publication No. 8

Gould, S.J. 1989. *Wonderful Life: The Burgess Shale and the Nature of History*. New York: W.W. Norton and Company

Ludvigsen, R. and G. Beard. 1994. *West Coast Fossils: A Guide to the Ancient Life of Vancouver Island*. Vancouver: Whitecap Books

Rudwick, M.J.S. 1972. *The Meaning of Fossils: Episodes in the History of Palaeontology*. New Haven: American Elsevier

Whittington, H.B. 1985. *The Burgess Shale*. New Haven: Yale University Press

3
The Origin and Evolution of Canada's Western Mountains

James W.H. Monger

This chapter introduces the concepts underlying our present understanding of the origin of the Canadian Cordillera – the mountains of western Canada. A general understanding of the geological evolution of the Cordillera is needed as background for subsequent chapters concerned with fossils.

The Cordillera lies along the boundary between the western North American continent and the flanking ocean basin. This boundary has been in existence for 750 million years, although processes leading directly to mountain building have only been active for the last 180 million years.

The rocks of the Cordillera contain a remarkably developed fossil record. The time spanned by these rocks is exceedingly long, longer than the span of Phanerozoic time, the interval for which evidence of life is abundant and obvious. Phanerozoic time extends from about 650 million years ago to the present. All systems from the Vendian to the Neogene are represented by strata in the Cordillera, and fossils from all these systems are known.

The Cordilleran fossil record preserves not only animals and plants that lived on the western edge of the North American continent but also some that probably lived and died in the ocean basin, in locations thousands of kilometres from the edge of the continent. The strata hosting these fossils were swept into their present locations within the Cordillera by enormous lateral movements of the earth's crust. These movements were responsible for Cordilleran mountain building.

Major Subdivisions of the Canadian Cordillera

Traditionally, the Canadian Cordillera has been divided into five longitudinal belts on the basis of the distinctive geology and landforms displayed by each belt. From east to west, they are the Foreland, Omineca, Intermontane, Coast, and Insular belts (Figure 3.1A and Table 3.1).

Table 3.1

Geological belts of the Canadian Cordillera

Foreland Belt: includes the Rocky, Mackenzie, and Franklin mountains
Cryptozoic through Middle Jurassic sedimentary rocks deposited along the passive, ancient continental margin and Upper Jurassic and Lower Cenozoic rocks derived from erosion of uplifted mountains to the west. Rocks were folded and thrust eastward on to the continental margin in Cretaceous and Early Cenozoic time.

Omineca Belt: includes the Purcell, Selkirk, Monashee, Cariboo, Omineca, Cassiar, and Selwyn mountains
Cryptozoic continental crust, overlain by Cryptozoic to Jurassic sedimentary and volcanic strata and Paleozoic to Cenozoic granitic rocks. All of these rocks were deformed in Middle Jurassic to Early Cenozoic time and metamorphosed to high grades.

Intermontane Belt: includes the Interior, Stikine, and parts of Yukon plateaus, and the Skeena Mountains
Unaltered sedimentary and volcanic rocks ranging in age from Devonian to Holocene, intruded by Mesozoic and Cenozoic granitic rocks and deformed in Mesozoic and Early Cenozoic time.

Coast Belt: includes the Coast and Cascade mountains
Jurassic through Cenozoic granitic rocks, and minor Paleozoic through Cenozoic rocks, variously metamorphosed to high grades and deformed in mid-Cretaceous through Early Cenozoic time.

Insular Belt: includes the Insular Mountains and the Coastal Depressions
Unaltered Paleozoic to Holocene volcanic and sedimentary rock, and Paleozoic through Cenozoic granitic rock, deformed mainly in Late Cretaceous to Holocene time.

These belts were identified by the earliest geological mappers in the Cordillera and are still used today, although some of the names have changed (Dawson 1881; Gabrielse and Yorath 1991). The belts reflect the sum of all processes that shaped the Cordilleran region from the Cryptozoic to the present. They are dominated, however, by features formed by mountain building in Middle Jurassic to Paleogene time, that is, a period lasting from 180 million years ago to fifty million years ago.

The geological character of the belts is most clearly defined on a metamorphic map (Figure 3.1B). High-grade metamorphic rocks that were buried to depths of ten to thirty kilometres and strongly altered by heat and pressure (indicated on the map as amphibolite and greenschist facies) or by high pressure and low heat (indicated as blueschist facies) are distinguished from those that were never buried very deeply. The Coast and

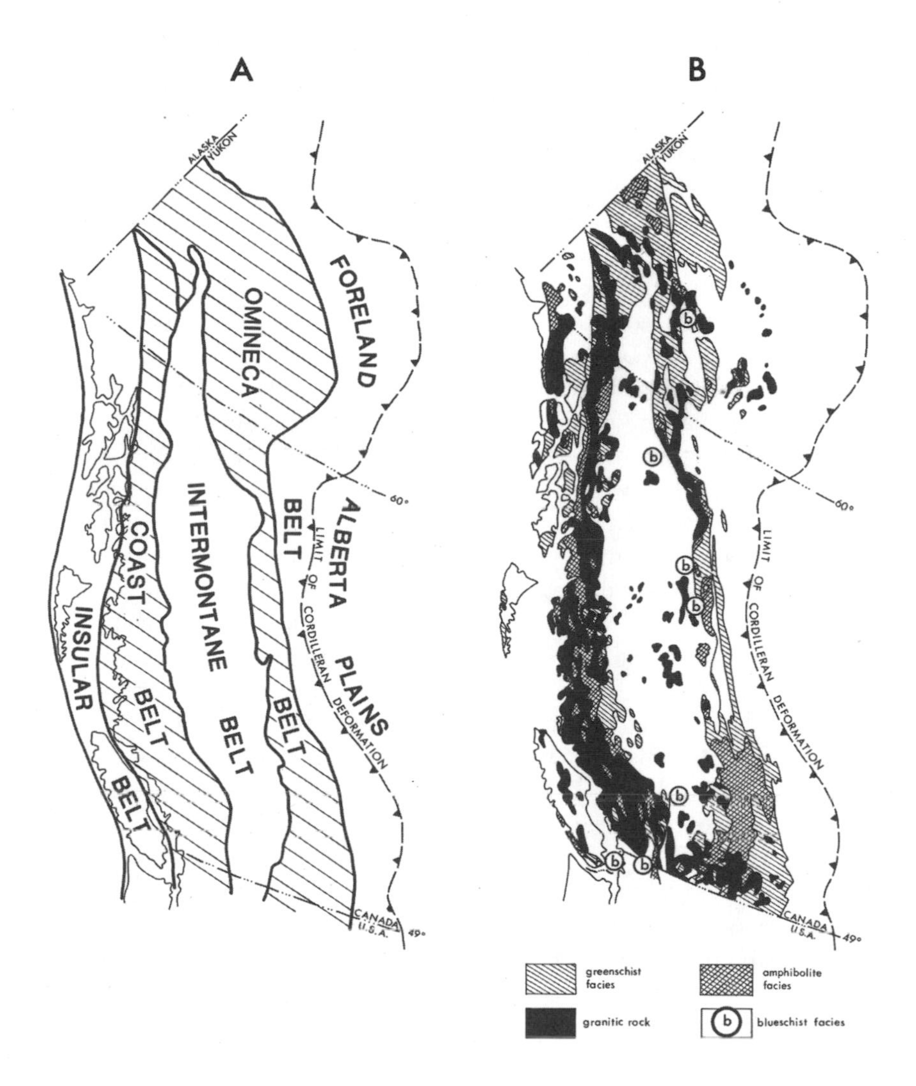

Figure 3.1 **Major divisions of the Canadian Cordillera.** (A) The five geological belts of the Canadian Cordillera. Belt descriptions are summarized in Table 3.1. (B) Simplified metamorphic map of the Canadian Cordillera.

Omineca belts consist of high-grade metamorphic and granitic rocks. Because these rocks formed deep in the crust but now are exposed at the surface, they represent the tracts of greatest uplift and erosion. These two belts separate the Foreland Belt, composed of sedimentary rocks, from the Intermontane and Insular belts, composed of volcanic, sedimentary, and lesser granitic rocks. To understand how this alternating pattern of deep-seated and near-surface rocks formed, we must examine the processes involved in Cordilleran mountain building.

Continental Drift and Plate Tectonics

Until the 1960s, many geologists thought that continents and oceans were geographically fixed in relation to one another. Others, however, believed that the continents moved across the face of the earth, a concept called 'continental drift.' Early support for this idea came from the southern hemisphere, where similarities in the geology and fossils of eastern South America, Africa, India, and Australia suggested to geologists that at one time these regions were joined to form a larger continent, 'Gondwana.' This continent later broke into the present continents, which are separated by intervening ocean basins.

Prior to the 1960s, the concept of continental drift lacked a generally accepted mechanism to explain how continents moved in this way. Eventually, however, the answer was found – on the floor of the oceans.

In the decades following the Second World War, the floors of deep ocean basins were thoroughly explored, and we learned that oceanic crust is very different from continental crust. Oceanic crust is relatively homogeneous, is composed mainly of dense basalt overlain by deep water sediments, and is geologically young. No crust older than 180 million years is known in the ocean basins; most is much younger, and some is forming today. By contrast, the continents are heterogenous, are composed mainly of less dense rocks including widespread granite, and in places are very old. The core of the North American continent, for example, is the Canadian Shield, which is composed of rocks from 900 to 3,700 million years old and surrounded by mountain belts containing mostly younger rocks.

In the early 1960s, the youngest parts of the ocean floors were found to be located at oceanic ridges such as the Mid-Atlantic Ridge. It was recognized that ocean floors grow continuously at these ridges by adding basalt to the oceanic crust from the deeper part of the earth, called the mantle. As the crust of the ocean floors is continually growing, and assuming the earth's surface area remains constant, crust elsewhere must be lost to compensate for the new oceanic crust. Old, cold, and dense oceanic crust returns to the mantle along linear features called 'subduction zones.' Thus there is continual recycling of the floors of the oceans, from their growth at oceanic ridges, their lateral movement across ocean basin floors, and their disappearance at subduction zones, a process probably driven by convective cooling of the earth. Because continental crust is less dense than oceanic crust it remains at the earth's surface, and the continents move around as passengers on the oceanic conveyer belts.

From these and related considerations, geoscientists in the 1960s formulated the hypothesis of plate tectonics, which was adopted by most geologists in the following decade. The hypothesis holds that the outer part of the earth, to a depth of 100 to 150 kilometres, consists of seven major rigid plates and some smaller plates, all of which move with respect to one another (Figure 3.2). The plates grow at oceanic ridges, which are called 'divergent plate boundaries'; disappear at the subduction zones, which mark 'convergent plate boundaries'; and, in places, slide laterally past one another on strike-slip faults, which mark 'transform plate boundaries' (Figure 3.3). Some plates, such as the Pacific Plate, consist almost entirely of oceanic crust. Others contain both continental crust and oceanic crust, in which case the within-plate boundary separating the two kinds of crust is called a 'passive margin' (Figure 3.3D). Thus the North American Plate comprises the North American continent and the western North Atlantic ocean floor, the two being separated by a passive margin along the eastern edge of the continent.

Tectonic Assemblages: Keys to Unravelling Cordilleran Evolution

Each category of plate boundary, or margin, is associated with a characteristic range of rock types, called a 'tectonic assemblage.' Tectonic assemblages are basic rock units shown on the current geological map of the Canadian Cordillera (Wheeler and McFeely 1991). By identifying these assemblages in the rock record, geologists are able to infer the record of past plate tectonics and to conclude that Cordilleran geology resulted

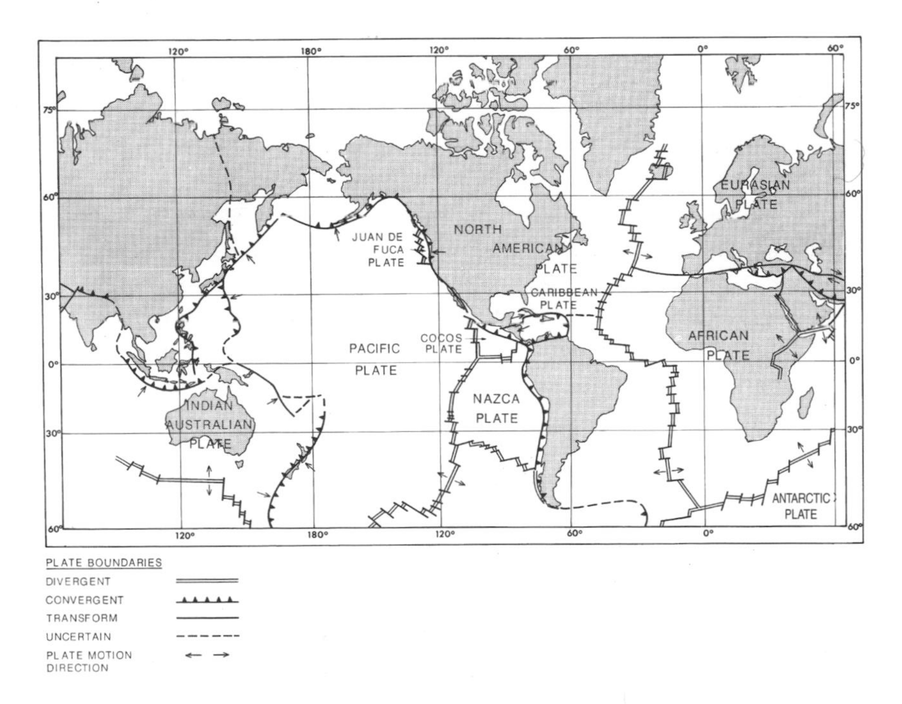

Figure 3.2 **Distribution of tectonic plates on the earth's surface and the nature of their margins**

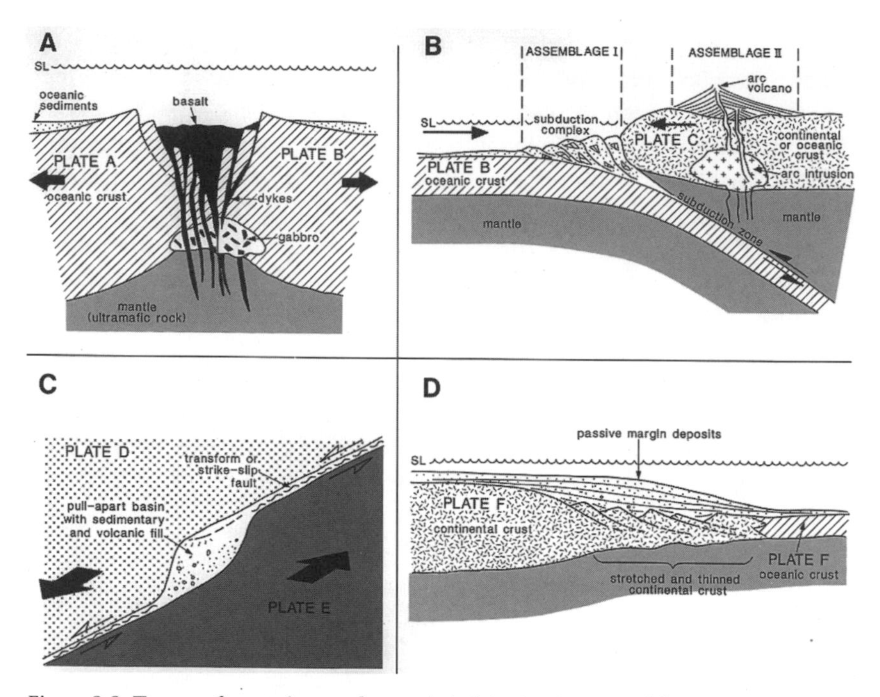

Figure 3.3 **Types of margins and associated tectonic assemblages.**
(A) Divergent plate margin. (B) Convergent plate margin. (C) Transform margin. (D) Passive margin. The heavy arrows indicate relative motions of plates; SL indicates sea level.

from complex, long-term interactions between the western edge of the North American Plate and various Pacific plates.

Divergent Plate Margin Assemblages

Divergent plate margin assemblages form at mid-ocean ridges where plates spread apart and grow by the addition of material from the mantle (Figure 3.3A). These assemblages are made up of the following: basalt that is overlain by deep-water sediments; dykes that run deep into the crust and once carried basalt from the mantle; and gabbro, a coarsely crystalline rock of basaltic composition that cooled slowly deep within the oceanic crust. At the base of the crust and in the underlying mantle are ultramafic rocks. Such an assemblage is forming today at the Juan de Fuca Ridge, located 200 to 400 kilometres west of Vancouver Island (Figure 3.4). This ridge separates the enormous Pacific Plate to the west from the small Juan de Fuca Plate to the east.

Convergent Plate Margin Assemblages

Convergent plate margin assemblages occur where oceanic plates return to the mantle along subduction zones. Two complementary assemblages

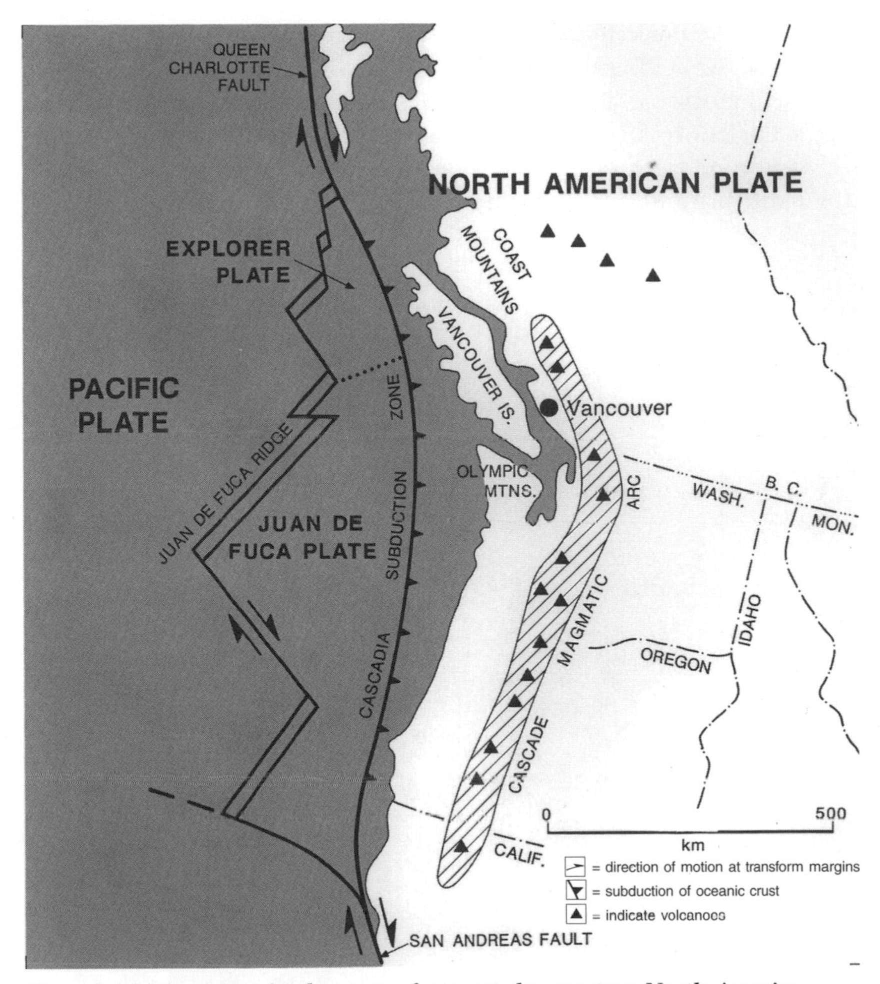

Figure 3.4 **Major tectonic elements of present-day western North America.** The Cascade magmatic arc lies east of the subducting small oceanic Juan de Fuca Plate. North and south of this arc, transform plate margins juxtapose the Pacific and North American plates. Plate boundaries are as indicated in Figure 3.2.

form in the upper or overriding plate: one at the boundary between the plates and the other within the margin of the overriding plate.

The first, called a 'subduction complex,' comprises oceanic crust (formed at a divergent margin) and possibly deep-water sediments. A subduction complex may or may not contain sands and muds eroded from the overriding plate and deposited on the ocean floor (Figure 3.3B). In either case, the complex consists mainly of rocks scraped off the lower oceanic plate during subduction and transferred to the edge of the upper plate, a process that characteristically breaks them up and disrupts them. An example of a modern subduction complex is found 100 to 200 kilometres west of Vancouver Island in water depths of up to 2,000 metres, where the westernmost part of the North American Plate overrides the subducting oceanic Juan de Fuca Plate (Figure 3.4). An ancient example is the Cache Creek terrane of the Intermontane Belt (Figure 3.5).

The second assemblage in a convergent margin is generally located in the upper plate, 100 to 200 kilometres from the subduction zone (Figure

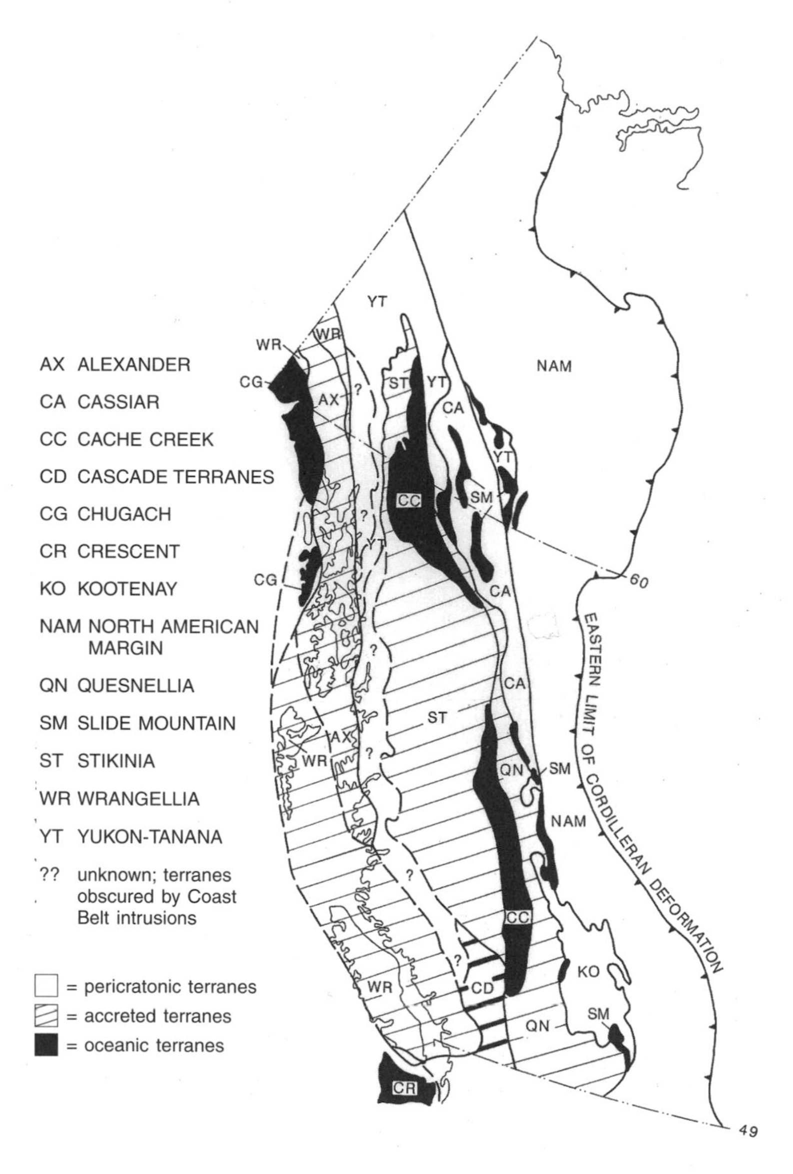

Figure 3.5 **Simplified terrane map of the Canadian Cordillera.** Terrane descriptions are summarized in Table 3.2 on pages 38-9.

3.3B). It comprises long chains of volcanoes and of granitic rocks that cooled deep within the earth's crust but which were subsequently uplifted and exposed at the surface by erosion. Because the granitic rocks are of the same general chemistry as the volcanics, and because in many cases they are the same age and occur next to one another, they probably originated from the same molten rock (or magma) source. Such chains of volcanoes, with or without the granitic intrusions, are called 'volcanic arcs' or 'magmatic arcs,' because of the arcuate geographic form they commonly display.

Mount Baker, Mount Garibaldi, Mount Cayley, and Mount Meager, all recently active volcanoes near Vancouver, are the northernmost volcanoes of the Cascade magmatic arc, which extends into northern California. This arc was built on the North American Plate margin above the subducting Juan de Fuca Plate (Figure 3.4). Elsewhere, for example in the southwestern Pacific Ocean, chains of volcanic arcs lie entirely within ocean basins. These assemblages are probably modern analogues of the ancient arc rocks that underlie much of the western Canadian Cordillera in the Alexander, Quesnellia, Stikinia, and Wrangellia terranes (Figure 3.5).

Transform Plate Margin Assemblages

Transform plate margin assemblages form where plates slide past one another (Figure 3.3C). In contrast to assemblages formed at divergent and convergent margins, transform plate margin assemblages are hard to identify, as they consist of narrow belts of thick, sedimentary deposits with or without volcanic rocks. They are associated with great linear zones of sheared and fractured rock that mark strike-slip (or transform) faults. The best-known of these faults is the active San Andreas Fault in California, which separates the North American Plate from a small continental fragment in southwestern California attached to the northward-moving Pacific Plate (Figure 3.4). An analogous structure is the submerged Queen Charlotte Fault north of Vancouver Island and west of the Queen Charlotte Islands (Figure 3.4). On land, ancient strike-slip faults that were moving between thirty-five and eighty-five million years ago include the Tintina and Northern Rocky Mountain trench faults in the Yukon and northeastern British Columbia, and the Fraser Fault, now highly eroded, along the Fraser Canyon, east of Vancouver (Figure 3.6A). These two faults formed when parts of the Cordillera west of them were temporarily wedded to northward-moving oceanic plates.

Passive Margin Assemblages

Passive margin assemblages form at the boundary separating continental and oceanic crust within a plate (Figure 3.3D). These assemblages feature thick deposits of sedimentary rocks that can be traced laterally into

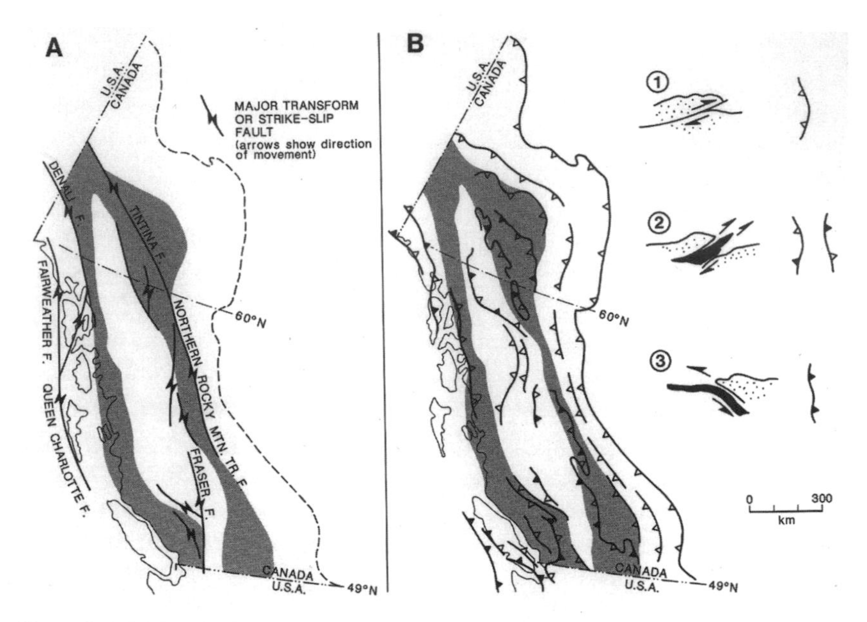

Figure 3.6 **Major faults of the Canadian Cordillera.** The shaded areas show distribution of Coast and Omineca belts for geographic reference. (A) Transform or strike-slip faults. These faults mostly moved late in the history of the Cordillera (Late Cretaceous-Holocene) and disrupted older features. (B) Thrust fault systems. In (1), the open barbs denote thrust faults within once-continuous stratigraphy, across which offsets can be measured. In (2), the open and solid barbs denote complex thrust faults that involve oceanic crust; some probably represent subduction zones involved in later thrusting. In (3), the solid barbs denote thrust faults that were, or are, subduction zones involving oceanic crust.

thinner shallow-water sequences on the continent and into deep-water strata on oceanic crust. A modern example of a passive margin is the eastern continental margin of North America. An ancient example is the thick succession of limestone, shale, and sandstone of late Cryptozoic to early Mesozoic age found in the Foreland Belt of the easternmost Canadian Cordillera (Figure 3.1A).

Evidence for Great Lateral Displacements

The concept of mobility is implicit in the plate tectonic hypothesis. As might be expected from the location of the Canadian Cordillera on or near a plate boundary throughout Phanerozoic time, the Cordillera contains evidence that many rocks within it have moved great distances across the earth's surface.

The geological structures called folds and faults record the stresses that were imposed on a rock after it formed and that changed its original shape. Structures that record shortening, or contracting, include folds and thrust faults (Figure 3.6B). Lateral movement of rocks past one another occurs on strike-slip faults (Figure 3.6A). Stretching, or extension, of rocks takes place on normal faults (Figure 3.3D). If a geological marker that records the original form of the rock, such as a layer or bed, is broken by a fault and offset across it, then the amount of displacement is measured by the offset. Similarly, a carefully mapped-out folded bed can be theoretically flattened to its original shape, and the difference between the original and the folded length of the bed is the amount of displacement that was caused by the folding.

Although the amount of displacement across a single fold or fault may be small, the cumulative offset of many folds and/or faults may be considerable. In the southern Foreland and Omineca belts (Figures 3.1A and 3.6B), the cumulative shortening on folds and thrust faults is about 300 kilometres. This means that rocks at the town of Revelstoke, 300 kilometres west of Calgary, were originally located before folding and thrusting at least 600 kilometres west of Calgary. The cumulative offsets across all strike-slip faults in the Cordillera amount to about 1,000 kilometres. This means that mid-Cretaceous and older rocks in the Queen Charlotte Islands were probably once located at the latitude of present-day Oregon.

Note, however, that the kinds of direct estimates given above can only be made when we find a geological marker on both sides of a fault or a continuous folded layer. In many cases, perhaps most, we are unable to do this, so that the amounts of displacement mentioned above are probably minimal. In addition, this kind of direct estimate cannot be made across a subduction zone, which is akin to an enormous thrust fault, because almost all the rocks in the lower plate have disappeared into the mantle. Other kinds of evidence must be sought.

West of Vancouver Island, the North American Plate overrides the subducting oceanic Juan de Fuca Plate at a rate of forty-five millimetres per year (Figure 3.4) (Riddihough and Hyndman, in Gabrielse and Yorath 1991). This rate seems slow – it's how fast fingernails grow. When continued for millions of years, however, the amount of displacement possible across subduction zones is stunningly large. Reconstructions of the past positions of oceanic plates suggest that a length of oceanic crust about 13,000 kilometres long has been subducted beneath western North America since the beginning of the Cretaceous (about 145 million years ago). This amount is almost equivalent to one-third the circumference of the earth! Furthermore, the displacement took place in only the last fifth of Cordilleran history.

The region across which this displacement occurred must lie west of the Coast Belt because Cretaceous and younger subduction complexes do not occur farther east. Older subduction complexes in the central part of the Cordillera likewise may record enormous amounts of displacement.

The evidence for this comes from the fossil record. The Cache Creek terrane of the Intermontane Belt (Figure 3.5) was active as a subduction complex in the early Mesozoic and contains fossils similar to those in Japan and China. Also, certain late Paleozoic and early Mesozoic faunas in other parts of the western Cordillera are more similar to faunas of the same age found in the southwestern United States than to faunas presently at the same latitude in the eastern Cordillera and continental interior. These observations can be explained by the northward movement on strike-slip faults of the rocks carrying those fossils.

The term 'paleolatitude' refers to the latitude at which a rock originally formed. Determining paleolatitudes involves measuring the inclination of the magnetic field preserved in a rock, for example, the field frozen into a lava flow as it cooled. We know that the present magnetic field is vertical at the poles, horizontal at the equator, and inclined in between. We also know that flows of lava cool as horizontal layers. Paleolatitudes are established by measuring the angle of the magnetic field and comparing it to the ancient horizontal plane. Paleomagnetic measurements like these show that the original latitudes of some rocks in the central and western Cordillera differ from those of the continental interior. Mid-Cretaceous rocks near Vancouver, for example, formed about 2,000 kilometres south of their present position with respect to the continental interior (Irving and Wynne, in Gabrielse and Yorath 1991).

Terranes: Cordilleran Building Blocks

For many years geologists working in the western Cordillera recognized that the succession of rock strata or the range of rock types within a region might remain constant and then change abruptly across a big fault, even though rocks on both sides of the fault were the same age. We call such regions of internally consistent geology 'terranes,' and we map terrane boundaries by finding discontinuities in the geology (Figure 3.5). 'Terrane' is used for a piece of the earth's crust with a geological record distinct from the records of adjacent terranes and from the record of the old continental margin. The juxtaposition of different geology across a fault can readily be explained once it is realized that parts of the Cordillera may have moved great distances with respect to other parts.

In 1977, David Jones and others proposed the name Wrangellia for a terrane featuring similar Triassic rock successions on Vancouver Island, the Queen Charlotte Islands, and southern Alaska. This stratigraphy not only differed from that of rocks of the same age to the east but the terrane also carried a paleomagnetic record different from that of the continental interior, which indicated that it was displaced from a more southerly location. Shortly after, terrane maps were produced for the entire North American Cordillera, from Mexico to Alaska (Coney et al. 1980; Jones et al. 1982).

On these maps, much of the Cordillera appears as a mosaic or collage of terranes that formed in different places and were later brought together by plate movements along the margin of North America.

Unlike the term 'tectonic assemblage,' the term 'terrane' implies nothing about its origin. In places, however, both terms are applied to the same body of rock. The Cache Creek *terrane,* for example, contains Carboniferous to Lower Jurassic strata that are very different from the strata of adjacent terranes of the same age. The Cache Creek *assemblage,* on the other hand, was the floor of a Late Paleozoic and Early Mesozoic ocean basin. Other terranes contain several stacked assemblages of different ages. The Stikinia terrane consists of Late Paleozoic, Late Triassic, and Early Jurassic arc assemblages separated by unconformities. Wrangellia consists of mid-Paleozoic arc, Late Triassic oceanic plateau, and Early Jurassic arc assemblages.

The rocks of the Cordillera can be grouped into three broad terrane categories, based on their composition and on the time that they became part of the North American continent.

First, buried below the Foreland Belt is the western edge of the Canadian Shield, made up of metamorphic rocks about two billion years old. These rocks have always been part of the North American continent and are referred to as the 'craton.' Deposited on the western edge of the craton and ranging in age from late Cryptozoic to Cenozoic are thick deposits of sedimentary strata laid down mainly in relatively shallow water. These are the rocks seen at the surface in the Foreland Belt.

The second category includes rocks in the Omineca Belt that were deposited not far from the craton. These rocks constitute the 'pericratonic terranes,' the terranes that formed around the craton. They consist mainly of sedimentary rocks, mostly deposited in deep water, but also of volcanic and granitic rocks. In age, they are mainly Paleozoic but range from late Cryptozoic to Jurassic. In many cases the rocks are so highly metamorphosed that their original nature is uncertain. The pericratonic terranes were thrust over the craton margin in the Jurassic to become part of the continent.

The third category is found in the Insular, Coast, and Intermontane belts. It includes rocks whose nature and chemistry suggest that they originated within an ocean basin, either as parts of the ocean floor (Cache Creek and Slide Mountain terranes) or as islands analogous to those in the present-day western Pacific Ocean (Alexander, Quesnellia, Stikinia, and Wrangellia terranes). These rocks were added, or 'accreted,' to the continental margin in Jurassic and Cretaceous time and so can be called 'accreted terranes' (Table 3.2).

The terranes are the building blocks of the Cordillera. About 750 million years ago, the ancient continental margin was located in the eastern Cordillera near the boundary of the Foreland and Omineca belts. Since

Table 3.2

Summary descriptions of the craton and the major Cordilleran terranes

The craton	
NAM	North American Margin A complete and thick succession of strata deposited along the ancient continental margin, comprising Cryptozoic to Upper Jurassic sandstone, shale, and carbonate
Pericratonic terranes	
CA	Cassiar Upper Cryptozoic to Devonian carbonate, sandstone, and shale. Rocks are similar to North American margin rocks, but sliced off and transported north on Tintina and Northern Rocky Mountain Trench strike-slip faults.
KO	Kootenay Lower Paleozoic argillite, sandstone, volcanic rocks, and Devonian intrusions, plus older rocks of continental origin. Rocks were deformed in the mid-Paleozoic and are overlain by Upper Paleozoic argillite, conglomerate, sandstone, and limestone.
YT	Yukon-Tanana Mainly metamorphosed sedimentary and volcanic strata and granitic intrusions of Paleozoic age
Accreted terranes	
AX	Alexander Pre-Ordovician metamorphics and Upper Cambrian to Triassic sedimentary and volcanic rocks and Paleozoic granites
CC	Cache Creek Disrupted Carboniferous through Lower Jurassic chert, argillite, basalt, ultramafic rocks, and carbonate. Some Permian and Triassic fossil groups resemble those from regions on the other side of the Pacific Ocean
CD	Cascade Several small Upper Paleozoic to Upper Mesozoic terranes of volcanic and sedimentary rocks
CG	Chugach Cretaceous and Upper Jurassic sandstone, argillite, minor volcanic rocks, chert, and ultramafic rocks
CR	Crescent Lower Cenozoic basalt, gabbro, and sandstone
QN	Quesnellia Upper Paleozoic sandstone, shale, chert, volcanic rock, and carbonate, overlain by Upper Triassic and Lower Jurassic volcanic and sedimentary rocks. Permian, Triassic, and Jurassic faunas differ from faunas of the same age at the same latitude on the continent, and more closely resemble those in the western United States.

[continued on next page]

Table 3.2 [continued]

Summary descriptions of the craton and the major Cordilleran terranes

SM	Slide Mountain Upper Paleozoic chert, argillite, sandstone, basalt, ultramafic rocks, and carbonate. Permian faunas are similar to those of the southwestern United States.
ST	Stikinia Upper Paleozoic to Lower Jurassic volcanic and sedimentary rocks and granitic intrusions. Permian, Triassic, and Jurassic faunas are similar to those in the western United States.
WR	Wrangellia Upper Paleozoic to Lower Jurassic volcanic rocks, granite, limestone, sandstone, and argillite. Paleomagnetic data from Triassic and Jurassic strata indicate origin from the latitude of the western United States.

that time the margin of the continent has been built 700 kilometres oceanward, mainly by the incremental addition of terranes to the old continental margin. The following section shows in more detail how this process occurred (Figure 3.8).

The Evolution of the Canadian Cordillera

The paleogeographic record of the earth shows that its continents at times cluster into single supercontinents. One such supercontinent, called 'Rodinia,' formed about one billion years ago and then broke up into smaller, continent-sized fragments about 750 million years ago. About 250 million years ago the fragments then re-clustered to form another supercontinent, called 'Pangaea.' Subsequently, Pangaea broke up into the continents we know today. In that break-up, North and South America separated from Europe and Africa, and the Atlantic Ocean was formed (Murphy and Nance 1992; Davidson 1992).

The western margin of ancient North American was formed by the rifting and separation of parts of Rodinia about 750 million years ago (Figure 3.7A). Today, this ancient continental margin is located in the eastern Canadian Cordillera near the boundary of the Foreland and Omineca belts (Figure 3.1A). For 600 million years, until latest Jurassic time, this margin was the boundary between continental crust to the east and oceanic crust to the west. The oldest rocks deposited along this margin are Late Cryptozoic coarse sandstones (grits), shales, and slates, deposited during the initial rifting. These deposits were covered by a succession of Cambrian to Jurassic strata, a thick passive margin (Figure 3.3D) made up of mainly limestone and shale. Today these strata constitute much of the Foreland Belt.

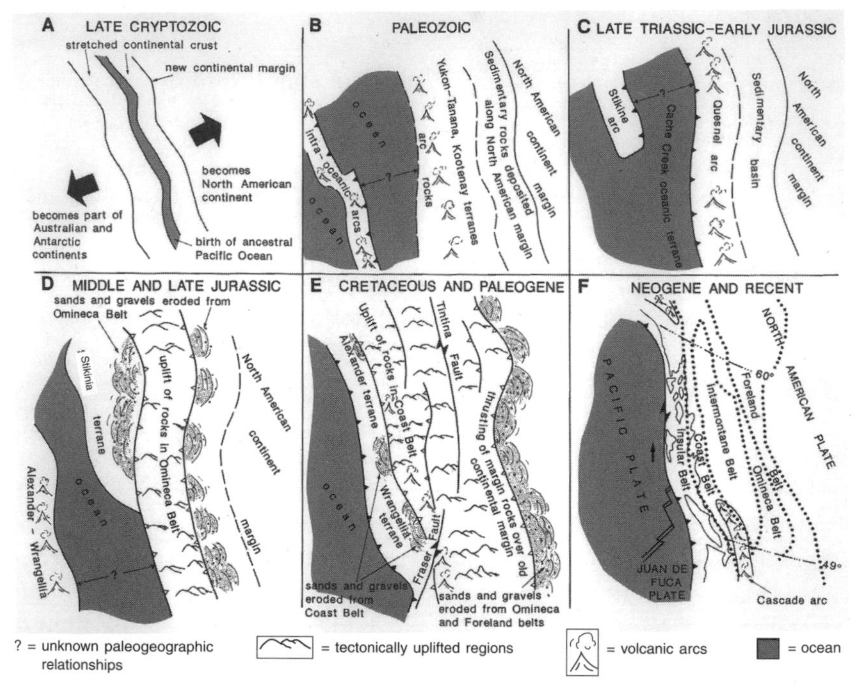

Figure 3.7 **Evolution of the Canadian Cordillera in six stages.** (A) Late Cryptozoic: the old continental margin formed by the rifting and separation of the supercontinent Rodinia, 750 million years ago. (B) Paleozoic: deposition of mainly limestones and shale on and near the ancient continental margin. Kootenay and Yukon-Tanana terranes include arc volcanics. Farther west was the ocean basin (the Slide Mountain terrane), west of which were intra-oceanic arc terranes (Alexander, Stikinia, and Wrangellia) in unknown relationship to North America, as indicated by arrows separated by a question mark. (C) Late Triassic-Early Jurassic: Quesnel arc on the western edge of North American Plate. Subducting oceanic rocks to west are represented by the Cache Creek terrane. (D) Middle and Late Jurassic: the start of mountain building. Uplift was caused by contraction (shown by symbols for flanking thrust faults). Erosion of rocks in Omineca Belt caused sand and gravel to be shed to adjacent regions. (E) Mid-Cretaceous through Paleogene: uplift across entire Cordillera followed accretion of Alexander and Wrangellia terranes on the site of the Coast Belt. Uplift was accompanied by thrust and transcurrent faulting. (F) Neogene and Recent: subduction limited to Vancouver Island and southern Alaska region. Continental arcs built on the North American plate margin. The remainder of the plate margin is a transform fault.

In the Paleozoic, the margin of the North American Plate lay well oceanward of the old continental margin, probably between the pericratonic terranes and the Slide Mountain terrane. Generally, the types of rocks deposited along the old continental margin were not strongly changed during the Paleozoic. West of the oceanic Slide Mountain terrane, Paleozoic arc-related volcanic and granitic rocks are common (Figure 3.7B).

By the Late Triassic (about 230 million years ago) a continuous west-facing arc was built on deformed oceanic basin deposits (Figures 3.5, 3.7C, 3.8). This arc forms the youngest part of the Quesnellia terrane. West of the arc, the subduction complex called Cache Creek terrane marks the western edge of the North American Plate boundary in Late Triassic and Early Jurassic time. It is not known how the Stikinia, Alexander, and Wrangellia terranes, which today lie west of the Cache Creek terrane, were related to the continental margin in the earliest Mesozoic. Quite possibly, the Cordilleran region at that time resembled the present western Pacific Ocean basin, with scattered chains of volcanic arcs and oceanic plateaus separated by basins floored by oceanic crust from Asian and Australian continents.

About 180 million years ago, in the Jurassic, a major change began that led to mountain building on the site of the former submerged continental margin. This change may be related to the opening of the Atlantic Ocean, which accompanied the break-up of Pangaea. The North American Plate moved northwest over the ancestral Pacific basin, sweeping up terranes on its leading edge. This process probably created the strongly contractional setting that built the Cordilleran mountains, a process recorded by the transition from marine rocks in the early Mesozoic to non-marine rocks in all places by the end of the Mesozoic. The crust was thickened by folding and thrust faulting of terranes over one another and over the old continental edge. The distinctive geometry of the Canadian Cordillera, namely high-grade metamorphic and granitic rocks in the Omineca and Coast belts, flanked by belts of rock that is only slightly metamorphosed, probably reflects accretion of terranes in Middle Jurassic time and later mid-Cretaceous time (Figures 3.7D and 3.7E) (Monger et al. 1982).

The deformation that led to the present Cordillera started in the Jurassic, about 170 to 180 million years ago. In the region centred on the present Omineca Belt, the Slide Mountain and Quesnellia terranes were thrust eastward over the pericratonic Kootenay terrane, and the Stikinia terrane was thrust beneath the Cache Creek terrane (Figure 3.7D). As these thrust faults stacked up, the crust thickened. Part of it was buried to great depths, metamorphosed, and intruded by granitic rock to form the Omineca Belt. Uplift accompanied the crustal thickening, and sand, gravel, and mud were eroded from the uplifted region and shed westward on to Stikinia and eastward on to the Foreland Belt. The initial marine deposits in these flanking basins were followed by non-marine deposits as the entire region was uplifted. Passive margin deposits and overlying clastic rocks derived from the Omineca Belt were incorporated in the growing stack of thrust fault slices that were moving east. In Late Cretaceous and earliest Cenozoic time (sixty-five million years ago), these rocks were thrust over the ancient continental margin, overlapping it by about 150 kilometres.

By Early Cretaceous time, the plate boundary was probably located

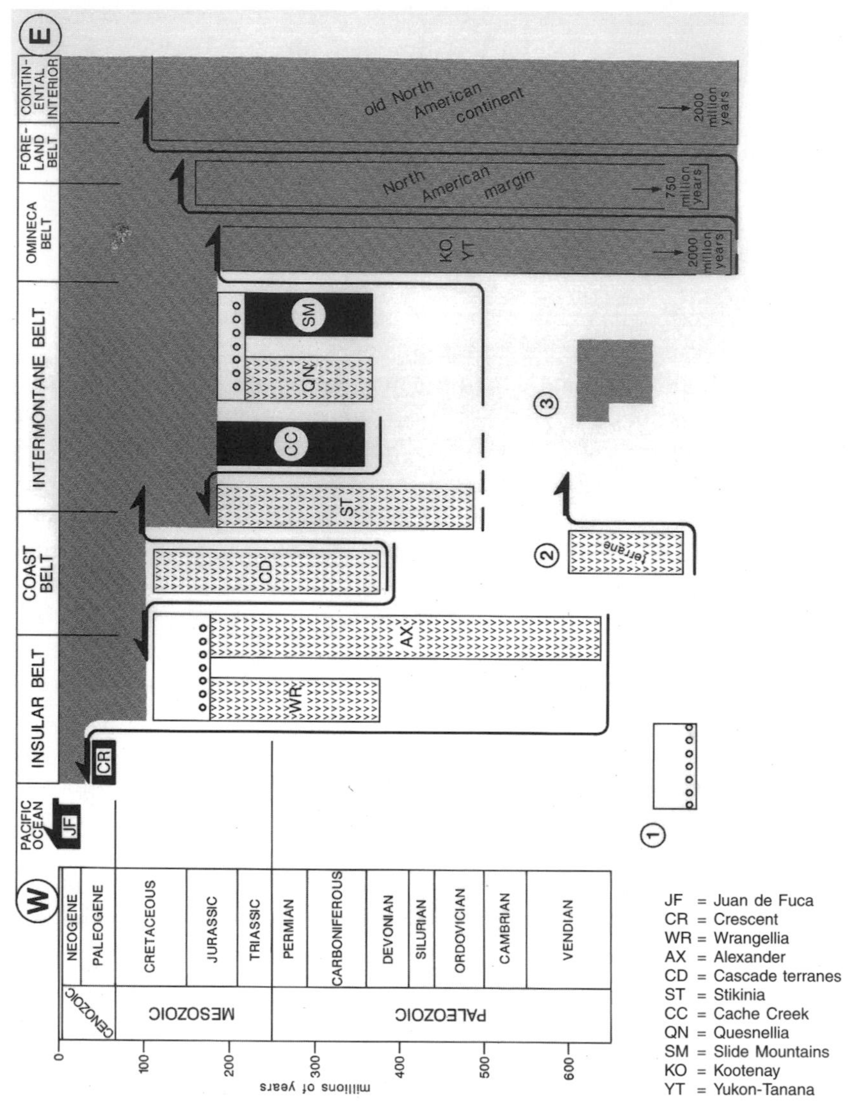

Figure 3.8 **The Cordilleran crust as built oceanward by accretion of terranes.** Time is represented vertically and space is represented horizontally. Accreted arc terranes are shown stippled and oceanic terranes as black. The gaps between vertical columns represent unknown paleogeographic relationships. (1), (2), and (3) explain patterns used. (1) Represents two terranes stratigraphically overlapped and linked by a single rock unit. (2) Represents a thrust fault (see Figure 3.6B), shown by a heavy line, which picks up a terrane and carries it over an adjacent terrane; the time when thrusting occurred is shown by the horizontal position of the heavy line and the barb shows the direction of thrusting. (3) Represents oceanward growth of the North American Plate by terrane accretion, as shown by shading.

along the Coast Belt, parts of which remained the site of marine deposition until the mid-Cretaceous (about 100 million years ago). Wrangellia and the Alexander terrane were accreted to the North American Plate by the mid-Cretaceous. This event was accompanied by a second episode of thrust faulting, metamorphism, granitic intrusion, and uplift centred in the Coast Belt (Figure 3.7E). As was the case for the Omineca Belt in the Jurassic, sand and gravel derived from the uplifted region were eroded and shed westward, this time into the Upper Cretaceous and Lower Cenozoic basins located along the west side of the Coast Belt.

After the mid-Cretaceous, the plate boundary moved to a position west of Vancouver Island and the Queen Charlotte Islands, close to its present position. The entire Cordilleran region, with the exception of places close to the present continental margin, was uplifted above sea level (Figure 3.7F). In the early Late Cretaceous, plate convergence changed from head-on to oblique with a strong northward component. Northward-moving plates in the Pacific Ocean basin sheared across the leading edge of the advancing North American Plate. The result was major strike-slip faulting that disrupted many earlier established relationships within the Cordillera (Figures 3.6A, 3.7E, and 3.7F). In the Early Paleogene, thrust faulting ceased in the Foreland and Omineca belts and was replaced by crustal extension and normal faulting, which formed small basins filled by sedimentary and volcanic rocks.

The landforms we see today in the Cordillera – the partly submerged Insular Belt, the rugged Coast Belt, the subdued plateau-like Intermontane Belt, and the high Omineca and Foreland belts – reflect uplift and erosion over the last ten million years. The details of these landforms – the rocks of which they are made – reflect events that have taken place over a time span of 700 million years.

Acknowledgments
The writer thanks Jim Roddick and Rolf Ludvigsen for their constructive reviews. This chapter is Contribution No. 21094 of the Geological Survey of Canada.

References

Burchfield, B.C. 1983. The Continental Crust. *Scientific American* (September 1983):130-42

Coney, P.J., D.L. Jones, and J.W.H. Monger. 1980. Cordilleran Suspect Terranes. *Nature* 288:329-33

Davidson, G. 1992. Piecing Together the Pacific. *New Scientist*, 18 January 1992, pp. 25-9

Dawson, G.M. 1881. Sketch of the Geology of British Columbia. *Geological Magazine*. New Series 8:156-62, 214-27

Francheteau, J. 1983. The Oceanic Crust. *Scientific American* (September 1983):114-29

Gabrielse, H., and C.J. Yorath, eds. 1991. *Geology of the Cordilleran Orogen in Canada*. Geological Survey of Canada, Geology of Canada, No. 4

Jones, D.L., A. Cox, P. Coney, and M. Beck. 1982. The Growth of Western North America. *Scientific American* (November 1982):70-84

Jones, D.L., N.J. Silberling, and J. Hillhouse. 1977. Wrangellia – A Displaced Terrane in Northwestern North America. *Canadian Journal of Earth Sciences* 14:2,565-77

Monger, J.W.H., R.A. Price, and D.J. Tempelman-Kluit. 1982. Tectonic Accretion and the Origin of the Two Major Metamorphic and Plutonic Welts in the Canadian Cordillera. *Geology* 10:70-5

Murphy, J.B., and R.D. Nance. 1992. Mountain Belts and the Supercontinent Cycle. *Scientific American* (April 1992):84-91

Wheeler, J.O., and P. McFeely. 1991. *Tectonic Assemblage Map of the Canadian Cordillera and Adjacent Parts of the United States of America.* Geological Survey of Canada, Map 1712A

4
Lower Cambrian Trilobites: Most Ancient Mariners

Rolf Ludvigsen and Lisa L. Bohach

The Reverend Adam Sedgwick, Professor of Geology at the University of Cambridge, defined the Cambrian System more than 150 years ago. Ever since, it has been subjected to intense scrutiny by paleontologists and biologists, because Cambrian rocks contain the oldest fossil shells. In the last century, these fossils – trilobites, mollusks, brachiopods, echinoderms, and sponges, as well as various caps, cones, and tubes of uncertain affinities – were accepted as documentary evidence of the appearance and diversification of the first multicellular organisms on earth, or even as evidence of the beginning of life. To the devout Reverend Professor, these fossils were proof of divine creation as set out in Genesis.

Our knowledge of the fossil record of Cambrian and older rocks has increased markedly over the past forty years, and these early interpretations have now been superseded. Well-preserved fossils of simple, minute unicellular algae and bacteria are now known from hundreds of localities in Cryptozoic rocks, some extending back as far as 3.5 billion years ago. And just below the Cambrian, a rich record of complex multicellular fossils has now been documented in Vendian rocks on most continents. These fossil specimens are peculiar impressions of soft-bodied organisms without shells or skeletons – animals that might, in life, have resembled quilts or air-mattresses. Some paleontologists consider them to represent true animals, ancestral to the more familiar metazoans found in Cambrian and younger rocks. Others, however, claim that these 'Vendozoa' were an unusual group of multicellular organisms with a unique body design, in essence, an evolutionary experiment in multicellularity that failed before the Cambrian began.

It is now clear that paleontologists must look farther back in time than the Cambrian to answer questions about the origin of life and the beginning of multicellularity. Still, some of the most exciting paleontological research carried on today involves fossils from Lower Cambrian rocks. These shells provide evidence for a critical juncture in the history of life, namely the appearance of the first unequivocal animals – the kingdom Metazoa. In other words, the Early Cambrian signals the beginning of life as we know it, based on the familiar structural designs that we call animals.

In British Columbia, Lower Cambrian fossils have been known for just over a century from localities in the eastern part of the province. In 1886, George M. Dawson, the pioneer geologist who was responsible for numerous firsts in BC geology and paleontology, collected an olenellid trilobite along the main Canadian Pacific Railway line in the Kicking Horse Pass. This find was the first documentation that Lower Cambrian rocks occur in British Columbia.

Lower Cambrian Rocks in British Columbia

Most people driving through the Rocky Mountains from British Columbia into Alberta marvel at the massive grey peaks that tower above the highway just west of the provincial boundary. Within this belt, the peaks are made entirely of marine sedimentary rocks of Cambrian age that were faulted and upthrust when the mountains were formed in the late Mesozoic. These rocks are extremely thick packages of mainly carbonate rocks that are cliff-forming and resistant to erosion. Located just north of the Yellowhead Highway in easternmost British Columbia, Mount Robson, the highest peak in the Rocky Mountains, is essentially a stack of Cambrian strata 3,000 metres thick. Farther to the south in Yoho National Park, home of the Burgess Shale, the Trans-Canada Highway passes beneath the twin peaks of Mount Field and Mount Stephen, also entirely made of Cambrian strata.

These rocks were originally deposited as sediment on a broad Cambrian continental shelf that formed after the ancient supercontinent Rodinia rifted into a number of smaller continents about 750 million years ago. One of these continents was Laurentia, the geological core of North America, then located south and east of an expanding ancient Pacific Ocean. The initial sediments on the margin of Laurentia are gritty sandstones deposited during the latest Cryptozoic, the Vendian, and the Early Cambrian, when the climate was cold, sea levels were low, and erosion rates were high. During those times, the shoreline lay close to the present border between British Columbia and Alberta. Sea levels rose through the Middle and Late Cambrian, pushing the shoreline as far east as present-day Saskatchewan. As the climate improved later in the Cambrian, carbonate sediments began to predominate across the broad shelf.

Lower Cambrian rocks are confined to a rather narrow belt in the Cordillera of British Columbia that corresponds to the margin of Laurentia and to the adjacent pericratonic Kootenay and Cassiar terranes. Lower Cambrian trilobites are common in the multicoloured shales of the Eager

Formation exposed in the Fort Steele-Cranbrook area (on the Kootenay Terrane) and in the mixed shale and limestone successions exposed through the Grainger and Hughes ranges east and south of Canal Flats (on the margin of Laurentia). Farther to the north, Lower Cambrian trilobites are found in limestones and shales exposed in the Main Ranges of the Rocky Mountains within Yoho National Park and Mount Robson Provincial Park and in the Cassiar Mountains, north and east of Williston Lake.

Demarcating a Base for the Cambrian

Traditionally, the base of the Cambrian was effectively defined by the appearance of the first trilobite above Cryptozoic or Vendian rocks. At virtually all localities across North America, these trilobites are olenellid trilobites. Olenellids are distinct and easily identified, even by first-year geology students and mining company geologists, and so this definition was widely accepted, being one that could readily be used in the field. Because olenellids became extinct before the Middle Cambrian, the stratigraphic range of these trilobites became synonymous with the entire Lower Cambrian interval. On other continents olenellids do not occur, and the first trilobites belong to different groups.

Over the years, however, a fundamental question plagued those stratigraphic paleontologists who were documenting the nature, extent, and timing of the proliferation of animal life in the Cambrian 'explosion.' Can we be sure that the first trilobites found at different localities and on different continents are the same age? This question also troubled those seeking an unambiguous biostratigraphic datum for the base of the Cambrian System. A clear-cut answer was elusive, but most Cambrian paleontologists suspected that the first appearance of Lower Cambrian trilobites was 'diachronous,' that is, the fossils in different areas are of different ages. So, if trilobites cannot provide an 'isochronous' datum at the base of the Cambrian, can other fossils provide it?

In 1972 an international working group of geologists and paleontologists was assigned the task of choosing a site that provided an assemblage of fossils suitable for defining the base of the Cambrian. The site would then serve as a stratigraphic type section – a stratotype. This procedure is now the favoured method of achieving international agreement in stratigraphic matters. After nearly twenty years of data collecting, site visits, and deliberations, the working group was ready to vote on the three best sites selected from more than a dozen. These three were the Aldan River in eastern Siberia, Jining in southern China, and the Burin Peninsula in eastern Newfoundland. At each site a somewhat different biostratigraphic level was selected, all of them well below the first trilobites.

Of the twenty-three votes cast in the first round, Newfoundland received twelve, China eight, and Siberia three. Because the plurality fell short of the recommended 60 per cent, another vote was necessary. As in political leadership races, the last was dropped off the ballot and, in the deciding vote, Newfoundland received fourteen, and China nine. The selection of the Burin Peninsula was ratified in 1992 by the International Commission on Stratigraphy. According to this decision, the base of the Cambrian was now defined by an assemblage of trace fossils dominated by the characteristic and widely distributed *Phycodes pedum* – a small trace fossil that looks like a length of twisted rope. This choice was a special triumph for paleoichnologists (paleontologists who study trace fossils) as trace fossils had always been considered second class by paleontologists who study body fossils. The stratotype near the tip of the Burin Peninsula now includes a permanent reference point for the base of the Cambrian System on all continents in the form of a steel spike driven into the sea cliff at the bottom of the *P. pedum* Zone.

On all continents, the base of the Cambrian is now located hundreds of metres below the first trilobite. In British Columbia the first trilobite – an olenellid trilobite – occurs more than 500 metres above the *P. pedum* Zone. Thus even though it is no longer the index fossil for the Lower Cambrian, olenellid trilobites are still the signature fossils for the upper part of that interval.

Arthropods

The vast majority of animals alive today belong to the phylum Arthropoda, which includes familiar groups such as insects, centipedes, millipedes, spiders, scorpions, crabs, lobsters, and barnacles. These animals are also among the oldest metazoans documented in the fossil record. Arthropods have always been in the evolutionary vanguard: in the Cambrian, arthropods diversified quickly to dominate all other metazoans in marine environments; in the Silurian, they were the first animals to develop lungs and follow the early plants onto land; in the Carboniferous, they were the first animals to acquire wings enabling flight; and, in the Mesozoic, they became the first social animals.

An arthropod possesses an exoskeleton, either rigid or flexible, that is periodically shed to allow for growth. This exoskeleton is composed of organic cuticle such as chitin or is mineralized, commonly as calcite. Its body is constructed of segments, each segment bearing a pair of appendages. Most arthropods possess a pair of compound eyes and, as their name implies, all have jointed limbs.

By the Early Cambrian, arthropods had already diversified into a number of discrete groups. Each group is primarily differentiated by the degree to which the limbs are specialized and the extent to which segments of the head are fused. Those easily recognized are the early ancestors of familiar extant groups such as crustaceans (e.g., crabs, lobsters, and barnacles) and chelicerates (e.g., spiders, scorpions, and horseshoe crabs). An extinct arthropod group is the trilobites.

Most arthropods, fossil or living, lack robust mineralized exoskeletons and fossilize only under exceptional circumstances, such as those responsible for the preservation of the Burgess Shale assemblage. Trilobites, however, possessed a rigid exoskeleton composed of calcite, and therefore these early arthropods have an excellent fossil record spanning almost the entire Paleozoic.

Trilobites

Second to dinosaurs, trilobites are probably the most familiar fossils to the layperson. Dinosaurs, however, are rare fossils that should be collected only by specialists, whereas trilobites are relatively common fossils in Lower Paleozoic rocks and are frequently collected by interested amateurs. For those not prepared to collect their own, most rock and fossil shops stock a good supply of entire trilobites – some even mounted as pendants or earrings. Most of these commercial trilobites belong to a single species, *Elrathia kingii,* which is quarried in huge numbers at one of the few trilobite quarries in the world. That quarry is located in Middle Cambrian rocks in the House Range west of Salt Lake City in Utah.

One of the best non-technical introductions to this intriguing group of fossils was written not by a paleontologist but by a physicist from the University of Chicago. Riccardo Levi-Setti (1993) collects, photographs, and studies trilobites 'as a passion.' Although these long-dead animals may seem an unlikely source for a physics lesson, optical studies of the compound eyes of trilobites allow Levi-Setti to combine his professional work with his intellectual passion. Whittington (1992) gives a more detailed introduction to trilobites.

With a few minor exceptions, trilobites are the only fossils found with their eyes preserved, or rather, their lenses preserved. Trilobites are unique among Metazoa in having lenses that are composed of calcite. These conspicuous structures on the trilobite head are evidence of an unbelievably ancient visual system. Recognizing these structures as eyes, early paleontologists began to study these fossils not as curiosities but as extinct animals comparable to those now living. In 1840, Timothy Conrad, the first State Paleontologist of New York, wrote 'Ode to a Trilobite,' which includes these lines:

> The race of man shall perish but the eyes
> Of trilobites eternal be in stone,
> And seem to stare about with wild surprise
> At changes greater than they yet have known.

Trilobites are divided into three lobes by two deep, lengthwise furrows. (Trilobite is Greek for 'three-lobed.') Their rigid exoskeleton is also divided into three parts: an anterior cephalon, commonly possessing a pair of compound eyes and crossed by sutures, which are lines of weakness that split during moulting; a thorax consisting of hinged segments; and a posterior plate-like pygidium. A raised middle portion of the cephalon known as the glabella enclosed the stomach, which was protected ventrally by a plate called the hypostome. The soft parts of trilobites are known for only about a dozen species. These appendages include antennae and paired jointed legs with gill branches. Unlike those of other arthropods, trilobite limbs are not specialized and differ only in size.

Trilobites lived in the seas for about 350 million years between the late part of the Early Cambrian and the Late Permian. Several thousand genera belonging to eight orders have now been identified. Although some trilobites look like today's horseshoe crabs (chelicerates) or isopods (crustaceans), these extinct arthropods were not ancestral to any other arthropod group.

Lower Cambrian Trilobites

Five major groups of trilobites are found in the upper portions of Lower Cambrian successions, including the olenellids and corynexochids that dominate the earliest trilobite associations in southeastern British Columbia (Figure 4.1). Even though these groups are the oldest, they comprise fully formed trilobites, which differ from younger trilobites but are not notably primitive.

Few of these five groups occur everywhere in the world. Clear differentiation of faunal provinces was already evident in the Early Cambrian. Many forms were found to be restricted to the shallow seas around a single continent. Olenellid trilobites are found in Laurentia and Siberia but not in China or Australia, where the earliest trilobites are redlichiids. These two groups of trilobites are similar – redlichiids differ primarily in having dorsal sutures. They occur together in only a few localities.

Other trilobite groups have a widespread distribution. Agnostids, small trilobites with only two body segments, are most common in deep-water sediments off all continents. They rarely occur with the other groups, which are best developed in shallow-water sediments.

Olenellids

In Laurentia, Siberia, and north Africa, olenellids are invariably the oldest trilobites found. They diversified rapidly through the rest of the Early Cambrian (possibly no longer than fifteen million years) into about fifty genera (Palmer and Repina 1993).

The brown, purple, and grey shales that make up the Eager Formation in the vicinity of Cranbrook and Fort Steele contain numerous fragments

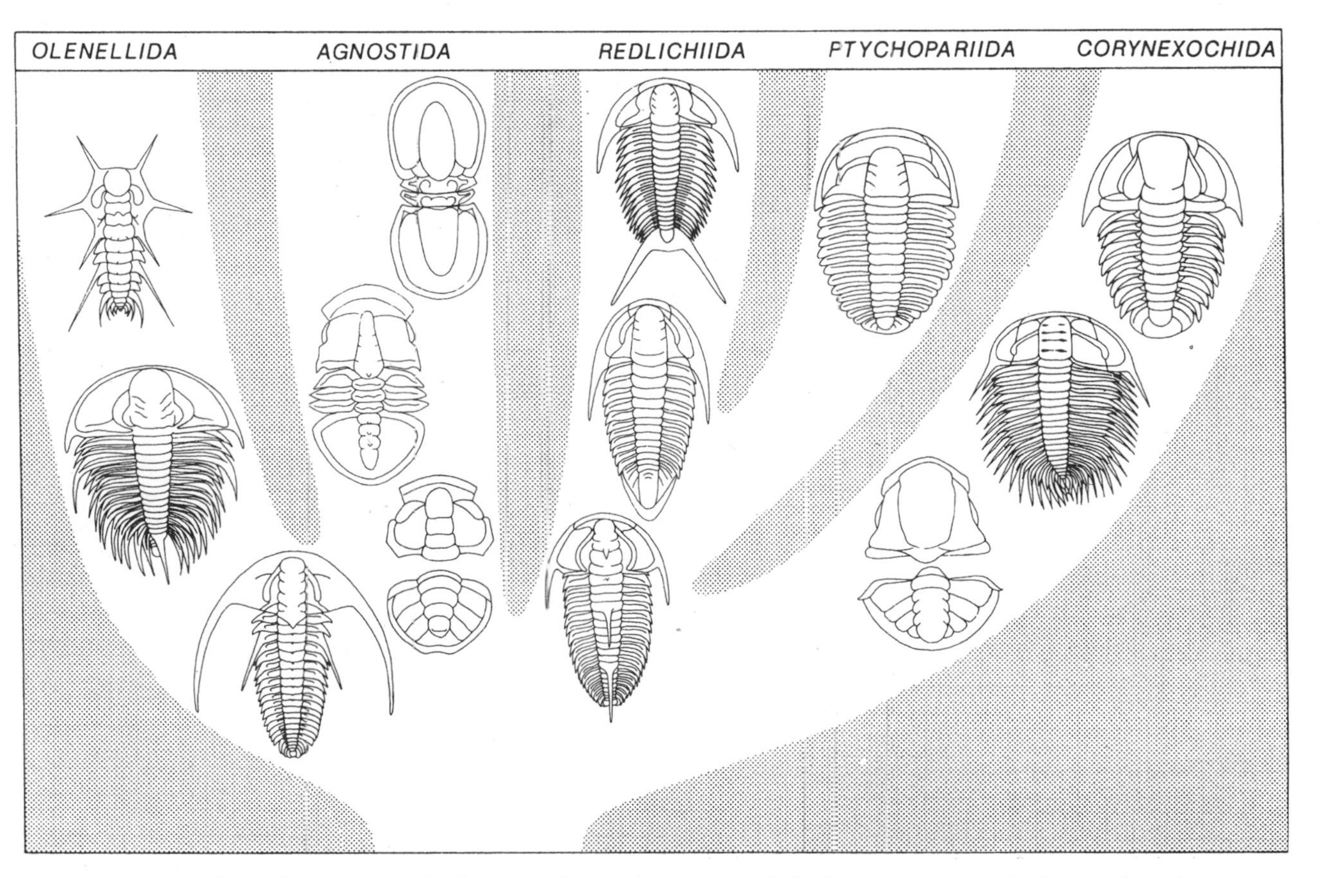

Figure 4.1 **Initial diversification of trilobites into five major groups.** Only the upper part of the Lower Cambrian is shown. *Olenellus* is the second olenellid from the bottom and *Bonnia* is the lowest corynexochid shown.
Source: From Briggs and Fortey (1992)

and many complete specimens of olenellid trilobites. These fossils have been avidly collected by amateurs and professionals since 1921, when they were first made known to paleontologists by Colonel C.H. Pollen of Cranbrook.

The most abundant olenellid in the Eager Formation, by far, is *Olenellus* sp. (Figure 4.2), which has a large, semi-circular cephalon; a wedge-shaped body of fifteen segments, of which the third is extended into long back-swept spines; a needle-like median spine on the fifteenth segment; and a minute pygidium that is rarely visible. The cephalon is characterized by a furrowed glabella flanked by a pair of large, curved eyes. Like all olenellids, *Olenellus* sp. lacks the dorsal sutures on the head.

A less common olenellid in the Eager Formation is *Wanneria* sp., which is characterized by a low, mushroom-shaped glabella without deep furrows and a broad thorax without an expanded third segment. Being an olenellid, *Wanneria* sp. has no dorsal sutures but, as shown in the specimen in Figure 4.3, it has a marginal suture that separated upon moulting. This specimen also shows a hypostome connected to the displaced lower rim of the cephalon.

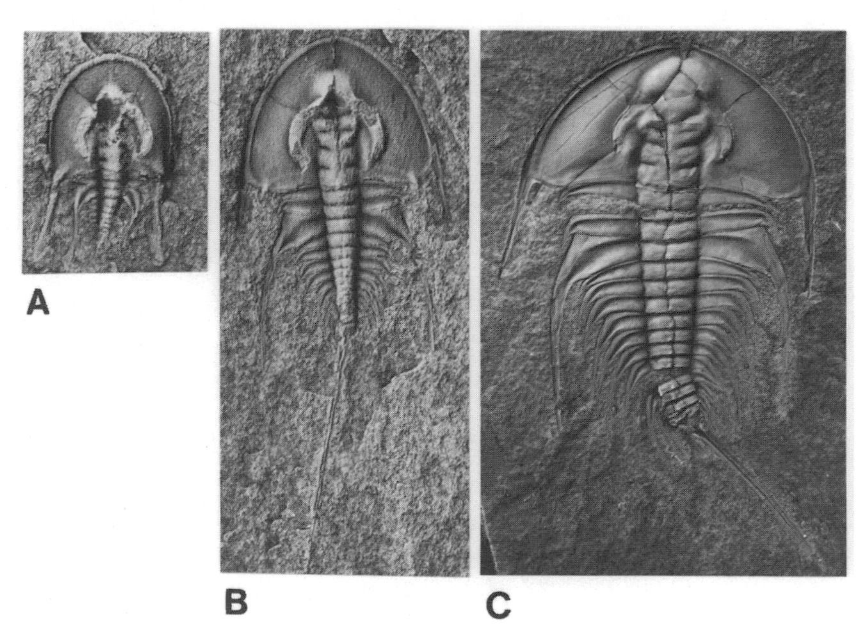

Figure 4.2 **Olenellus sp. from the Eager Formation near Cranbrook.**
An abbreviated growth series of three complete specimens. (A) A larval trilobite 5 mm long. (B) A small trilobite 10 mm long (excluding spine). (C) An adult trilobite 5 cm long (excluding spine).
Source: Collections of the Department of Geological Sciences, University of British Columbia

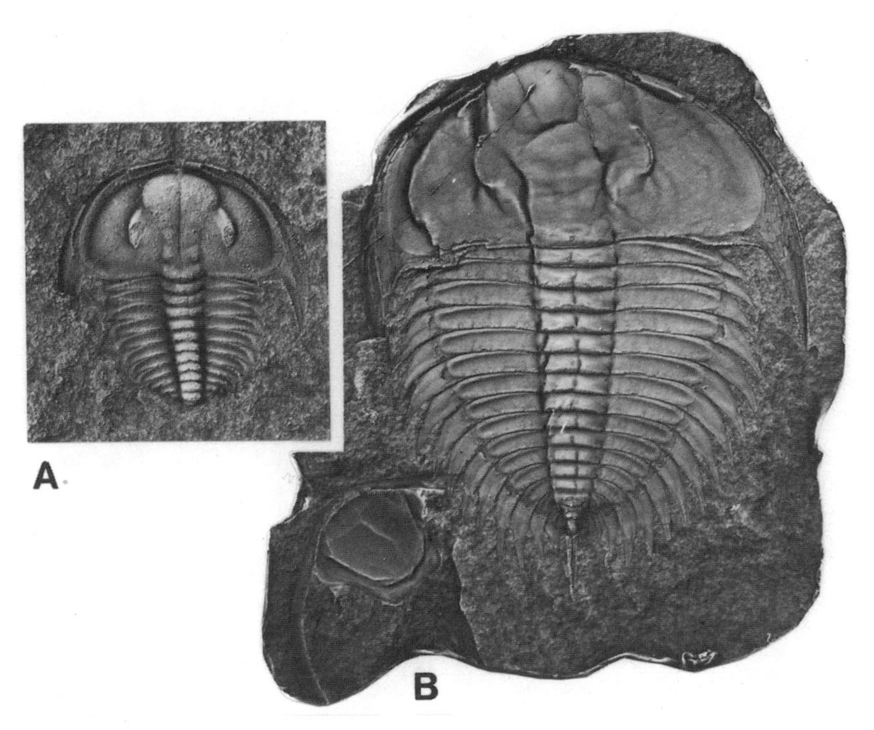

Figure 4.3 ***Wanneria*** **sp. from the Eager Formation near Cranbrook.**
(A) A small trilobite 8 mm long. (B) An adult trilobite 4.5 cm long. The hypostome and ventral rim of the cephalon displaced on the lower left indicate that this specimen is a moult.
Source: Collections of the Department of Geological Sciences, University of British Columbia

All of the trilobites collected from the Eager Formation shales are compressed and crinkled, making it difficult to reconstruct the original convexity of these fossils and, consequently, the original shape of the living animals. Fortunately *Olenellus* and *Wanneria* are also found in limestone units of the same age, exposed in the Grainger and Hughes ranges to the north. These small specimens retain their original convexity (Figure 4.4) due to the nature of limestone, which resists compaction. In *Olenellus,* the glabella has deep, narrow furrows, and is flanked by swollen, banana-shaped eyes that rise steeply above the flanks of the cephalon. In *Wanneria,* the cephalon is gently arched and the mushroom-shaped glabella is slightly swollen.

Corynexochids

A few small trilobites are so squashed in the Eager Formation shales near Cranbrook that they cannot be identified. Even so, they are clearly

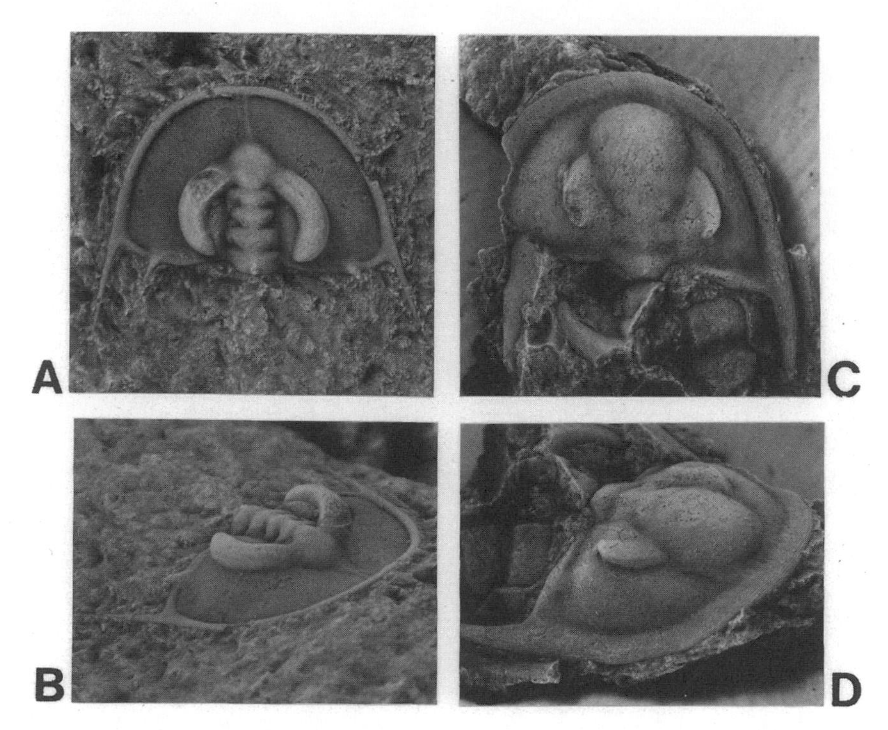

Figure 4.4 **Uncompressed cephala of two olenellid trilobites from Lower Cambrian limestones, Ram Creek area southeast of Canal Flats.** (A) and (B) *Olenellus* sp., 4 mm long. (C) and (D) *Wanneria* sp. 5 mm long.
Source: Collections of the Geological Survey of Canada

different from olenellids. Fortunately, well-preserved and uncompressed specimens are common in limestone beds of the same age, exposed in the Grainger and Hughes ranges to the north. In these beds, they are so abundant that the rock is almost a shell-hash of trilobites. The abundance of these trilobites in limestone compared to their rarity in shale is a graphic example of a preference for one type of habitat.

One of the most common of these small trilobites is the corynexochid *Bonnia* sp. (Figure 4.5). Unlike olenellid trilobites, corynexochids possess a pair of sutures that, on moulting, separate the cephalon into a cranidium and a pair of free cheeks – the lateral parts of the cephalon that carry the lens surface of the eyes. These sutures pass precisely across the top of the visual surface of the eye, a path that presumably served to protect the delicate lenses during the precarious moulting procedure.

Named for Bonne Bay in western Newfoundland, *Bonnia* has a thick exoskeleton marked by fine pits and granules. Its swollen glabella is bald, and stout spines occupy the corners of the cephalon. The hypostome is an inflated rectangle. No complete specimen of *Bonnia* sp. with an entire

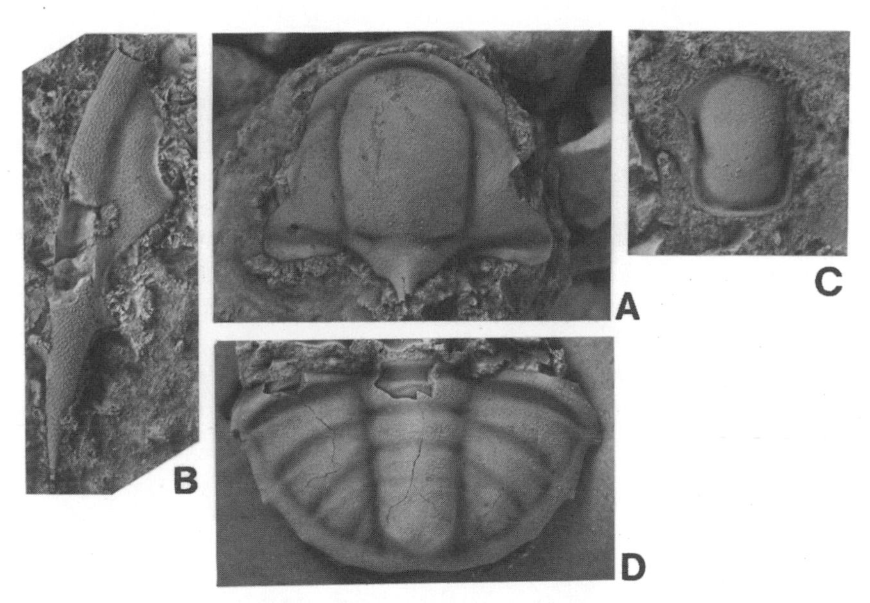

Figure 4.5 ***Bonnia* sp. from Lower Cambrian limestones exposed on Mount Watson, east of Canal Flats.** (A) Cranidium. (B) Free cheek. (C) Hypostome. (D) Pygidium. (All X 6)
Source: Collections of the Geological Survey of Canada

thorax has been found in British Columbia, but species discovered elsewhere have only eight segments in the thorax. The pygidium is a large, slightly vaulted, ribbed shield that carries minute spines and nodes around the rim.

Growing up a Trilobite

Like any arthropod encased in a rigid exoskeleton, a trilobite must periodically moult, or exuviate, in order to grow. As a result of this method of growth, many fossil specimens of trilobites preserved in rock are not the remains of dead animals. Rather, they are moult remains left behind by living animals. It is conceivable that a single trilobite could have left behind twenty or more moults as it grew, each with the potential to become a fossil trilobite. The missing free cheeks or gaping dorsal sutures on the cephala of moulted trilobites serve to distinguish moults from carcasses. It is more of a challenge to distinguish the two for olenellids, as these trilobites lack dorsal sutures. In *Wanneria* sp., however, the displaced rim with attached hypostome (Figure 4.3B) identifies a moult.

The successive moults discarded by a trilobite clearly disclose its growth history, or ontogeny – that is, the increase in size, the addition of segments, and the change of shape as a trilobite grew and matured from larva to adult (Chatterton and Speyer 1990). The shales of the Eager

Formation contain numerous complete specimens of *Olenellus* sp., including many larvae smaller than five millimetres and some adults as long as fifteen centimetres (Figure 4.6). The ontogeny is expressed by the addition of parts and changes in shape that occurred as this trilobite experienced a thirtyfold increase in length. The cephalon became broader, the crescent-shaped eyes grew proportionately smaller, and the inner cephalic spines became shorter, disappearing entirely in adults, while the outer cephalic spines became longer. With the addition of segments, the thorax became longer and broader. A trilobite with the full complement of thoracic segments characteristic for that species is considered an adult (Figure 4.6). This state, however, is more akin to a teenager – sexually mature but with a lot of growth yet to come. The needle-like spine on the fifteenth segment increased dramatically in length until, in large adults, it was nearly as long as the rest of the trilobite.

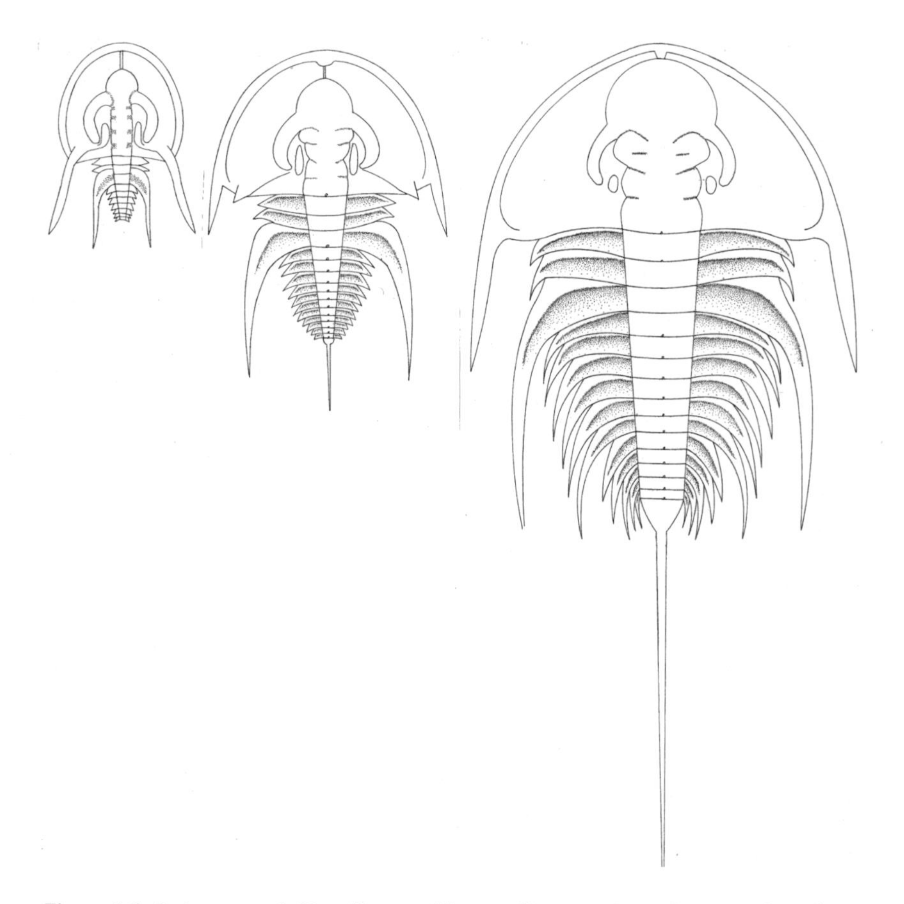

Figure 4.6 **Ontogeny of *Olenellus* sp.** The smallest specimen here is a larval trilobite 5 mm long; the largest is an adult trilobite 5 cm long (excluding spine).

Telling Time with Trilobites

Although the entire geological range of trilobites is extensive, their heyday was short-lived. Only in Cambrian rocks, where three out of four identified fossils are trilobites, did they reign supreme. In Ordovician and younger Paleozoic marine rocks, trilobites are crowded out by abundant fossils of other rapidly diversifying groups such as brachiopods, mollusks, and corals.

Because of their abundance, trilobites are the biostratigraphic indices of choice in Cambrian rocks everywhere in the world. Much of the biostratigraphic research on trilobites in the Cambrian rocks of Laurentia is being done here in Canada. Over the past decade, Steve Westrop of Brock University and Brian Pratt of the University of Saskatchewan have established detailed successions of trilobite zones in Upper Cambrian rocks of the Rocky Mountains and the Mackenzie Mountains. These zones and subzones, based on *species,* make possible correlations as accurate as those based on ammonoid zones in Mesozoic rocks. By contrast, Lower Cambrian rocks have been divided into only three long trilobite zones that are based on *genera.* Thus only crude correlations are possible using genus-based zones.

The *Bonnia-Olenellus* Zone is the youngest Early Cambrian trilobite zone in North America (Fritz 1972). At some localities, the two name-bearing genera are found through several hundreds of metres of strata; at others, through close to a thousand. As demonstrated by trilobite occurrences in southeastern British Columbia, however, only rarely are the two found in the same rocks. *Bonnia* mainly occurs in limestones, whereas *Olenellus* dominates the shales. Both the Eager Formation near Cranbrook and the stratigraphically equivalent but unnamed rocks of the Hughes and Grainger ranges clearly belong to the thick *Bonnia-Olenellus* Zone, but at present it is difficult to specify where in the zone they belong. We are now gathering data that should allow us soon to replace the existing Lower Cambrian trilobite genus-based zones of western North America with zones defined by species.

Once considered the oldest Cambrian fossils, trilobites are now known to have been tardy arrivals that made their first appearance rather late in the Early Cambrian. Nonetheless, the trilobite remains the most familiar example of an extinct and ancient animal group – a compact icon of eternity.

Trilobites are also natural time capsules packed with information about ancient life and habitats. To West Coast poet Earle Birney, they were dispatches from the deep past. In 'David,' his quintessentially Canadian narrative poem about mountains, friendship, and tragedy set in the Rockies, Birney wrote:

And then Inglismaldie. Now I remember only
The long ascent of the lonely valley, the live
Pine spirally scarred by lightning, the slicing pipe
Of invisible pika, and the great prints, by the lowest

Snow, of a grizzly. There it was too that David
Taught me to read the scroll of coral in limestone
And the beetle-seal in the shale of ghostly trilobites,
Letters delivered to man from the Cambrian waves.

References

Briggs, D.E.G., and R.A. Fortey. 1992. The Early Cambrian Radiation of Arthropods. In *Origin and Evolution of the Metazoa*, ed. J.H. Lipps and P.W. Signor, 336-73. New York: Plenum Press

Chatterton, B.D.E., and S.E. Speyer. 1990. Applications of the Study of Trilobite Ontogeny. In *Arthropod Paleobiology*, ed. S.J. Culver, 116-36. Paleontological Society, Short Courses in Paleontology, No. 3

Fritz, W.H. 1972. *Lower Cambrian Trilobites from the Sekwi Formation Type Section, Mackenzie Mountains, Northwestern Canada*. Geological Survey of Canada, Bulletin 212

Levi-Setti, R. 1993. *Trilobites*. 2nd ed. Chicago: University of Chicago Press

Palmer, A.R., and L.N. Repina. 1993. *Through a Glass Darkly: Taxonomy, Phylogeny and Biostratigraphy of the Olenellina*. The University of Kansas Paleontological Contributions, New Series No. 3

Whittington, H.B. 1992. *Fossils Illustrated: Trilobites*. Woodbridge: Boydell and Brewer

5
The Trilobite Beds of Mount Stephen, Yoho National Park

David M. Rudkin

Several places in Canada might rightfully qualify as hallowed ground for students of ancient armoured arthropods. The site with the most legitimate claim to such exalted status, however, is probably the Middle Cambrian 'Trilobite Beds' of Mount Stephen in Yoho National Park, British Columbia.

These beds formed from muddy sediments deposited on an ancient subtropical seabed some 515 million years ago in the Middle Cambrian. Although eclipsed in the recent popular and scientific media by the Burgess Shale, a near neighbour and stratigraphic cousin, the Trilobite Beds continue to hold a place of prominence in Canada's impressive inventory of paleontological treasures. It is difficult to find a geology text, fossil field guide, or children's handbook of prehistoric creatures that does not contain a photograph or drawing of at least one of the common trilobites from this sequence of shales and siltstones full of fossils.

The Trilobite Beds: History and Description

The Trilobite Beds are located on the northwest slopes of Mount Stephen about 800 metres above the town of Field. They were first brought to widespread attention in 1886 during construction of the Canadian Pacific Railway through the Kicking Horse River valley in southeastern British Columbia. Over the next three decades, several pre-eminent researchers undertook pioneering descriptive work on the abundant and conspicuous trilobite fossils that give the locality its name.

The most notable of the earliest investigators was Charles Doolittle Walcott, geologist and Cambrian paleontologist *extraordinaire*. His report on this area (Walcott 1908), which was published in the *Canadian Alpine Journal* the year before his far more celebrated discovery of the Burgess Shale, provided the most comprehensive description of Trilobite Bed fossils of the time, and indeed to date. Walcott included the Beds in his newly named Stephen Formation. When the Burgess Shale was later discovered about six kilometres to the north on the ridge between Mount Field and Wapta Mountain, he regarded this site as a stratigraphic equivalent of the Trilobite Beds.

In the years immediately following their discovery, fossils in the Trilobite Beds were eagerly collected by geologists and tourists alike, with the result that the locality achieved widespread fame, and the fossils worldwide distribution. When it later became apparent that continued unrestricted access would seriously degrade the site, officials of Yoho National Park, in a series of moves over some fifty years, first prohibited collecting, then discouraged visits, and ultimately declared the Trilobite Beds a 'Closed Area.' Nowadays, public access to the locality is restricted to the lower portion of the beds, and that, by permit only. Nevertheless, guided tours of this area operated under Park regulation afford the public a unique opportunity to view one of Canada's premiere paleontological attractions.

The Mount Stephen Trilobite Beds constitute a singularly impressive occurrence, though not necessarily at first sight. Because they are located on steep treeless slopes at the end of a demanding three-hour alpine trail, they are usually viewed initially through sweat-stung eyes. It sometimes takes the first-time visitor a few minutes to realize that virtually every piece of rock littering the extensive talus accumulation contains at least part of a trilobite fossil. While the sheer number of the remains is in itself truly astounding, another remarkable characteristic of the deposit is the large proportion of the trilobite fossils representing complete or nearly complete individuals. Although we are not yet able to fully explain these characteristics, it seems certain that one factor was the same as that contributing to the spectacular preservation in the Burgess Shale, namely, rapid burial in an oxygen-poor environment.

In terms of other measures of preservation, fossils from the Trilobite Beds do not rank particularly high. Compaction has greatly reduced the original convexity of the trilobites. This conclusion arises from examinations of the same or similar species in carbonate rocks (such as limestones) elsewhere, in which little or no flattening has occurred. Also, the trilobites seldom retain much of the outer mineralized body covering. Instead, most consist primarily of impressions of inner or outer surfaces of the original exoskeleton in the shales. Despite these preservational shortcomings, there are rare surprises in the Trilobite Beds, including three known specimens of trilobites that show remains of anatomical features such as antennae, gills, and legs in addition to the mineralized shell.

The name 'Trilobite Beds' belies the fact that many different kinds of fossils are found at this site, including typical Middle Cambrian hard-shelled organisms like brachiopods and hyoliths, prodigious numbers of the unmineralized claws of the large predatory arthropod *Anomalocaris*, and a variety of lesser-known fossils. Given the stunning preponderance of trilobite remains, however, it is not surprising that 'Trilobite Beds' became the accepted label for this unique occurrence.

The Trilobites

A single species with the glottally challenging name *Ogygopsis klotzi* (Figure 5.1A) overwhelmingly dominates the fauna of the Trilobite Beds, a fact that prompted Walcott (1908) originally to name the unit the '*Ogygopsis* Shales.' This trilobite is so conspicuously large – often more than twelve centimetres long – and occurs here in such vast numbers that complete specimens were easily acquired in the early days of unrestricted collecting. As a result, *Ogygopsis* found its way into museums, universities, and private collections around the world. Because it appears in a great many illustrated books on fossils, this trilobite has come to be regarded as something of an archetype for Cambrian trilobites. Curiously, it is virtually unknown in Middle Cambrian faunas elsewhere.

In addition to the ubiquitous *Ogygopsis,* at least three other kinds of trilobites are present in sufficient numbers of complete individuals that they can be found after just a few minutes of searching the talus of the Trilobite Beds. Two of these, the species of *Bathyuriscus* and *Elrathina* (Figures 5.1B and 5.1C), are common enough, both in this site and elsewhere in western North America, to have prompted another distinguished early researcher, Charles Deiss, to use their occurrence to define the *Bathyuriscus-Elrathina* Zone, a biostratigraphic time interval. A third, *Olenoides* (Figure 5.1E), is also well known, primarily through the remarkable appendage-bearing specimens from the nearby Burgess Shale but also as a typical Cambrian trilobite of wider geographic distribution. A fourth form, *Oryctocephalus* (Figure 5.1D), is a much less common but exceptionally widespread trilobite. Representatives have been found in Middle Cambrian rocks, not only of western North America but also Europe, Asia, Australia, and South America.

The story does not end with these relatively well-known and occasionally spectacular components of the fauna. Another five kinds of trilobite occur infrequently to rarely along with their more common companions. It is among these unsung denizens that we find some of the most fascinating tales of the Trilobite Beds.

Rominger's Very Thorny Beast

While *Ogygopsis* and its more common cohorts *Bathyuriscus* and *Olenoides* dominate the fauna of the Trilobite Beds, at least in terms of sheer numbers and often in size, otherwise they are not particularly striking trilobites. The prize for most ornate appearance might easily go to a less frequently encountered creature bearing the delightful name of *Zacanthoides*

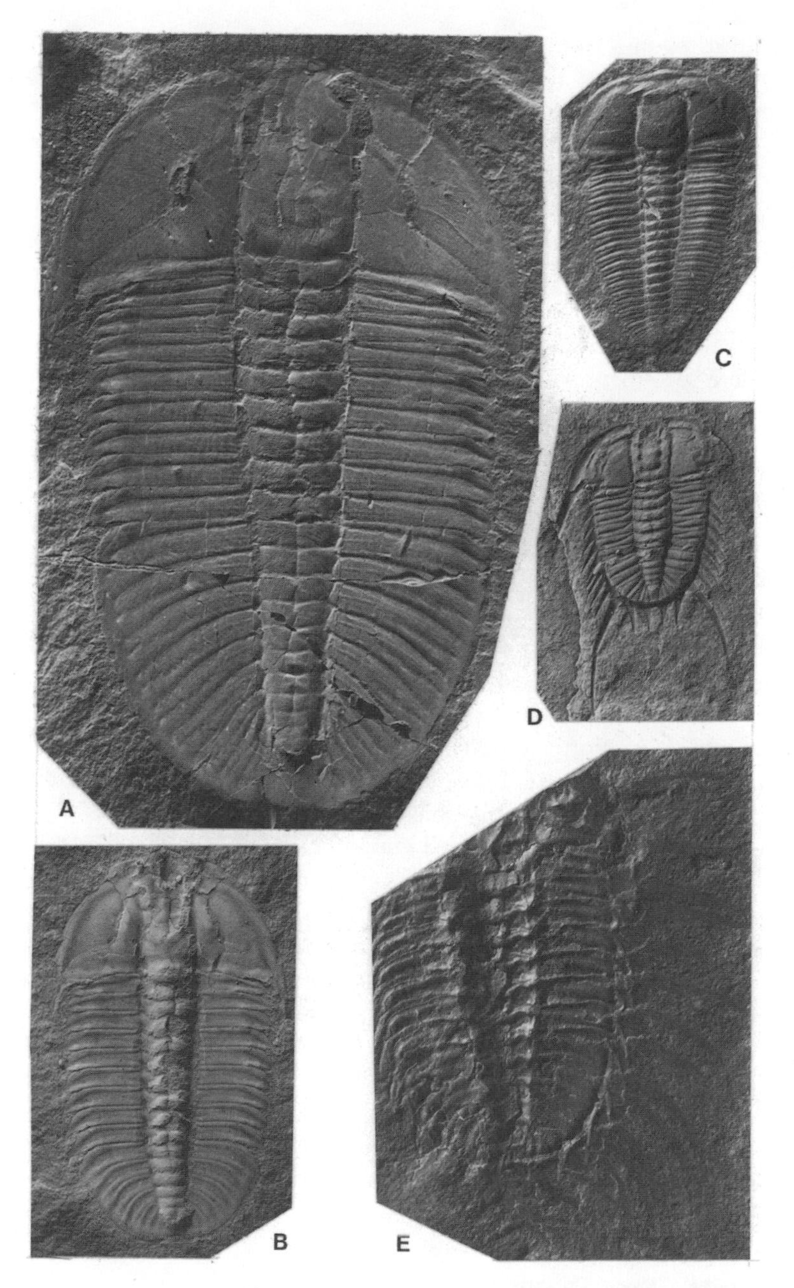

Figure 5.1 **Common trilobites of the Trilobite Beds on Mount Stephen.** (A) *Ogygopsis klotzi.* (B) *Bathyuriscus rotundatus.* (C) *Elrathina cordillerae.* (D) *Oryctocephalus reynoldsi.* (E) *Olenoides serratus,* negative impression showing preserved appendages. (All X 1.5)
Source: Collections of the Department of Invertebrate Palaeontology, Royal Ontario Museum

romingeri, a veritable porcupine of a trilobite sporting an impressive array of spines.

Like many of the fossils from Mount Stephen described in the latter half of the last century, *Zacanthoides* has a complex history of name changes. Taxonomy – the science of naming organisms following an explicit set of internationally recognized rules – is more than just a formal exercise in labelling. It is one aspect of a larger effort directed toward understanding and communicating the genealogical relationships among living and once-living things. Thus the long series of name changes associated with *Zacanthoides* reflects both our expanding knowledge of trilobite relationships in general and our appreciation of where this particular form fits.

The trilobite that we now know as *Zacanthoides romingeri* was first named in 1887 by Carl Rominger, at the time a paleontologist and retired Michigan State Geologist. Rominger characterized the trilobite on the basis of material sent to him by his long-time friend, Otto Klotz, who was engaged on a geodetic survey along the newly opened CPR line. While working in the vicinity of Field, Klotz had acquired one of the first collections of Mount Stephen fossils. From the sample sent by his friend, Rominger described and named six new species of trilobites, following the very basic rules of taxonomy of the 1880s. Each name he proposed was a binomen, consisting of two parts, the first being the genus (the generic name) and the second, the species (the specific name). The largest and most numerous specimens in the collection, for example, he called *'Ogygia Klotzi.'* The genus name *'Ogygia'* was already in use for similar looking British trilobites; the species name *'Klotzi'* was in honour of his friend.

Rominger chose the binomen *'Embolimus spinosa'* for a single, more or less complete specimen of a distinct and markedly spiny form, about 3.5 centimetres long, which he depicted like the other described forms by a rough schematic sketch. The specific epithet *spinosa* may have been appropriate, but because the generic name *Embolimus* was already in use for an insect, Rominger erred according to the rules of taxonomy. It has long been a convention that two different groups of organisms cannot share a common taxonomic name above the level of species.

At that time, Walcott was an up-and-coming geologist with the US Geological Survey. In 1888, he revised much of Rominger's somewhat shaky taxonomy and assigned a Middle Cambrian age to the previously undated collection from Mount Stephen. Walcott coined the name *Zacanthoides* for the new genus to include specimens that he had collected in Nevada and described in 1884 as *Olenoides spinosus,* as well as Rominger's *'Embolimus spinosa,'* which he regarded as belonging to the same species. Thus Walcott's revisions resulted in the binomen

Zacanthoides spinosus being applied to both the spiny form from Mount Stephen and the Nevada specimens.

The slightly different species names *'spinosa'* and *'spinosus'* were used by these workers to achieve gender agreement in names derived from classical Latin and Greek roots. Their use of these names was in obvious allusion to the needle-like projections of various parts of the exoskeleton. But how did Walcott choose his generic name? Although he did not publish the reasons for selecting *Zacanthoides,* we can work backward through the classical derivations of the word to arrive at a pretty good guess. The Greek root *akantha* means a 'thorn or prickle': the word *akanthion* means 'thistle,' 'porcupine,' or 'hedgehog.' The last part of the name, *-oides,* is based upon a contraction of the Greek term suggesting 'likeness.' It is probable that the first letter, *Z,* is a contraction of the Greek *za* meaning 'very.' Thus *Zacanthoides* can be translated literally as 'very thorn-like' or, more euphoniously as 'very thorny.' Either way it is an apt label for this particular animal. Subsequently, in his 1908 article in the *Canadian Alpine Journal,* Walcott provided a considerably more accurate drawing of a specimen of *Zacanthoides* from Mount Stephen.

The next chapter in this tale was written in 1942 when Charles Resser, a paleontologist at the Smithsonian Institution, recognized that the Mount Stephen and the Nevada forms of *Zacanthoides* were sufficiently different from one another to warrant being placed in separate species. Walcott's original species *'spinosus'* was left to represent the forms described earlier from Nevada, and Resser established the new specific name *romingeri* for the material from the Trilobite Beds, honouring the man who first named this trilobite fifty-five years earlier.

Finally, in 1951, Franco Rasetti, a renowned physicist who became an authority on Cambrian trilobites, published a summary based on two years of field work studying rocks and fossils of the southern Canadian Rockies (Rasetti 1951). Now a classic, this volume provides photographs of three specimens of *Zacanthoides romingeri* from Mount Stephen. Also included are descriptions and figures of five new species and one undetermined form of the genus discovered at sites elsewhere in the Rockies.

While the taxonomic history of *Zacanthoides romingeri* is all very interesting and instructive, it tells us relatively little about the true nature of this beast. For this we must turn to the fossils themselves. Fortunately, we now have more to go on than Rominger's single incomplete specimen and crude sketch.

During a major reinvestigation of the Burgess Shale and other fossil localities in Yoho National Park, the Royal Ontario Museum (ROM) was granted permission to collect selectively from the Mount Stephen Trilobite Beds on a number of occasions between 1975 and 1989. The work, under the direction of Dr. Desmond Collins, was primarily directed toward finding and documenting new occurrences of 'soft-bodied' fossils. Nevertheless, representative collections of trilobites were made wherever possible. This work was undertaken not only because these trilobites are aesthetically appealing or because they occasionally include preserved traces of unmineralized anatomy but also because they are the consummate indices of time for the Cambrian System. Thus these collections can be of help in deciphering the confusing sequences of rocks and fossils in this geologically complex area. Among the recent ROM trilobite collections from Mount Stephen are specimens of *Zacanthoides romingeri* that expand our knowledge of the species beyond that presented by Rominger, Walcott, Resser, and Rasetti.

The single specimen upon which Rominger based his figure and original description apparently lacked free cheeks. The hard upper exoskeletal covering of the head of most trilobites was crossed from front to back by a pair of symmetrical lines of weakness called facial sutures. The sutures allowed the free cheeks to split away from the cranidium when the animal moulted its exoskeleton to accommodate growth. The lack of free cheeks is typical of many specimens from the Trilobite Beds, suggesting that a large number of these trilobites are partly disarticulated remains representing moulted exoskeletons.

Walcott's figure of *Zacanthoides* in the *Canadian Alpine Journal* (Walcott 1908) showed the free cheeks in place, a restoration presumably based on reconstruction from disarticulated elements. Resser did not include an illustration to accompany his taxonomic revision and description, and Rasetti's photographs were of partial specimens only, none showing the free cheeks intact. But the specimens of *Zacanthoides* in the ROM collections include an almost complete exoskeleton (Figure 5.2A), in which the left free cheek is still attached to the cranidium, thus confirming the presence of a long, slender projection – the genal spine – of the back corner of the cheek. This specimen also shows the elongated bow shape of the left eye, with the lens surface in place.

In *Zacanthoides,* a small thorn-like spine is borne on each of the outer posterior corners of the fixed cheeks – the portions of the head flanking the central raised region (glabella) of the cranidium. Fixed cheeks are located inside the facial sutures and remain fixed in place during moulting. These spines can be seen on a specimen (Figure 5.2B) that lacks both free cheeks. Cranidia of disarticulated specimens sometimes reveal a short, strong spine extending backward along the centre-line from the occipital ring (the 'neck,' or rear part of the glabella). More complete exoskeletons, such as the one shown in Figure 5.3B with the cranidium still attached to the articulated thorax, seldom have the occipital area preserved intact. Such specimens are usually thoroughly flattened and

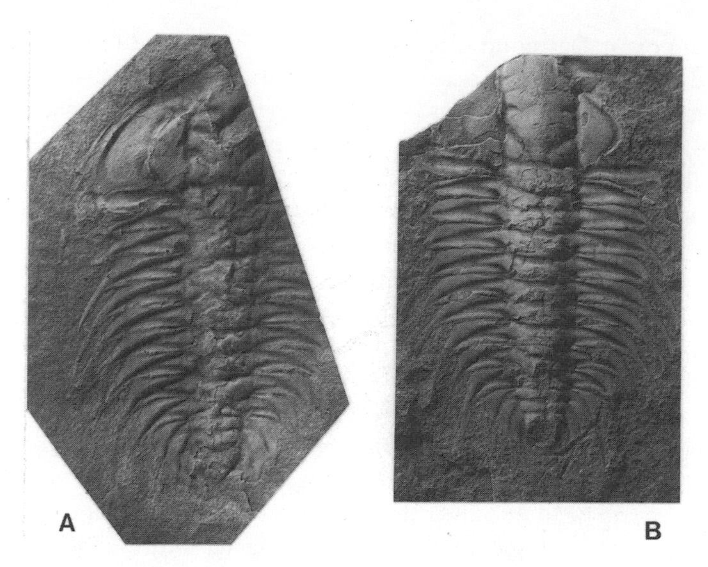

Figure 5.2 **Zacanthoides romingeri.** (A) Latex mould of nearly complete individual showing left free cheek and genal spine in place. (X 1.2) (B) Specimen showing tiny spines at posterior corners of fixed cheeks. (X 1.4)
Source: Collections of the Department of Invertebrate Palaeontology, Royal Ontario Museum

Figure 5.3 **Zacanthoides romingeri.** (A) Negative impression of partially disarticulated thorax and pygidium, showing needle-like median thoracic spine. (Actual size) (B) Specimen lacking free cheeks, showing broadly triangular occipital area and spinose pygidium. (X 1.3)
Source: Collections of the Department of Invertebrate Palaeontology, Royal Ontario Museum

fractured, but they may show evidence of a bluntly pointed node, rather than the spine, on the occipital ring.

The lateral extremities of each segment of the articulated nine-part thorax are extended into backwardly directed spines, which become longer and more blade-shaped toward the back of the animal (Figure 5.2A). A long, needle-like spine that runs back from the centre of the eighth segment of the thorax can sometimes be seen on smaller, or partly disarticulated, individuals like that in Figure 5.3A. Although the occipital and median thoracic spines lie flat against the thoracic segments in these compressed specimens, they very likely were directed up at a moderate angle in the living animal.

The rear-most element of the trilobite exoskeleton is called the pygidium. In *Zacanthoides,* it is a relatively small, bluntly triangular plate carrying four pairs of backwardly directed marginal spines. Unlike the thoracic spines, each pair of pygidial spines is progressively shorter than the one in front of it (Figure 5.3B).

Thus the overall impression of *Zacanthoides* is one of a mass of spines with a trilobite hidden somewhere inside. *Zacanthoides* is not unique in this regard, however, as many different kinds of trilobites evolved elaborately spinose exoskeletons over the 300-million-year history of the group. A variety of reasons have been suggested for the development of spinosity, the most obvious being that spines functioned to discourage or prevent predatory attack. Certainly, in this particular instance, there is ample justification for considering a protective role, because a formidable predator lurked in the Middle Cambrian seas inhabited by *Zacanthoides.* In the rocks of the Trilobite Beds, occurring with nearly the same frequency as trilobite remains, are the fossilized feeding claws of *Anomalocaris.* With an estimated maximum body length approaching half a metre, this bizarre creature is the largest known Cambrian predator (Collins 1986).

There is good evidence to implicate *Anomalocaris* in the attack and injury of at least one other kind of Mount Stephen trilobite. Several specimens of *Ogygopsis* collected by the ROM display distinctive healed wounds on the margins of the exoskeleton, suggesting that the trilobites had survived a traumatic injury (Rudkin 1979). The wounds are of such a configuration that it is entirely possible that they were generated by the peculiar circular biting apparatus of *Anomalocaris.* So far, however, no complete specimens of *Zacanthoides* have been found with signs of healed injuries that might have been inflicted in a failed attack. Could it be argued that its spines protected *Zacanthoides* from this attack? Perhaps, though we do occasionally encounter clusters of broken spines that might conceivably represent the results of a successful attack by a predator.

Another function of trilobite spines might be to obscure the outline shape of the animal in such a way as to help it blend in with its surroundings. This proposition seems reasonable in the case of another infrequently detected form of the Mount Stephen trilobites, one that is also comparatively tiny, being about twenty millimetres long. *Oryctocephalus* is arguably just as 'thorny' as *Zacanthoides,* having a pair of slender genal spines that sweep back from the cephalon to beyond the full length of the exoskeleton, a number of sharp lateral pleural spines, and a series of marginal pygidial spines, of which the fourth pair is greatly elongated. Complete or partially disarticulated specimens of *Oryctocephalus* are sometimes found in association with small fossil sponges called *Pirania,* which have a distinct cactus-like appearance. In such instances it is difficult to distinguish between the spines of the trilobite and the needle-shaped spicules protruding from the body of the sponge. It is easy to imagine a diminutive *Oryctocephalus* hiding out among a sea-floor thicket of *Pirania,* while larger trilobites and the dreaded *Anomalocaris* crawled and cruised nearby. *Zacanthoides* might well have sought similar refuge.

After more than a century of increasingly detailed investigation, taxonomic refinements, and paleontological pondering, we still lack answers to the many intriguing questions about the half-billion-year-old creatures entombed within the Mount Stephen Trilobite Beds. We certainly know more about the evolution and diversity of trilobites in general, thanks in no small part to the principal players in this particular tale (Whittington 1992). And we are now in a much better position to engage in informed speculation and supposition about aspects of trilobite existence not preserved in rock. The work continues. Someday we may be able to decipher successfully the origin of this unique cache of Cambrian history and describe accurately the lives of all its cryptic citizens.

References

Collins, D.H. 1986. The Great *Anomalocaris* Mystery: How a Shrimp Became the World's First Monster. *Rotunda* 19(3):51-7

Rasetti, F. 1951. Middle Cambrian Stratigraphy and Faunas of the Canadian Rocky Mountains. *Smithsonian Miscellaneous Collections* 116(5):1-277

Rudkin, D.M. 1979. *Healed Injuries in* Ogygopsis klotzi *(Trilobita) from the Middle Cambrian of British Columbia.* Royal Ontario Museum Life Sciences, Occasional Paper 32

Walcott, C.D. 1908. Mount Stephen Rocks and Fossils. *Canadian Alpine Journal* 1:232-48

Whittington, H.B. 1992. *Fossils Illustrated: Trilobites.* Woodbridge: Boydell and Brewer

6
The Burgess Shale: A Spectacular Cambrian Bestiary

Desmond H. Collins

The fossils of the Burgess Shale of eastern British Columbia are famous for two reasons: their exquisite preservation and their great age of about half a billion years. The significance of the age is that these fossils document life as it was just after the great evolutionary 'explosion' early in the Cambrian, when most animal groups first appeared on earth. Because of their extraordinary preservation (Figure 6.1), these fossils present a vivid and compelling view of early life in the Cambrian seas.

These fossils indicate which early animal groups originated 530 million years ago. Depending on how they have been classified, they also disclose how well different animal groups have survived on earth since the Cambrian. Instead of focusing on individual fossils, this chapter reviews the history of discovery of the Burgess Shale and then explores how these fossils have influenced our view of animal life on earth.

The discovery site and main quarry yielding Burgess Shale fossils is located about 2,300 metres above sea level on the west side of the ridge between Wapta Mountain and Mount Field, about eight kilometres west of the Alberta border in Yoho National Park. The Trans-Canada Highway passes below Mount Field so that, weather permitting, the quarry site can be seen by anyone looking east from the highway west of Field, British Columbia.

Discovery and First Impact

The first Burgess Shale fossils were found loose on the west slope of Mount Field on Monday, 30 August 1909, by Charles Doolittle Walcott. Walcott was Secretary of the Smithsonian Institution at the time and, as such, was one of the top government scientists in the United States. He was widely regarded as the world authority on Cambrian rocks and fossils. So, another remarkable aspect of the Burgess Shale story is that the most important Cambrian locality in the world was discovered by the one person most able to recognize its significance.

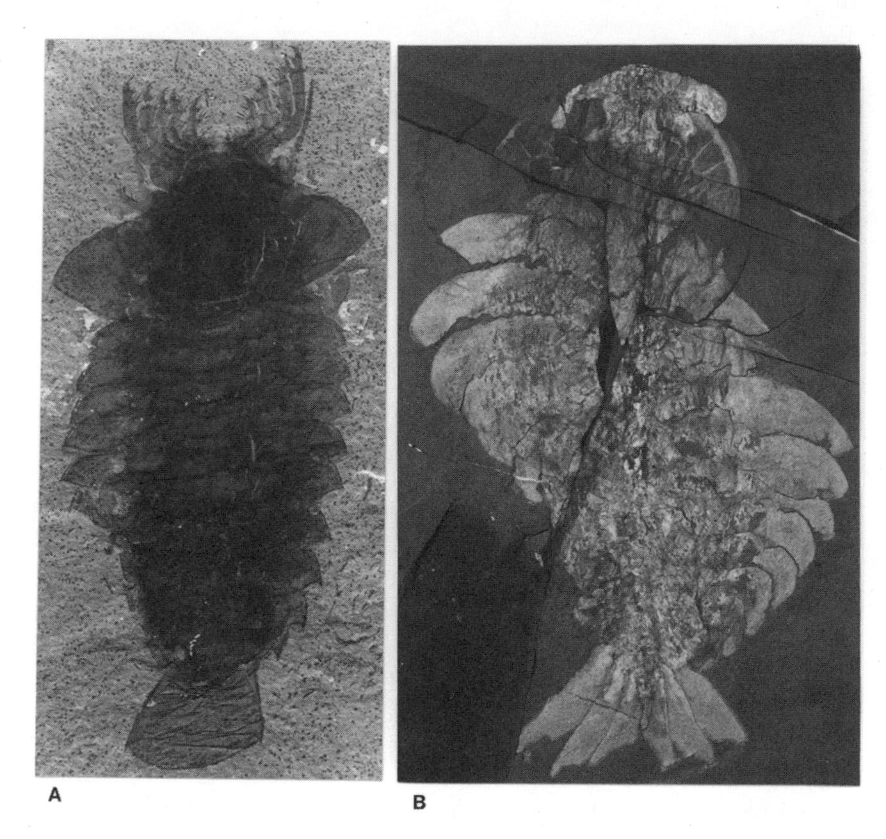

Figure 6.1 **Two spectacular Burgess Shale fossils.** (A) The chelicerate *Sanctacaris uncata,* 7 cm long. (B) The world's first monster *Anomalocaris canadensis,* 20 cm long.
Source: Collections of the Department of Invertebrate Palaeontology, Royal Ontario Museum

Walcott first came to Yoho National Park in the summer of 1907 to visit and collect from the famous Trilobite Beds on Mount Stephen, the mountain just across the Kicking Horse River valley from Mount Field. Walcott had described some of the Mount Stephen trilobites in 1888 and 1889, and had assigned a Middle Cambrian age to them. However, 1907 was the first time Walcott had actually visited the Trilobite Beds. On that visit, he described the rocks on Mount Stephen and measured other geologic sections in the area as part of a long-term project studying Cambrian rocks in western North America.

A continuation of this search for good Cambrian rock sections with fossils led Walcott along the Burgess Pass trail on Mount Field in late August 1909 to his discovery of the Burgess Shale fossils in talus blocks. Walcott and his family collected Burgess Shale specimens here for five days before moving across the valley to the Mount Stephen Trilobite Beds, where they collected for the last three days of the season. In a letter dated 27 November 1909, to William Arthur Parks of the University of Toronto, Walcott observed, 'I had a few days collecting in the Stephen formation in the vicinity of Field in September, and found some very interesting things.'

Walcott came back the following August, again with his family (wife Helena, sons Sidney and Stuart, and daughter Helen). By working his way layer by layer up the west slope of Mount Field, he discovered the 'lace crab beds' *(Marrella),* the main source of the Burgess Shale fossils. Walcott spent six weeks in 1910 measuring the stratigraphic section, excavating a quarry, and collecting fossils. By this time, he knew what he had found. In his diary dated 31 December 1910, Walcott exulted, 'Great find of Cambrian fossils at Burgess Pass.'

Walcott didn't waste any time in describing the Burgess Shale fossils collected in 1910. From April 1911 to March 1912, he produced four Smithsonian publications giving 'preliminary' descriptions of jellyfish, sea-cucumbers, annelid worms, and arthropods. These papers had an immediate impact on other paleontologists because they demonstrated the presence in Cambrian seas of a variety of animals previously unknown and unsuspected. Some had been known previously as fossils in much younger rocks but not in Cambrian rocks; others such as the marine 'velvet worm,' *Aysheaia,* had no known fossil record.

The main reason the remains of such animals had not been found in Cambrian rocks elsewhere was that they were soft-bodied. Virtually all fossils are the remains of animals or plants with hard parts – shells, bones, tree stems, etc. The Burgess Shale is remarkable because it includes the remains of a great many soft-bodied animals such as jellyfish, sea-cucumbers, and worms. In addition, the fossils of the Burgess Shale often have soft appendages or organs preserved, for example, the legs and antennae of trilobites.

In a twelve-month period, therefore, Walcott showed that the animals living in Middle Cambrian seas were much like those living today, with the exception of fishes and other vertebrate animals that had yet to evolve. Of course, some of these animals, like *Wiwaxia* or *Opabinia,* look strange compared to their living descendants. Yet Walcott and the colleagues he consulted attributed these differences to the fossils representing early members of living groups and thus exhibiting primitive characteristics.

The Burgess Shale site thus became enshrined as a locality where the life of the Cambrian could be seen to a much fuller extent than anywhere else. Subsequently, all paintings, dioramas, and descriptions of Cambrian life showed Burgess Shale animals almost exclusively.

Although he visited the locality in 1919 and 1924, Walcott's last excavation of the Burgess Shale was in 1917. He wrote lengthy papers on

Burgess Shale trilobite appendages, algae, and sponges, but he never did complete the detailed descriptions of the soft-bodied animals that he had intended to prepare. Walcott died in 1927, in his seventy-seventh year. After his death, his assistant at the Smithsonian, Charles E. Resser, published some additional descriptions of Burgess Shale fossils compiled from the photographs and descriptions that Walcott had prepared from his post-1910 collections. With the exception of these specimens, the rest of Walcott's collections of 1911, 1912, 1913, and 1917, representing about 80 per cent of the total, remained unstudied and unknown for almost fifty years.

The main Burgess Shale work of note after Walcott's death was that of Percy Raymond, who led a number of parties from Harvard University summer school to the Canadian Rockies, beginning in 1924. These visits to the Burgess Shale culminated in 1930 with a fifteen-day excavation in Walcott's quarry and at a second level that is twenty-three metres stratigraphically higher, now known as Raymond's quarry.

For the next thirty-five years no work was done on the Burgess Shale fossils. Walcott thought that his 1917 excavation had essentially exhausted the Burgess Shale quarry and, even though disproved by Raymond's subsequent excavation, this view persisted. Also, Walcott's influential widow, Mary Vaux Walcott, discouraged research on her husband's collection at the Smithsonian Institution. During these years, it was fully accepted that the Burgess Shale animals were primitive ancestors of invertebrate animals living today. Dioramas at most of the major museums in North America showed these animals as the epitome of life in the Cambrian.

Revision and Second Impact

The second phase of research on the Burgess Shale began as the Canadian Centennial project of the Institute of Sedimentary and Petroleum Geology (ISPG) in Calgary. ISPG is a branch of the Geological Survey of Canada (GSC). In 1966 and 1967, GSC field parties under Jim Aitken and Bill Fritz extended Walcott's quarry to the north and enlarged Raymond's quarry. The Director of the ISPG, Digby McLaren, invited Harry Whittington, a well-known trilobite paleontologist from the University of Cambridge, to oversee the redescription of the Burgess Shale animals. This work was to be based on both Walcott's collections in Washington and the specimens newly excavated by the GSC. When Whittington began, it was with the expectation of completing the detailed descriptions of Burgess Shale animals that Walcott had been unable to finish. Whittington focused on the arthropods (Figure 6.2), beginning with *Marrella,* the most common fossil in the Walcott quarry. At the time, *Marrella* was classified within the Trilobitomorpha, a group of fossils thought to be related to trilobites.

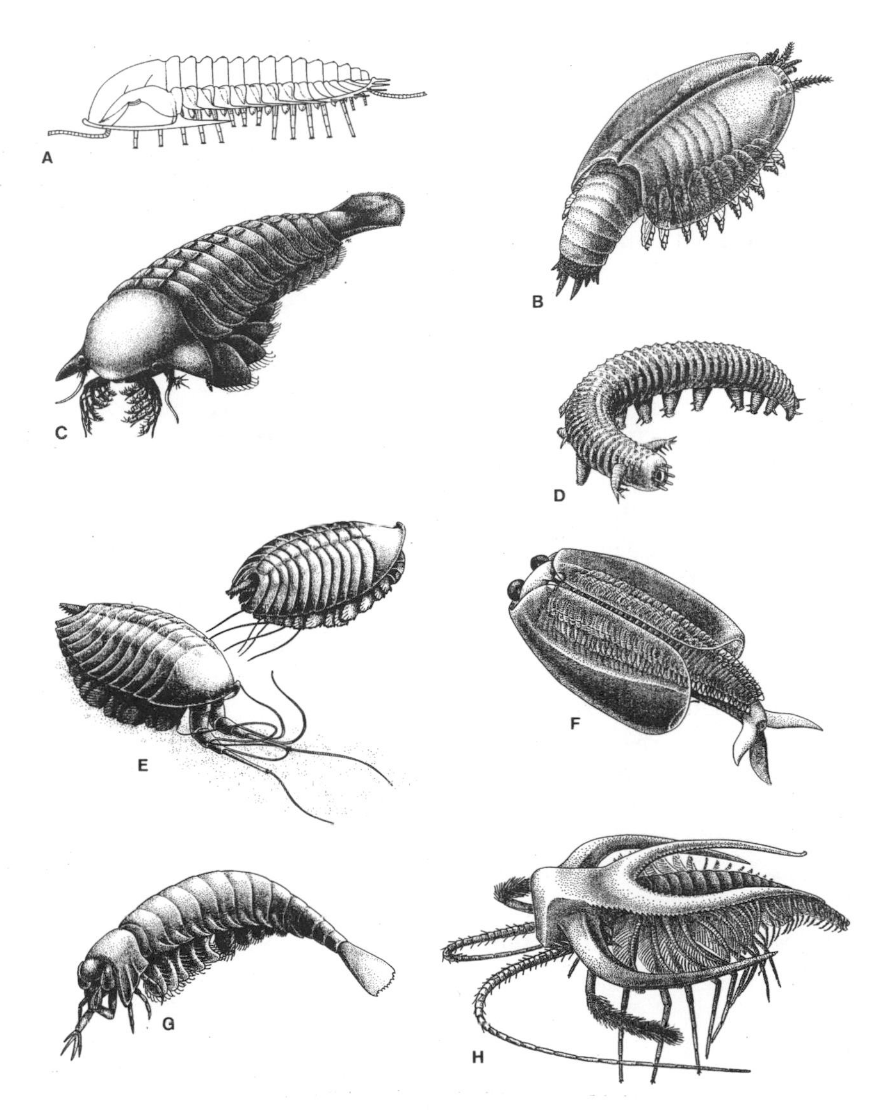

Figure 6.2 **Burgess Shale arthropods.** (A) Trilobite *Olenoides* with appendages. (B) Crustacean *Canadaspis.* (C) Chelicerate *Sanctacaris.* (D) Onychophoran *Aysheaia.* (E) Unclassifiable *Leanchoilia.* (F) Unclassifiable *Odaraia.* (G) Unclassifiable *Yohoia.* (H) Unclassifiable *Marrella.*
Sources: (A) From Whittington (1985); (C) From Briggs and Collins (1988); all others from Gould (1989)

Through careful preparation of many specimens of this arthropod, Whittington discovered new anatomical details. With this information, he found that *Marrella* certainly was not related to trilobites and could not be assigned to any arthropod class.

Whittington had similar difficulties with the classification of *Yohoia* and *Plenocaris,* two arthropods also represented by exquisitely preserved specimens. The next animal he studied possessed five eyes, a frontal digitate trunk, and a tail with fins directed obliquely upward and outward like a Beechcraft Bonanza airplane, and it apparently lacked jointed limbs. This animal, *Opabinia,* did not appear to fit in any known category of arthropod, or in any known phylum.

As Whittington continued, it became obvious that other arthropods, such as *Emeraldella* and *Leanchoilia,* also failed to fit into existing classes. His student, Derek Briggs, was able to assign some arthropods, such as *Canadaspis,* to the class Crustacea but not others, like *Odaraia.* Lastly, Whittington and Briggs together described an almost complete *Anomalocaris,* identifying it as a member of a 'hitherto unknown phylum.'

Parts of *Anomalocaris* had previously been described as separate animals: a 'shrimp' body, a jellyfish, a sea-cucumber, a bristleworm, and a jellyfish superimposed on a sponge. Whittington's paleo-dissection of an *Anomalocaris* specimen revealed the shrimp body to be one of a pair of claws, and the jellyfish to be a circlet of radiating teeth. Whittington and Briggs then reconstructed *Anomalocaris* as an extraordinary half-metre-long carnivore – truly the world's first monster (Collins 1986). Recently, field parties from the Royal Ontario Museum have collected more complete specimens (Figure 6.1B), showing *Anomalocaris* to look even more fearsome than before (Figure 6.3H).

Another of Whittington's students, Simon Conway Morris, who had been given Walcott's 'worms' to study, found his task even more challenging. He discovered that two worms, namely *Canadia* and *Burgessochaeta,* were indeed polychaetes as Walcott had said. Several others, such as *Ottoia,* were priapulids. Six others – *Amiskwia, Dinomischus, Hallucigenia, Nectocaris, Odontogriphus,* and *Wiwaxia* – could not be classified by Conway Morris, even to phylum.

In 1989, Stephen Jay Gould, the well-known paleontologist and science writer from Harvard University, published a synthesis of the reinterpretations of the Burgess Shale fossils in his best-selling book *Wonderful Life.* Gould took the reinterpretations one step further. Whereas Whittington and his co-workers suggested the animals were of 'unknown affinities' or belonged to a 'hitherto unknown phylum,' Gould proposed that they belonged to animal groups that became extinct after the Cambrian. Furthermore, he observed that most Burgess Shale arthropods could not be assigned to any one of the three classes of Cambrian arthropods – trilobites or crustaceans, both of which were already known; or chelicerates, the first of which, *Sanctacaris,* had recently been described from a new Burgess Shale locality on Mount Stephen (Briggs and Collins 1988) (Figure 6.1A). Gould therefore concluded that more Cambrian

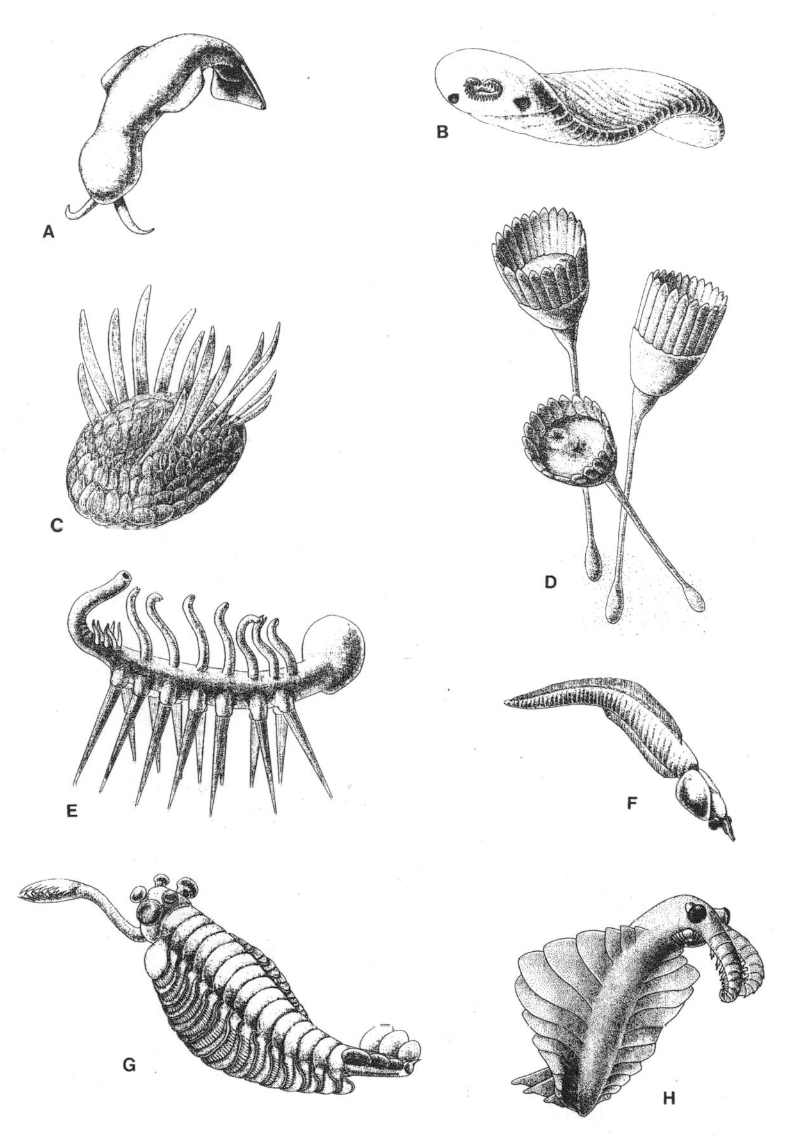

Figure 6.3 **Burgess Shale 'weird wonders.'** (A) *Amiskwia.* (B) *Odontogriphus.* (C) *Wiwaxia.* (D) *Dinomischus.* (E) *Hallucigenia.* (F) *Nectocaris.* (G) *Opabinia.* (H) *Anomalocaris.*

In (E) *Hallucigenia* is shown walking on spike-like spines and having a single row of slender tentacles along its back. This reconstruction has recently been challenged by the Swedish paleontologist Lars Ramsköld. He demonstrated the presence of two rows of tentacles, not one, and quite reasonably claimed that these were the paired legs of the animal. The spines, therefore, must have been located on the back and had a protective function. Ramsköld's flip of *Hallucigenia* has been accepted by most paleontologists. *Editor*

Sources: (H) From Collins (In press); all others from Gould (1989)

arthropod classes had become extinct than are alive today. Then Gould turned his attention to the eight 'weird wonders' that could not be assigned to living phyla – Conway Morris's six 'worm' genera listed in the preceding paragraph plus *Opabinia* and *Anomalocaris* (Figure 6.3). And he further concluded that as many animal phyla became extinct after the Cambrian as survive today.

The cause of the extinctions was what Gould called 'contingency,' referring principally to the five mass extinctions known to have occurred during the 500 million years since the Cambrian. Gould proposed that mass extinctions had had much more drastic effects on the evolution of life through geologic time than previously realized. Indeed, we can now add mass extinction as a third factor to variation (from mutation) and natural selection in the process of evolution. An example of immediate relevance to us is the mass extinction sixty-five million years ago, when the dinosaurs, pterosaurs, and large marine reptiles became extinct. This extinction allowed the surviving mammals to diversify and become dominant on land and in the sea, and, incidentally, to produce primates – a branch of which evolved into humans.

So, what began as an esoteric exercise in describing and classifying a number of strange fossils from the Burgess Shale of eastern British Columbia led to a profound change in our understanding of how life has evolved on earth.

The Current Perception

Gould's provocative proposals have not gone unchallenged. Derek Briggs and two colleagues have demonstrated that Burgess Shale arthropods show no more overall variation in form and structure than do arthropods living today. As well, attempts have been made, with varying success, to find a home for some of Gould's 'weird wonders.' Nick Butterfield, a paleontologist now at the University of Western Ontario, has demonstrated convincingly that *Wiwaxia* is a polychaete worm, as Walcott had said. My own research on *Anomalocaris* and *Opabinia* indicates that they are arthropods, probably belonging to an extinct class. And recent work by Swedish and Chinese paleontologists suggests that *Hallucigenia* may be a lobe-footed animal related to onychophorans (velvet worms) like *Aysheaia*. Certainly, any reduction in the number of unclassified animals lessens the possible effect of contingency on animal life.

The concerns relating to the purported effect of contingency on the evolution of animal life and to the classification of the animal groups that appeared in the Cambrian 'explosion' are far from resolved. Moreover, although the Burgess Shale is still the main source of new interpretations using both Walcott's collections and new ones being made by Royal Ontario Museum parties, it is no longer the only source. The

Chengjiang fauna discovered in Lower Cambrian rocks of southwestern China in 1984 includes both Burgess Shale fossils and new forms. This locality seems to have the potential to eventually yield as great a variety of animals as the Burgess Shale. A third source is the Buen Formation of northernmost Greenland. This locality is producing a more restricted fauna than the other two but has yielded two quite extraordinary animals so far. Coloured illustrations of these remarkable Cambrian animals – from the Burgess Shale fauna and the Chengjiang and Greenland faunas – were published in the October 1993 issue of *National Geographic.*

The Burgess Shale story continues. The combination of great age, the exquisite preservation of extraordinary animals, and the message to us about evolution of life on earth makes the Burgess Shale in eastern British Columbia the most important fossil locality in the world. As well, its setting above Emerald Lake, high in the Canadian Rockies, makes it one of the most beautiful.

Acknowledgments

The drawings in this chapter were done by Marianne Collins of the Royal Ontario Museum and are reprinted here with her permission. The drawings from *Wonderful Life* are reprinted with permission from Stephen Jay Gould. The plates were assembled by Peter Fenton, and the manuscript was processed by Joan Burke.

References

Briggs, D.E.G., and D. Collins. 1988. A Middle Cambrian Chelicerate from Mount Stephen, British Columbia. *Palaeontology* 31:779-98

Collins, D. 1986. The Great *Anomalocaris* Mystery: How a Shrimp Became the World's First Monster. *Rotunda* 19(3):51-7

–. In press. The 'Evolution' of *Anomalocaris*: A Paradigm for Understanding Cambrian Life. *Journal of Research and Exploration,* National Geographic Society

Gould, S.J. 1989. *Wonderful Life: The Burgess Shale and the Nature of History.* New York: W.W. Norton

Whittington, H.B. 1985. *The Burgess Shale.* New Haven: Yale University Press

7
The Microscopic World of Conodonts

Michael J. Orchard

Wherever Cambrian to Triassic sedimentary rock of marine origin is encountered in British Columbia, there is a good chance that it contains the tiny fossils called conodonts. The name of these enigmatic microfossils derives from Greek, and brings to mind the conical, tooth-like shape of the first described specimens. Many other conodonts have extraordinarily elaborate shapes resembling miniature combs and leaves.

Conodonts are often described as simple cones (or coniforms), bars and blades (sometimes combined as ramiforms), or platforms (Figure 7.1). Most conodonts are less than one millimetre in any dimension and therefore are rarely identifiable in the field. Collections of conodonts may nevertheless be extracted from rock through relatively simple laboratory procedures. It is probably true to say that these microfossils are one of the most common types of fossils found in the province.

Conodonts provide a considerable amount of geological information about the rocks in which they occur. They can be used to determine the age of the host rock, the maximum temperature to which it has been heated, the environment in which the constituent sediments originally accumulated, and the geographic region of their origin (that is, their location before continental drift). Of course, the nature of the extinct animal whose remains the conodonts represent, and the function of the conodonts within the animal, are also matters of considerable paleobiological interest.

Finding Conodonts in the Cordillera

The recovery of conodonts begins with the somewhat speculative collecting of rock samples, usually limestone but sometimes shale or chert. A carbonate-rich sample is cleaned and broken into walnut-size pieces that are then immersed in a pail of 10-per-cent acetic acid (essentially, strong vinegar). After about a week, the undissolved residue is passed through sieves to concentrate the fine-grained fraction, which is then

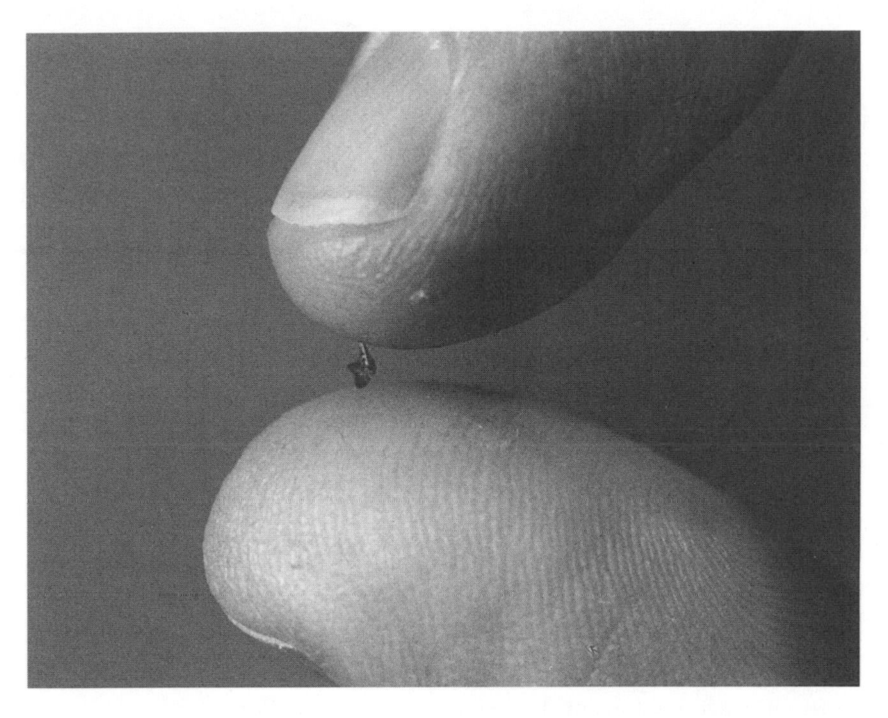

Figure 7.1 **Specimen of Late Devonian platform conodont *Palmatolepis*.** This specimen is relatively large. Those illustrated in Figures 7.2 to 7.8 are smaller, with those in Figure 7.5 being the smallest.

dried and, if necessary, concentrated further by a combination of different separation methods. Using a binocular microscope (about X 25 magnification), the final concentrate (Figure 7.2) is examined for conodonts and associated acid-insoluble microfossils, which are placed in special containers for study. Similar methods are currently used routinely in several laboratories in western Canada, where the study of conodonts has become an indispensable component of regional geological studies.

The first conodonts were discovered in British Columbia only about thirty years ago. At that time, conodonts had just begun to make a worldwide impact as index fossils – that is, fossils useful in dating and correlating rock strata. The first discovery was of Ordovician conodonts northeast of McMurdo, near Golden. A 1964 report on this conodont collection is the first to appear in the archives of the Geological Survey of Canada (GSC) in Calgary. It was written by Tom Uyeno who, a few years later, also reported the first conodonts from the accreted terranes of the western Cordillera, a collection of Triassic age from the Taseko Lakes area. In the 1970s, Bruce Cameron began processing conodont samples in the Vancouver laboratory of the GSC, and thereby set the stage for what might be called a 'conodont revolution' in Cordilleran geology. The process

Figure 7.2 **Exceptionally rich, acid-insoluble conodont concentrate from Upper Triassic limestone of the Peace River region**

moved into high gear in 1979 when I set up a large-scale conodont processing facility in Vancouver. To date, over 6,000 collections of conodonts and associated microfossils from western Canada have been assembled in Vancouver, about two-thirds of which come from British Columbia (Orchard 1991a). These collections have provided a 'time-framework' that has been fundamental to the mapping and interpretation of the geological history of Paleozoic and Triassic rocks.

What Are Conodonts?

The term conodont is applied to the mineralized elements that represent the only fossilized remains of otherwise soft-bodied creatures that inhabited the earth's oceans for over 300 million years. Conodonts first appeared in the Upper Cambrian (about 510 million years ago), and disappeared at the end of the Triassic, about 210 million years ago. The seas must have teemed with conodont-animals, since their teeth (composed, as are ours, of calcium phosphate) are globally distributed and often very common. The most abundant collections I have found in British Columbia contain more than one thousand specimens in a kilogram of limestone. Rocks in this province are often metamorphosed (changed by heat and pressure) to the point where fossils composed of calcium and silica are destroyed. Conodonts may still be recovered, however, since they are made of phosphates, making them particularly resilient – hence their pre-eminent role in complementing geological mapping.

'Conodonts: Those Fascinating Little Whatzits' is the title used by Walter Sweet for his presidential address to the Paleontological Society in the mid-1980s. These 'whatzits' have both baffled and intrigued paleontologists for some 140 years. They were discovered by Christian Pander, a Latvian, who described Lower Paleozoic specimens found in Estonia and Russia in an 1856 monograph on fossil fishes. (I should note here that researchers in this area honour the discoverer of conodonts by naming their association the 'Pander Society,' although they avoid the derivative 'Panderers'!) The original interpretation of conodonts was that they were the remains of an ancient group of fishes, a concept that was challenged by subsequent investigators, many of whom believed that conodonts represented the jaws of annelid worms. As will be outlined, however, it seems that Pander was right.

In 1879, George Hinde reported conodont species from North America, in which he included several different types of conodont elements, that is, both platform and ramiform elements. Hinde's ideas gave rise to the concept of conodont apparatuses or 'multi-element species.' This taxonomic approach was contrary to that of Pander, who named each conodont element as a separate species. Although Hinde's approach is now understood to have been zoologically correct, just about all people studying conodonts ignored or rejected it at the time. It was not until the middle of the present century that the multi-element nature of conodont species became widely accepted, when more 'natural assemblages' of different conodont elements were discovered on rock surfaces. At about the same time, it was also discovered that discrete conodonts could easily be extracted from carbonate-rich rocks by dissolving the rocks in acid. The flourish of studies that followed resulted in the discovery of many new conodont elements, and the single-element species became both the basis for classification and the effective index fossil. Nowadays, researchers attempt to combine the two systems, one based on single elements and the other based on multi-elements, to arrive at a classification that is biologically valid and a sequence of zones that expresses evolutionary steps.

What Were These Animals?

The zoological affinity of conodont-animals has been one of the classic paleontological uncertainties. Even when conodonts were first studied, Pander wrote that their classification would always be 'precarious and arbitrary.' Comparison of conodont elements with similar structures in the animal kingdom has led to all sorts of interpretations of both the function of the conodont-elements and the affinity of the conodont-animal. In the 1970s, several unusual soft-bodied fossils were presented as candidates for the whole conodont-animal, but none survived scrutiny. Then, in 1982, a rock slab bearing the impression of a single worm-shaped animal was discovered in a drawer containing fossil shrimp at the British Geological Survey collections in Edinburgh, Scotland.

This specimen was about four centimetres long and two millimetres wide, with fin-like posterior structures. Amazingly, conodont elements were preserved, probably in their original positions, close to a pair of lobe-like structures on the head. The slabs from the fossil shrimp bed had, in fact, been collected many years before from Carboniferous rocks along the nearby seashore. Fortunately, the source locality was well known and further collecting soon produced additional specimens of the conodont-animal. Subsequently, in South Africa, important Ordovician specimens were discovered, in which the anterior lobes were interpreted to have been eyes. The study of all these specimens has involved modern analysis of the chemistry and structure of conodont elements. The results suggest that conodonts are the remains of an ancient group of elongate, jawless fishes – perhaps akin to hagfishes and lampreys – that bore a toothed feeding apparatus composed of several different conodont elements (Figure 7.3).

The Conodont Apparatus

When conodont-animals died, the integrity of the conodont apparatus was usually lost as a result of predators and scavengers or of sedimentary processes. Present-day laboratory procedures may have a similar effect. The most revealing data describing the structure of conodont apparatuses come from natural assemblages such as those first described from Germany and the American midcontinent. These assemblages reveal that, within a single apparatus, six types of elements commonly occur, and most of these elements form mirror-image pairs arranged symmetrically about a midline. Although no natural assemblage is known from British Columbia, conodont elements that are fused together are found occasionally. Fused elements also provide insights into the kinds of conodont elements naturally associated in a single animal and into their orientation and distribution within that animal.

In northern British Columbia, west of Muncho Lake on the Alaska Highway, examples of fused conodonts have been found in collections from limestones of Late Devonian age (370 million years ago). The host rocks were originally deposited as muds in deep water off the western edge of the ancient North American landmass. Nowadays, the region is explored for mineral deposits such as lead, zinc, and associated precious metals that occur interlayered within the sedimentary rocks. These 'stratiform' mineral deposits probably originated as metallic brines periodically deposited on the sea-floor, as is the case today around deep-sea vents.

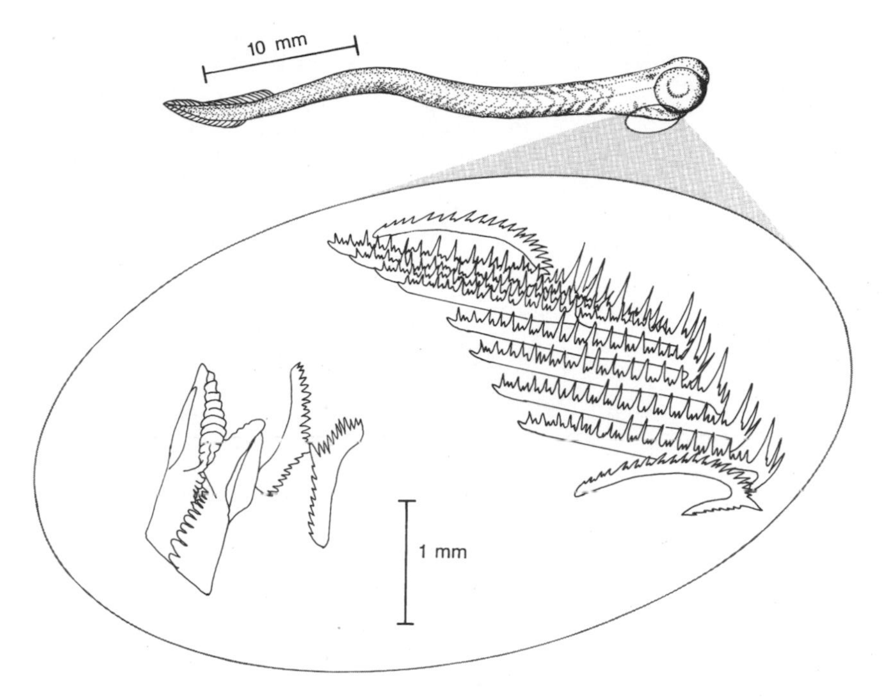

Figure 7.3 **Restoration of conodont-animal and location and architecture of typical Carboniferous apparatus.** At the left (posterior) are paired platforms and blades, and at the right (anterior) is a series of aligned bars.
Sources: Adapted from Aldridge et al. (1993) and Purnell (1993)

When corpses of conodont-animals came to rest in these sediments and if the conodont apparatus remained undisturbed, adjacent elements became fused by insoluble mineral material (Irwin and Orchard 1991). Fused specimens show us mirror-image pairs of platforms and blades and various bar elements, including a medially located symmetrical form, that were closely aligned within the conodont-animal (Figure 7.4).

From fused conodont elements such as these and from natural assemblages, paleontologists have established 'templates' of conodont apparatuses. By comparing collections of discrete conodont elements with these templates, we are able to reconstruct multi-element conodont apparatuses. Reconstruction is relatively straightforward when the sample yields a collection representing one conodont-animal. A good example of a reconstructed conodont apparatus comes from a Late Triassic sample from the Queen Charlotte Islands. This conodont collection is composed entirely of six kinds of similar conodont elements that are a good fit with a common template for multi-element apparatuses (Figure 7.5). This example illustrates how reconstructed conodont apparatuses can be used as a basis for a natural classification of the conodont group.

Figure 7.4 **Fused conodont elements of Late Devonian age from the ancient North American continental margin.** The original organization within apparatuses is revealed by pairs of opposed platforms of *Palmatolepis,* seen here in lateral view (left) and in upper and lower views (top), and by a pair of blades (right) and a cluster of bars (bottom).
Source: Collections of the Geological Survey of Canada

Cooking Conodonts

Conodonts change colour when heated. This characteristic was demonstrated experimentally by Anita Harris of the US Geological Survey in laboratory studies designed to reveal the physical conditions that might have produced the array of colours shown by conodonts (Epstein et al. 1977). Naturally occurring colours range from pale yellow through shades of brown to black and, in more metamorphosed rocks, through grey to white. For purposes of documentation, colours are identified by a scale of 'colour alteration indices' (or CAI), which range from 1 to 8. We now know that changes in CAI result from progressive, temperature-induced alteration of organic material that is interlayered with calcium phosphate in the conodont elements. Accordingly, the colour of conodonts may be used to indicate how hot the host rock was in the past, either because it was buried deep in the earth's crust or because it came in contact with molten igneous rock. This information has important application in the hydrocarbon industry, because the generation and preservation of oil and gas depend on the temperature to which organic material within sedimentary rocks has been heated. The 'oil window,' that is, the tempera-

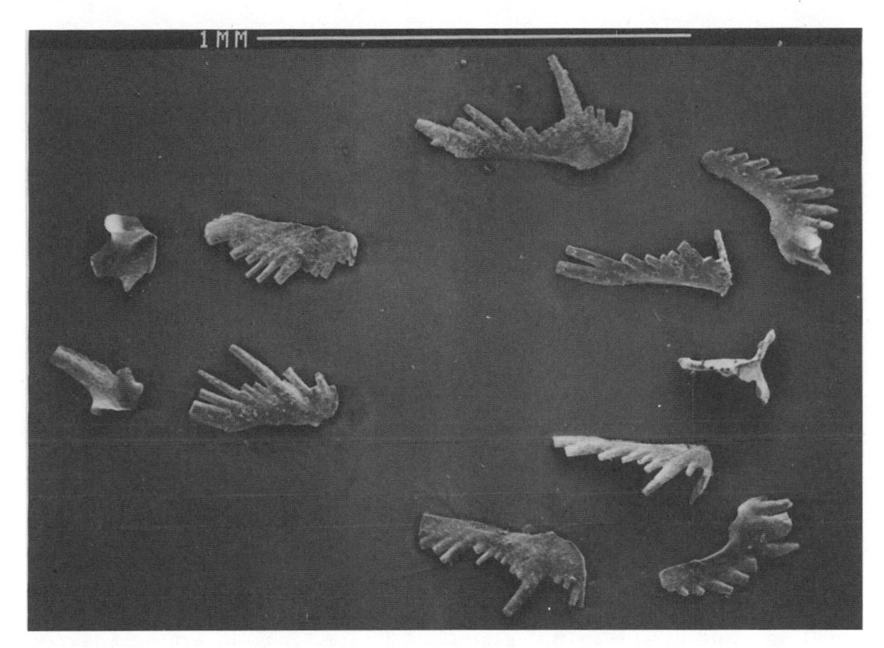

Figure 7.5 **Reconstructed multi-element apparatus of a Late Triassic conodont species showing relative position of six element types.** All elements are paired except for the bilaterally symmetrical specimen at centre right, which occupied the mid-plane of the apparatus. Note scale bar of 1 mm.
Source: Collections of the Geological Survey of Canada

ture range within which oil may be both generated and preserved, corresponds to a conodont CAI of between 1.5 and 3 (about 50 to 200 degrees Celsius).

The CAI of conodont collections often show regional trends. In British Columbia, the 'hottest' conodonts (CAI of 6-8, or 360 to 720 degrees Celsius) occur in the Coast and Omineca mountains, although there are many examples of 'cooked' collections adjacent to local igneous intrusions elsewhere. Relatively 'cool' conodonts (CAI of 2-4, or 60 to 300 degrees Celsius) occur throughout the Rocky Mountain foothills and in pockets of relatively unmetamorphosed strata to the west, for example, in rocks of the Taseko Lakes and Iskut River areas. A complete spectrum of conodont CAI is preserved within Triassic rocks of the Queen Charlotte Islands (Orchard and Forster 1991). CAI is generally low in northern Graham Island, but in southern Moresby Island it is higher and shows progressive increase toward the east coast, where there are large igneous intrusions. A regional trend also exists in the CAI of Triassic conodont collections in northeastern British Columbia, which once lay at the edge of the ancient North American continent. CAI increase toward the west of the region, both in the direction of increasing sediment accumulation and approaching the Rocky Mountain Trench, beyond which lie the accreted terranes.

Conodonts in a Mobile World

Conodonts remain essentially unchanged except in colour, and so they can be identified accurately even in regions of moderate metamorphism and structural disruption. The geology of much of British Columbia is now envisaged as a gigantic collage assembled from pieces of crust that formed far away and were then moved to their present location through the motion of plates ('plate tectonics'). As a result of the large-scale transportation and accretion of crustal fragments, or terranes, fossil conodont faunas now found in British Columbia may have originally lived, and died, in faraway places and may be quite unlike faunas now preserved nearby. Studies of past and present distributions of conodonts (and other fossils) help to identify the origins of these 'exotic terranes' and, therefore, the course of continental drift.

The study of 'conodont geography' is in its infancy compared with those of other fossil groups. Nevertheless, British Columbia provides some startling examples of 'exotic' conodont faunas. A good example comes from the Cache Creek terrane, a crustal segment that extends discontinuously from Cache Creek in the south through Fort St. James to the Atlin area in northern British Columbia. The best-known conodonts are from the southern part of the terrane, where the first conodonts in the Cache Creek terrane were discovered in 1980. This 'discovery fauna,' of Early Triassic age, was collected in Marble Canyon, northwest of Cache Creek near Pavilion Lake. Rocks of this age are rare in the Cordillera and, at that time, were unknown elsewhere in the Canadian western Cordillera. The discovery was all the more intriguing because of the presence of a few anomalous conodont specimens resembling Late Permian species formerly unknown in North America. In fact, the only known collections of these particular species are from China and Iran. Although the specimens from Cache Creek are thought to be reworked from older rock strata, they nevertheless pointed to an 'exotic origin' for those rocks.

Several years passed before Late Permian conodonts were found in Upper Permian rocks of central British Columbia (Figure 7.6). Now they are known from several localities north of Cache Creek (Beyers and Orchard 1991). These conodonts do not prove that the Cache Creek terrane was originally a piece of China. They do, however, provide linkage with distant places, as do other Permian fossil groups at Cache Creek. These Permian conodonts are far more common in the 'Tethys' region, the site of an ancient ocean that encompassed the Mediterranean region and southern Asia.

Figure 7.6 **Exotic Late Permian conodonts from Cache Creek terrane.** From left, specimens of *Iranognathus, Neogondolella,* and *Sweetognathus*. None of these conodonts has been described from elsewhere in North America. See Beyers and Orchard (1991) for details.
Source: Collections of the Geological Survey of Canada

Although conodonts are rarely observed in the field, they may sometimes be spotted in chert – a very fine-grained siliceous rock thought to be an ancient equivalent of sediments forming today through the accumulation of siliceous oozes in the deep oceans. Many of the Cordilleran terranes represent, or include, remnants of former ocean-floor that became uplifted as plates collided. Within these terranes, chert is a common rock type. Because it originally formed very slowly, probably far from land, and at depths where all carbonate material dissolves, chert contains only phosphatic fossils such as conodonts and siliceous fossils such as radiolarians. Conodonts can be abundant in chert. When they are dark, they may be clearly visible within a pale rock. Again, Cache Creek provides examples where chert occupies a belt adjacent to the 'exotic' limestones of Marble Canyon. This belt is considered part of the remnants of a closed ocean that formerly separated the Tethyan rocks of the Cache Creek terrane from those to the east in the Quesnellia terrane. Other good examples of conodonts preserved in chert have been collected on Vancouver Island (Figure 7.7).

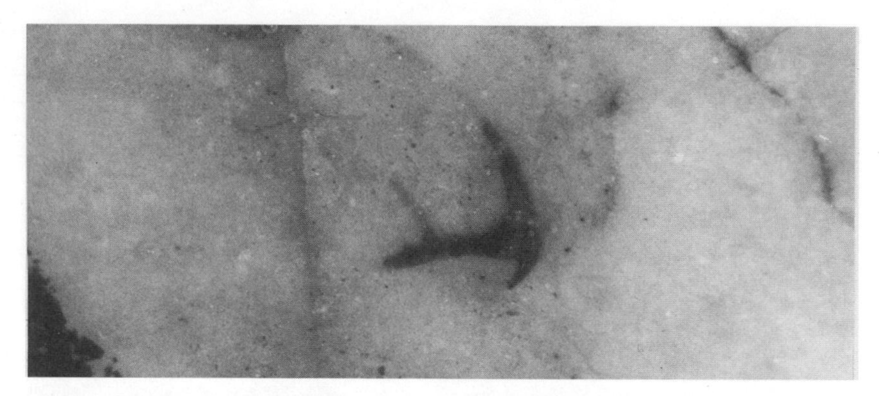

Figure 7.7 **Ramiform conodont element preserved on surface of chert sample from the Lower Carboniferous of Vancouver Island**
Source: Collections of the Geological Survey of Canada

Conodont Twilight: Dating Triassic Rocks

The primary use of conodonts in British Columbia is in determining the relative age of rocks (Orchard 1991c). This is accomplished by fossil zonation, that is, by establishing time scales divided into increments defined by the presence of particular index species. Zonations have been developed by studying well-developed fossil successions in rock sequences wherever they are found. British Columbia has outstanding successions of fossils. Triassic ammonoids and conodonts are examples. For these fossils, the province provides an unrivalled paleontological laboratory for the field of study called biochronology, which involves defining and studying fossil zones.

Paleontologists have long known that British Columbia has significant occurrences of Triassic fossils, particularly in the Peace River area in the Rocky Mountain foothills. Until recently, however, little was known about conodonts of this age, although Late Triassic time was known to encompass the decline and ultimate extinction of this group, a group that had endured for about 300 million years. Then, in 1979, Tim Tozer flew by helicopter over the newly dammed Peace River. Tozer's earlier work on ammonoids had established a standard time scale for the Triassic. What he saw on this trip was spectacular. Although Williston Lake had been enlarged, it was seasonally lowered, and the perimeter was stripped of vegetation and soil cover, exposing new rock outcrop along much of the shoreline. In 1980, Tozer and I began a collaboration that took us to the log-strewn edge of Williston Lake, where Upper Triassic strata full of fossils were laid out before us in continuous outcrop, the like of which was formerly unknown in the region. What we saw was an inspiration for any biostratigrapher. But did the rocks contain conodonts? And, if so, would they be useful for precision dating?

Although not obvious until I was back in the laboratory, virtually all the samples collected contained abundant well-preserved conodonts. These samples fit into the time scale indicated by the abundant ammonoid faunas for virtually the entire Norian, the final stage of the Triassic. For the first time, a complete succession of conodonts was available, mostly belonging to the genus *Epigondolella.* The Norian had a duration of about fifteen million years. Initially it was divided into eight zones on the basis of these conodont faunas; later, fourteen subdivisions were recognized (Figure 7.8), each covering a period of about one million years. Conodont-animals begat index fossils to the very end.

Two Final Problems

Two notable gaps remained in the Upper Triassic conodont zonation, and both were to be filled by further study of Triassic outcrops in British Columbia. First, the older conodont succession of the Carnian, the penultimate stage of the Triassic, was poorly understood. Second, the record of the very youngest Triassic conodonts was uncertain. For both intervals, work in the Peace River area had failed to produce significant conodont collections. In the first case, unfavourable rock types beneath Norian strata precluded the collection of a complete conodont succession. In the second case, the youngest Triassic strata were eroded beneath the earliest Jurassic strata. Enter the Queen Charlotte Islands.

In 1987, a multidisciplinary study of the geology of the Queen Charlotte Islands was undertaken by the GSC and university collaborators. My assigned task was to investigate the Triassic conodonts of the Kunga Group. During two summer seasons all large outcrops of Triassic sedimentary rocks on the Queen Charlotte Islands were visited and systematically sampled. Some of this work on south Moresby Island around Rose Harbour and Houston Stewart Inlet covered outcrops visited by G.M. Dawson during his 1878 expedition. Over a century later, our mother ship, the *Beatrice,* probably anchored close to where Dawson's boat, the *Wanderer,* sheltered. During Dawson's pioneering work, he collected a Triassic ammonoid later called *Acrochordiceras charlottense* from this area. His notes are a little ambiguous about the exact place. Tim Tozer was anxious to relocate it because of uncertainty about its relative age and because only four other specimens of the ammonoid were known in the world. I was interested in collecting conodonts at this locality because the ammonoid was suspected to be Late Carnian in age and, therefore, slightly older than the oldest *Epigondolella* faunas of Norian age that I had recovered from Peace River country. And so the search was on.

After two unsuccessful traverses without any sign of Late Carnian fossils, we found ourselves walking through a succession of Triassic carbonates on the west coast of Kunghit Island. After some time, we found a

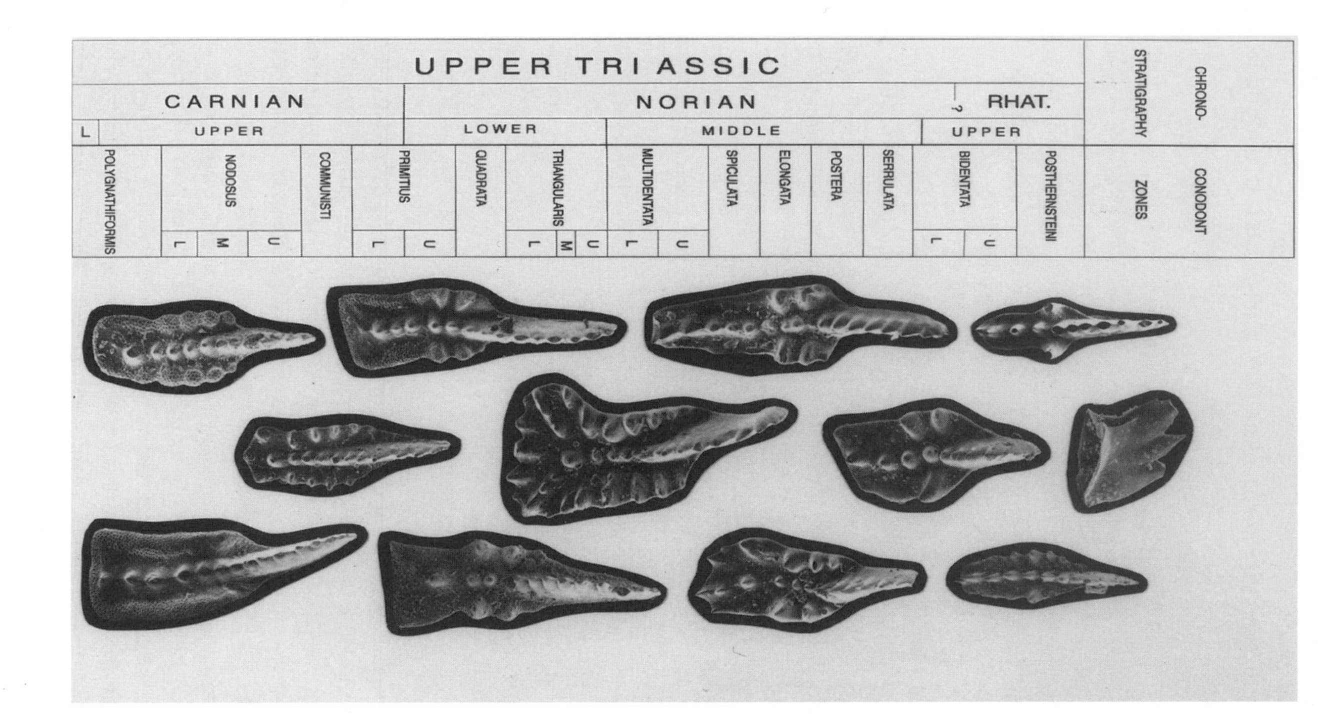

Figure 7.8 **Upper Triassic subdivisions, conodont zones, and representative conodont index fossils from the Peace River and Queen Charlotte Islands.** The scope of the Rhaetian ('Rhat.') is uncertain, and is shown here as equivalent to part of the Upper Norian. The 'last conodont on Earth,' *Misikella,* is shown centre top. Other illustrations are of species of Norian *Epigondolella* and Carnian *Metapolygnathus*. See Orchard (1991b) for details.
Source: Collections of the Geological Survey of Canada

bed containing Late Carnian ammonoids, but there was no sign of Dawson's ammonoid. We headed south, collecting conodont samples and looking for further fossils. It was late in the afternoon when we ran out of good outcrop. Beyond the final small embayment where we were, the rocks were faulted and metamorphosed: there was no more fossil potential. We sat and hammered the highest limestone beds. Eureka! The fifth known specimen of 'Dawson's ammonoid' fell to my hammer blow.

The expedition had been at least partly successful, but several months were to elapse before conodont samples from this section were fully processed. It then became clear that the section was important – for both ammonoid and conodont biochronology. My samples yielded a good succession of Late Carnian conodonts, which could be assigned to the six subdivisions of the Upper Carnian, based on the genus *Metapolygnathus* (Figure 7.8). At last the relationship between Carnian and Norian conodonts was established. The rock stratum from which Dawson and I collected our ammonoids produced conodonts indicative of the second youngest of the six conodont zones of the Upper Carnian.

One more gap remained to be filled. In 1988, we visited the other end of the Queen Charlotte Islands – Kennecott Point on northeast Graham Island. In this area, the Triassic-Jurassic boundary had been loosely constrained by fossil mollusk collections indicative of both latest Triassic and earliest Jurassic age. Although the succession was largely siliceous, we collected carbonate samples from throughout the section. Again, the resulting conodont collections fully complemented the existing Peace River data and filled much of the gap present at the top of the Triassic. One sample from this interval yielded *Misikella,* a conodont that is minute, even by conodont standards, and structurally simple. With this find, essentially the last conodont on earth, the Late Triassic conodont record in British Columbia was largely complete (Orchard 1991b).

Acknowledgments

I thank S. Irwin, R. Ludvigsen, G.S. Nowlan, and G.K. Woodsworth for reading drafts of the manuscript and, through their comments, improving it. P. Krauss is thanked for photography and T. Oliveiric for drafting. This chapter is Contribution No. 25794 of the Geological Survey of Canada.

References

Aldridge, R.J., D.E.G. Briggs, M.P. Smith, E.N.K. Clarkson, and N.D.L. Clark. 1993. The Anatomy of Conodonts. *Philosophical Transactions of the Royal Society of London,* Series B, 340:405-21

Beyers, J.M., and M.J. Orchard. 1991. Upper Permian and Triassic Conodont Faunas from the Type Area of the Cache Creek Complex, South-Central British Columbia. In *Ordovician to Triassic Conodont Paleontology of the Canadian Cordillera,* ed. M.J. Orchard and A.D. McCracken, 269-98. Geological Survey of Canada, Bulletin 417

Epstein, A.G., J.B. Epstein, and L.D. Harris. 1977. *Conodont Color Alteration – An Index to Organic Metamorphism.* United States Geological Survey, Professional Paper 995
Irwin, S.E.B., and M.J. Orchard. 1991. Upper Devonian-Lower Carboniferous Conodont Biostratigraphy of the Earn Group and Overlying Units, Northern Canadian Cordillera. Geological Survey of Canada, Bulletin 417, pp. 185-214
Orchard, M.J. 1991a. Conodonts, Time and Terranes: An Overview of the Biostratigraphic Record in the Western Canadian Cordillera. Geological Survey of Canada, Bulletin 417, pp. 1-26
–. 1991b. Upper Triassic Conodont Biochronology and New Index Species from the Canadian Cordillera. Geological Survey of Canada, Bulletin 417, pp. 299-335
–. 1991c. Dating Otherwise Undatable Rocks. Part 1. Conodonts Put Teeth into Cordilleran Geological Mapping. *Geos* 20(3):28-34
Orchard, M.J., and P.J.L. Forster. 1991. Conodont Colour Alteration as a Guide to Maturation of Triassic Strata, Queen Charlotte Islands, British Columbia. Geological Survey of Canada, Paper 1990-10, pp. 453-64
Purnell, M.A. 1993. Feeding Mechanisms in Conodonts and the Function of the Earliest Vertebrate Hard Tissues. *Geology* 21:375-7

8
Mesozoic Radiolarians of Haida Gwaii

Elizabeth S. Carter

Mesozoic fossils of Haida Gwaii, or the Queen Charlotte Islands, are crucial to biostratigraphy because at some levels ammonoids, clams, conodonts, foraminiferans, and radiolarians occur together in the same bed. As a result, we know that they lived at the same time and can begin to develop integrated time scales based on different fossil groups. The ammonoids and bivalves of the Queen Charlotte Islands were first collected by G.M. Dawson of the Geological Survey of Canada (GSC) in the late 1800s, but the study of radiolarians (affectionately known as 'rads') in this paleontological paradise is comparatively new.

What are radiolarians? How big are they? Where did they live? Where are they found? How are they extracted from rocks? How are they studied? And how can they be used? These are some of the questions that people ask when first coming into contact with these exquisitely beautiful microfossils. The discussion that follows addresses these queries. It then moves to the Queen Charlotte Islands to look at the rocks, collecting techniques, and some Mesozoic radiolarians.

Basic Radiolarian Facts

Radiolarians are microscopic, single-celled, floating organisms that have drifted in the world's oceans since the beginning of Phanerozoic time – about 600 million years ago. They are characterized by a radiating network of slender pseudopodia, or food-gathering arms, and a skeleton composed of transparent glass (opal). Only the skeletons are preserved. Radiolarians are best known to paleontologists for the variety of minute, exquisitely structured skeletons of remarkable delicacy and complexity.

Most radiolarians are solitary. Their size ranges from 50 to 200 micrometres (that is, less than 0.2 millimetres). Paleozoic and Mesozoic forms are usually large and quite robust, whereas Cenozoic and living forms tend to be much smaller, falling in the lower end of the size range.

The radiolarians most commonly preserved as fossils are divided into two main groups: spumellarians and nassellarians. Spumellarians are usually spherical or disc-shaped, and display the greatest variation in skeletal development. They grow outward from the centre, and possess an outer shell and frequently one or more internal shells. The shells are composed of a latticed or spongy meshwork of bars and spines. Nassellarians are typically cone-shaped and have multiple cavities, or chambers. The initial chamber is called the 'cephalis'; additional chambers are the thorax and abdomen. Nassellarians grow in one direction along a single axis by the accretion of chambers. Since Ernest Haeckel's landmark study of radiolarians from the expedition of the HMS *Challenger* (1887), paleontologists studying radiolarians have believed that spumellarians, with their larger radii and extended arms, have greater buoyancy and generally prefer shallower waters, whereas the larger, heavier-shelled nassellarians are more abundant at greater depths.

Radiolarians are exclusively marine. They appear to be well adapted to open waters ranging from polar to equatorial, and to depths from the surface to the abyss. Their maximum abundance is at depths of about 50 to 200 metres. They are particularly prevalent in areas of offshore upwelling that occur off the western coasts of the continents, but they are not commonly found in shallow coastal waters.

Fossil radiolarians are preserved in a variety of rocks that originally were laid down in ocean basins and now are found within mountain belts formed by the interaction of great oceanic and continental plates. In North America, Mesozoic radiolarians are most abundant in Alaska, the Queen Charlotte Islands, central Oregon, the California Coast Ranges and Great Valley sequence, and Baja California. Radiolarians are found in many kinds of sedimentary rocks, but the best are in concretions (hard, calcareous nodules formed in shale). Limestone concretions tend to yield well-preserved fossils because the rocks were cemented early, before the rads were affected by dissolution or mechanical erosion.

Radiolarians can be removed from Cenozoic or Holocene bottom sediments (marls, deep-sea clays, silts, and oozes) by a simple washing procedure. Mesozoic rocks require more vigorous techniques, involving acid digestion to disaggregate rock samples – acetic acid dissolves limestone, and hydrofluoric acid dissolves chert and siliceous siltstone. Acid treatment leaves a residue of grains, mud, and radiolarians. Shale and mudstone can usually be broken down by mixing the crushed sample with water in a high-speed blender or by using various industrial detergents. Following any of these procedures, the residue is washed and dried and then the radiolarians are picked from the residue with a very fine paintbrush and placed on cardboard slides. The diversity and abundance of radiolarians in a single sample generally varies directly with the quality of preservation; in well-preserved samples it is not uncommon to find well upward of 100 different species.

Mesozoic radiolarians are normally viewed with a binocular microscope in reflected light. Smaller Cenozoic or Holocene forms are mounted on glass slides using a translucent medium and are viewed with a transmitted-light optical microscope. This technique is also used by Mesozoic paleontologists to study the inner structure of well-preserved specimens.

In the early years of radiolarian study, specimens were illustrated with drawings, some very simple and others extremely artistic and detailed. The foremost examples of the latter are the intricate drawings in the monograph of Haeckel (1887) (Figure 8.1). Since the advent of the scanning electron microscope (SEM) in the late 1960s, SEM photographs have been the preferred method for illustrating Mesozoic radiolarians. The SEM captures at very high magnifications the details of inner structure and ornamentation that could never be visible to the eye, even with the highest-powered microscope.

Because of their cosmopolitan nature, their great diversity of form, and their rapid evolution, radiolarians are proving to be ideally suited for developing detailed time scales for dating Mesozoic marine strata. These time scales consist of 'zonations' that interpret the successive faunas

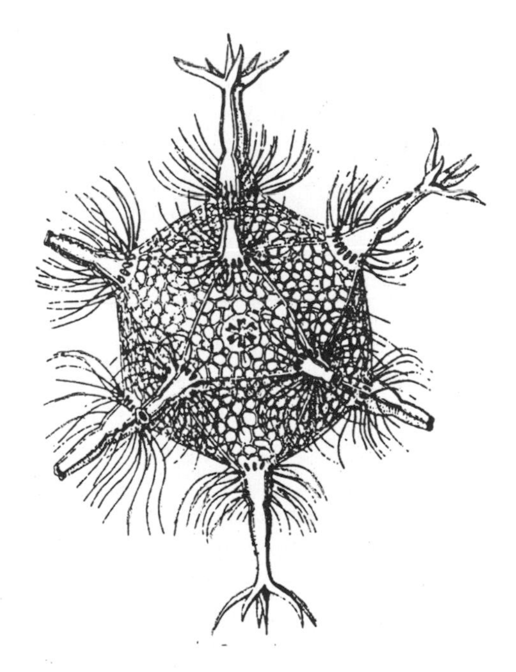

Figure 8.1 **Skeleton of the living radiolarian *Circogonia icosahedra***
Source: From Haeckel (1887)

preserved in rocks. Time scales are based on the law of superposition, which states that, in an undisturbed sequence of beds, the oldest beds occur at the base and the youngest at the top. The succession of faunas in such sequences records the evolution of the fossil group through time. The time scale thus assembled should be repeatable wherever rocks of this age are found. These time scales have wide application in mapping the complex rocks of the Canadian Cordillera and in deciphering the timing of accretion and successive movement of the various terranes in western North America.

Field Work in the Queen Charlotte Islands

In 'Islands at the Edge,' Jacques Cousteau described the Queen Charlotte Islands as the Misty Isles. This epithet is appropriate because, for most of the year, Haida Gwaii is densely shrouded in mist – the mist that makes the cedar, spruce, and hemlock grow tall, creating the wondrous northern rain forest that we treasure; the mist that promotes the growth of the dense, almost impenetrable undergrowth beneath the canopy; and the mist that feeds the moss blanketing the giant cedar beams of the ancient Haida longhouses that now lie prostrate in the silent villages. As truly wondrous as this mist is, it is a considerable impediment to field work. Even during the summer season, one must be prepared for light or heavy mist at all times.

Collecting fossils in the Queen Charlottes is always an adventure. Some roadcuts, quarries, and beach exposures on southern Graham Island, Skidegate Inlet, and northern Moresby Island can be reached by logging road, and shoreline localities in Skidegate Inlet are accessible by boat. The vast remote areas of the Charlottes are much more difficult to get to and can only be reached by float plane, helicopter, or sizeable boat.

The Queen Charlotte Islands are believed to have been part of an ancient island arc located somewhere in the eastern Panthalassic Ocean (the ancient Pacific Ocean). The Late Triassic and Early Jurassic fossil groups studied thus far are similar to faunas that originated in the Tethys, an ancient seaway that extended across Eurasia throughout the Paleozoic and Mesozoic. These features suggest that the Queen Charlotte archipelago originated far south of its present position, but the exact latitude is unknown.

The volcanic rocks of the Karmutsen Formation are the oldest rocks in the islands, and form the backbone upon which all Mesozoic sediments were deposited. The oldest sedimentary rocks (limestones) are Late Triassic in age. These were probably deposited on a warm, shallow platform similar to that found in parts of the Caribbean today, whereas the overlying ones were deposited in much deeper water. The younger Mesozoic rocks consist of sandstone, siltstone, shale or mudstone, and minor limestone, and are intermixed with occasional volcanic rock and sediments derived from them.

All radiolarians collected in the Queen Charlotte Islands are from limestone, as these rocks usually contain the best-preserved fossils. There is no guarantee that a certain limestone will contain radiolarians, but a sample having a fine-grained limestone matrix with added clay often yields good results. Coarse-grained limestones are not usually productive. Wherever possible, samples are collected from continuous sequences of sedimentary beds called 'sections'; at more limited outcrops, spot samples are collected.

Where continuous sequences of rocks are identified, one first walks over the section to see whether the limestones look as if they might contain radiolarians. Where the possibility seems good, samples are broken out of the rocks at intervals of approximately three to five metres (usually about 0.5 kilograms per sample). The samples are cracked into smaller pieces, bagged, and labelled with the locality, section number, and sample number; then the bags are left in place. Next, beginning at the base, the section is measured to the nearest tenth of a metre. Detailed notes are made, recording the position of the sample in the section, the composition of the sample (the size of concretion, colour, distinguishing features, etc.), the rock type and sedimentary structures, and any small faults or disruptions in the sequence (for example, gaps or repetitions) that might account for discrepancies in the faunal succession. This procedure is repeated upward through the section. When the collection is complete, the samples are shipped to Vancouver to be processed at the Cordilleran Division of the GSC.

Mesozoic Radiolarians of the Queen Charlotte Islands

The study of radiolarians in the Queen Charlotte Islands began in the late 1970s when Emile Pessagno and some of his students at the University of Texas in Dallas collected samples of Late Triassic and Early Jurassic age at Kunga Island and in the Skidegate Inlet area. At about the same time, Bruce Cameron of the GSC began collecting Lower and Middle Jurassic limestones on the Yakoun River in central Graham Island and on Maude Island in Skidegate Inlet, and processed them for radiolarians. I began studying these collections in 1982 and completed this work for a Master of Science degree at the University of British Columbia in 1985. Since then I have continued to study radiolarians in association with the GSC, and the scope of study has expanded to include all Mesozoic faunas of the Queen Charlotte Islands. The remainder of this chapter discusses characteristics of radiolarians of the Triassic, Jurassic, and Cretaceous systems, and some of the localities where they are found. Figures are provided to illustrate a few of the beautiful microfossils that make this work so interesting and enjoyable.

Triassic Radiolarians

In the worldwide extinction that occurred at the end of the Permian, nearly all Paleozoic radiolarians disappeared. Very little is known about radiolarians of the earliest Triassic except that some had very simple skeletons made up entirely of spicules and that others had large, very primitive shells with strong, heavy spines. By the Middle Triassic, however, a diverse assortment of structurally complex spumellarian radiolarians inhabited the seas, and the first cone-shaped nassellarians began to appear. The feature that distinguishes Middle and Late Triassic radiolarians from all others is the tendency for the radial spines to become strongly twisted (Figure 8.2C). Both spumellarians and nassellarians continued to diversify right up until the very end of the Triassic. Then, quite suddenly, nearly all were gone.

In the Queen Charlotte Islands abundant and diverse assemblages of beautifully preserved radiolarians are found in Upper Triassic limestones of the Kunga Group. Collections record the almost complete history of

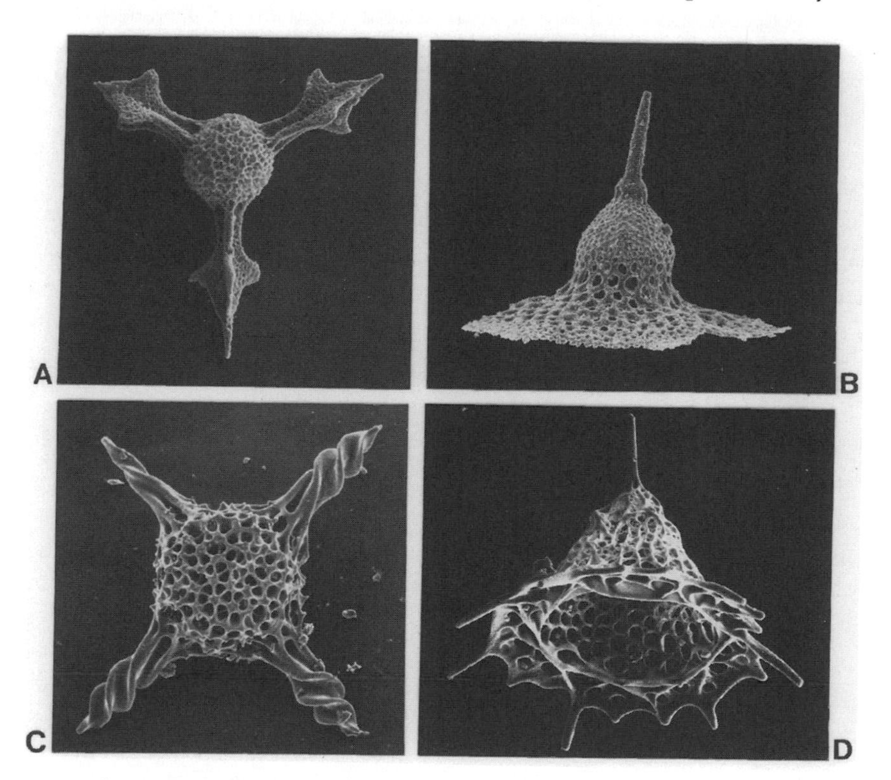

Figure 8.2 **Late Triassic radiolarians.** (A) *Sarla* sp., Frederick Island. (X 100) (B) *Haeckelicyrtium* sp., Burnaby Island. (X 90) (C) *Plafkerium fidicularium,* Kunga Island. (X 140) (D) *Squinabolella* sp., Louise Island. (X 140)
Source: Collections of the Geological Survey of Canada

Late Triassic radiolarians (Carter et al. 1989). Much work remains to be done on the older forms, but an extensive study of the youngest Triassic faunas has recently been completed (Carter 1993).

Examples of Late Triassic species are shown in Figure 8.2. The first is a new species of the spumellarian genus *Sarla*. This species has a double-walled outer shell and three large tri-bladed radial spines that are widely flared into barbs at the tips. The second is a species of the genus *Haeckelicyrtium,* a two-chambered nassellarian with a hat-shaped cephalis with a long apical horn and a rounded thorax with a widely expanded perforate skirt. Forms structurally similar to this species are known from Middle Triassic rocks in the alpine regions of southern Europe. This species is from the oldest assemblage in the Queen Charlotte Islands. The third is *Plafkerium fidicularium,* a spumellarian with a square box-like shell and four strongly twisted spines, one at each corner. And the fourth is a species of the genus *Squinabolella,* a hat-shaped nassellarian having a complex spicular network inside the cephalis and a widely expanded thorax composed of latticed and perforate meshwork edged with long spines.

The youngest Triassic radiolarians come from a long succession of hard, black rocks that line the southern shore of Kunga Island. I have searched this area for fossils several times and, except for radiolarians and rare conodonts in the limestone concretions, the rocks seem to be completely barren of life. But the ancient seas represented at this locality evidently were rife with radiolarians. Their abundant remains, piled layer upon layer, make this possibly the best locality in the world to study the evolution of this microfossil group across the Triassic-Jurassic boundary.

Jurassic Radiolarians

Jurassic radiolarians have been studied since David Rüst first began work in the Jura Mountains of Switzerland in the late 1800s. Following almost a century of little activity, the pace picked up in North America in the late 1970s, and it has been increasing ever since. Late Jurassic radiolarians are best known in the Coast Ranges of California and in Baja California; and Middle Jurassic radiolarians are most prevalent in eastern Oregon and Alaska. The Queen Charlotte Islands is possibly the best place in the world for studying Early to Middle Jurassic radiolarians because the specimens are abundant and excellently preserved. In addition, the rocks are almost continuously exposed, assuring a complete evolutionary succession, and they are replete with ammonoids to date the rads precisely.

Jurassic radiolarians differ considerably from Triassic radiolarians. The Jurassic ones tend to be more conservative in form with much less complex shapes and less ornamentation. Few major groups survived the Triassic extinction but some of these expand to their greatest diversity in the Early

Jurassic. Many stratigraphically useful groups arise in the latest Early Jurassic and dominate the fauna through the remainder of the system.

The rads I have chosen to illustrate are of earliest Middle Jurassic age from the uppermost part of the Maude Group. Besides their aesthetic beauty, they are important stratigraphically because only a few localities with radiolarians of this age are known worldwide. These faunas were found by Giselle Jakobs on the Yakoun River in central Graham Island during low water in the summer of 1989. The Yakoun River is a wonderful fishing river and the Haida have caught trout and salmon from its shimmering waters for years. The river is not easily accessible but, to the field worker with a good machete and compass, the rewards on its banks are well worth the difficult traverse. Long sequences of shale and sandstone containing abundant fossil ammonoids line both shores of the river. The limestone concretions within these sequences have yielded excellently preserved radiolarians (Carter and Jakobs 1991).

The two radiolarians illustrated in Figure 8.3 are *Hagiastrum* cf. *egregium* and *Napora nipponica*. The first is a four-armed spumellarian with a regular pattern of bars and nodes on its long delicate arms. The tips of the arms originally had very small spines around the periphery but only remnants of these are visible. The second is a small, two-chambered nassellarian with a robust tri-bladed apical horn extending past the cephalis and three inwardly curving, tri-bladed feet.

It is not known exactly how these radiolarians floated, but it seems likely that forms such as *Hagiastrum* with long radial arms would have floated in the water column with their arms or spines in the horizontal plane. This position probably would give stability to the organism as it drifted with the currents and would also provide maximum buoyancy for those forms preferring near-surface waters. In modern oceans, small nassellarians shaped like *Napora* are also common in the upper layers.

Cretaceous Radiolarians

Cretaceous radiolarians, like Triassic ones, are very diverse, and many unusually complex forms are known. Two of the most prominent early investigators were Karl Zittel, who worked in northern Germany, and Senofonte Squinabol, who conducted studies in the Venetian Alps in northern Italy. Their work has proven to be quite accurate and many of the species' names proposed then are still valid today. Cretaceous radiolarian studies experienced a lull of several decades until the late 1960s when William Riedel, Helen Foreman, and others began studying cores recovered from the Deep Sea Drilling Project. Combined with Emile Pessagno's work on radiolarians from land-based sections in the Coast Ranges of California, these studies have provided much of the essential paleontological framework in effect today. Several sequences of zones have been advanced that

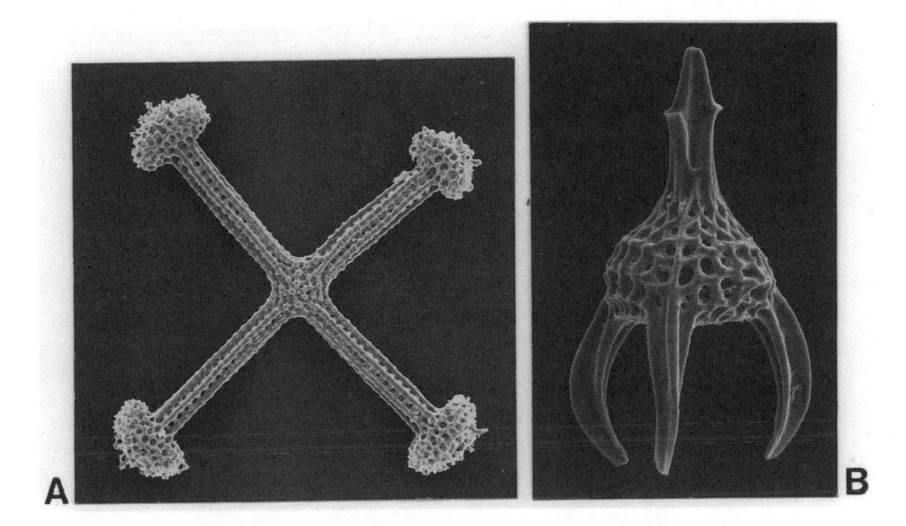

Figure 8.3 **Middle Jurassic radiolarians.** (A) *Hagiastrum* cf. *egregium,* Yakoun River, central Graham Island. (X 70) (B) *Napora nipponica,* Yakoun River, central Graham Island. (X 140)
Source: Collections of the Geological Survey of Canada

are widely applicable for low latitude Tethyan forms. As less is known of radiolarians from higher latitudes, these faunas are an important area for new investigations, especially those involving paleobiogeography.

The study of Cretaceous radiolarians is relatively new in the Queen Charlotte Islands so that the extent of faunal coverage for all stages of the Cretaceous is not yet known. Well-preserved Early Cretaceous faunas have been collected from limestone concretions in mudstone sequences of the Vancouver Group at localities on Moresby Island. As well, a few Late Cretaceous radiolarians are known from Skidegate Inlet and northwest Graham Island.

The Early Cretaceous radiolarians illustrated in Figure 8.4 are from Hotspring Island, a small island located off the east coast of Moresby. The island is well known to sailors and kayakers, and the hotspring pools are operated year-round by Haida watchmen. To soak in these waters after a long day of field work is wonderfully relaxing. The view west over Juan Perez Sound toward the San Christoval Mountains is truly magnificent at sunset.

The two radiolarians shown in Figure 8.4 are *Crucella* cf. *messina* and *Xitus plenus*. The first, a spumellarian, has a cruciform (four-armed) shape and long spines that extend from the tips of its arms. The shell is made up of parallel layers of spongy meshwork, and the external layer is covered with many small, raised nodes. The second, a nassellarian, has an elongate conical shell. The cephalis is smooth, and a rather sharply pointed

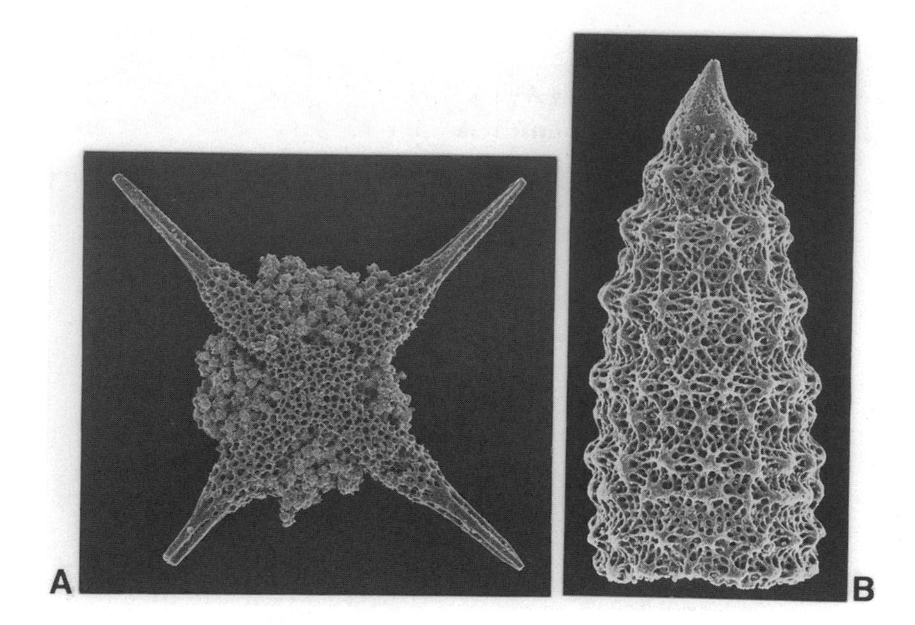

Figure 8.4 **Early Cretaceous radiolarians.** (A) *Crucella* cf. *messina,* Hotspring Island. (X 100) (B) *Xitus plenus,* Hotspring Island. (X 140)
Source: Collections of the Geological Survey of Canada

apical horn extends from its tip. The remaining shell displays raised ridges lined with massive knobs, which are connected by bars forming criss-cross ornamentation on the exterior of the shell. This feature is well developed on this species and is a distinguishing characteristic of the genus *Xitus.*

This chapter has been designed to give a taste of radiolarians – what they are, what they look like, and how we use them. In the Queen Charlotte Islands, the land of the Haidas, radiolarians occur through long, continuous sequences of rock and are exceptionally well dated by other co-occurring fossils. The rocks on these islands provide a unique opportunity for creating detailed zonations that intermingle with those of other fossil groups – a field guide, if you will – that can be used in areas of complex geology around the world to date the rocks and interpret their ancient history.

References

Carter, E.S. 1993. *Biochronology and Paleontology of Uppermost Triassic (Rhaetian) Radiolarians, Queen Charlotte Islands, British Columbia, Canada.* Mémoires de Géologie (Lausanne), No. 11

Carter, E.S., and G.K. Jakobs. 1991. New Aalenian Radiolarians from Queen Charlotte Islands, British Columbia: Implications for Biostratigraphic Correlation. In *Current Research,* Part A, Geological Survey of Canada, Paper 91-1A, pp. 337-51

Carter, E.S., M.J. Orchard, and E.T. Tozer. 1989. Integrated Ammonoid-Conodont-Radiolarian Biostratigraphy, Late Triassic Kunga Group, Queen Charlotte Islands, British Columbia. In *Current Research,* Part H, Geological Survey of Canada, Paper 89-1H, pp. 23-30

Haeckel, E. 1887. Report on the Radiolaria Collected by H.M.S. *Challenger* during the Years 1873-1878. In *The Voyage of H.M.S. Challenger,* Vol. 18, ed. C.W. Thompson and J. Murray, 1-1,803. N.p.

Haggart, J.W., and E.S. Carter. 1993. Cretaceous (Barremian-Aptian) Radiolarians from Queen Charlotte Islands, British Columbia: Newly Recognized Faunas and Stratigraphic Implications. In *Current Research,* Part E, Geological Survey of Canada, Paper 91-1E, pp. 55-65

9
Fishes of the Triassic: Trawling off Pangaea

Andrew G. Neuman

My first visit to the fossil-bearing rocks at Wapiti Lake was as a graduate student in 1984. Our group from the University of Alberta had planned a long time for this trip, but for several days it looked as if bad weather was going to prevent us from visiting the site. Finally the weather broke for a few hours and, after a hair-raising helicopter flight and some interesting hours setting up camp in the howling wind, I stopped to gaze out over the rocks that I knew represented a significant part of my future.

Fossil fish remains from the Early Triassic occur at a number of localities in British Columbia, extending from the Fernie area in the south almost to the Yukon border. Sites generally occur on low ridges and in valleys between the main ranges of the Rocky Mountains. The best-known sites are located on Ganoid Ridge near Fossil Fish Lake in the vicinity of Wapiti Lake, about 150 kilometres northeast of Prince George. The area is south of Dawson Creek and approximately sixty kilometres south of Tumbler Ridge. Ganoid Ridge is one of the northwest-southeast oriented ridges comprising the Canadian Rocky Mountains.

The mountains around Wapiti Lake hold the remains of animals that inhabited the seas approximately 210 to 250 million years ago – before the time when dinosaurs first appeared on the land. Fishes, sea-going reptiles, and small marine invertebrates are preserved in the black shales and siltstones.

Our 1984 trip to Wapiti turned out to be short and unrewarding, but the following year was much more productive. The weather was excellent, and we spent many sunny days climbing the ridges and bringing down packloads of good fossils. Ganoid Ridge comprises a number of cirques, each with a small lake in its centre (Figure 9.1). Standing on the top of Ganoid Ridge, we could look across Fossil Fish Lake, and on to Wapiti Lake to the northeast. To the west, the icefields atop Mount Bulley are visible at the east end of Monkman Provincial Park. To the south and southeast, Mount Becker and the front ranges of the Rocky Mountains stand tall.

Figure 9.1 **Looking south along Ganoid Ridge across a cirque exposing Lower Triassic rocks**

Collectors of the Fossil Fishes

Fossil fishes have been known from the Wapiti Lake sites since they were first discovered in 1947 by Lowell R. Laudon and a group of students from the University of Kansas. They reported finding 'excellently preserved specimens of ganoid fishes' on the steep slopes of the ridge above Fossil Fish Lake, which they named Ganoid Ridge. Ganoid fishes had thick, heavy scales that were usually dark and shiny and that had a type of enamel called ganoine. Because of the resistant nature of these scales, the fossil fishes are often well preserved.

Although Laudon's group was actually looking for potentially promising areas for petroleum exploration, it discovered many exceptional fossil localities. The field crew was composed of his wife Florence, their three sons – Thomas, aged fifteen, Richard, thirteen, and Robert, four – and nine ex-servicemen studying to be geologists through the retraining programs that followed the Second World War (Figure 9.2). Several of the students and at least two of the sons later became geologists in their own right. In preparation for the expedition, the Laudons mortgaged their home in Wisconsin and bought an amphibious Seabee airplane. A seaplane base was set up near Grande Prairie, Alberta, and the party and supplies were flown to the Wapiti Lake base camp. It was while the party was on a nine-day pack trip that the original fossil fish locality was discovered on Ganoid Ridge.

Following Laudon's lead, Charles M. Sternberg of the National Museums of Canada visited the Wapiti Lake site in 1948 and collected a

Figure 9.2 **Laudon's field crew, 1947.** Lowell Laudon is in the front row, far right.
Source: Photograph courtesy of the Laudon family

number of specimens. Several dozen additional specimens were collected during the 1950s and 1960s by representatives of oil companies doing exploration in the area and by expeditions from the Geological Survey of Canada, the University of Calgary, and the University of Alberta.

The first major expeditions to the Wapiti Lake sites were in 1961 and 1962. They were led by Harvey Champagne of the National Museums of Canada and Gilbert Stucker of the American Museum of Natural History. More than a hundred fish fossils were collected, as well as many other specimens. The expeditions travelled by truck to nearby Stony Lake, then by pack train for fifty kilometres. The campsite was at Fossil Fish Lake, near the site where the first specimens were discovered. This campsite was also used by many of the later expeditions.

The first major report on this material was published in 1976 by Bobb Schaeffer and Marilyn Mangus of the American Museum of Natural History in New York. This publication heightened interest in the Triassic fossil fish sites at Wapiti Lake, and in 1983 a joint expedition from the University of Alberta and the newly developed Tyrrell Museum of Palaeontology visited the area to collect display and teaching specimens. This expedition was led by Mark Wilson and Don Brinkman. Some 800 specimens were collected that year (Wilson 1984).

My own involvement with fossil fishes from the Wapiti Lake area began in 1983 while working as a curatorial assistant to Mark Wilson at the University of Alberta. This material became the subject of my graduate work in 1984. My colleagues and I collected in the area during seven field seasons between 1984 and 1995 (Figure 9.3).

Figure 9.3 **Field party from University of Alberta loading pack horses for the trip to Ganoid Ridge, 1985**

The Rocks at Wapiti Lake

The Lower Triassic section at Wapiti Lake consists of the Sulphur Mountain Formation, a sequence of siltstones, silty limestones, shales, and very fine-grained sandstones that weather to a characteristic dark grey to orange-brown colour. Triassic rocks along the front ranges and foothills of western Canada form a westward-thickening wedge of marine sediments, deposited at the margin of the supercontinent Pangaea. These sediments were laid down in relatively deep-water, open-shelf marine environments, similar in some respects to that now existing off the Atlantic coast of the United States (Neuman 1992).

Triassic Fishes

The fishes of the Triassic form a link between the ancient fishes of the Paleozoic and the modern teleost fishes that began their evolutionary expansion later in the Mesozoic. Teleosts are the most common fishes today. They are ray-finned fishes with light scales, tails with equal upper and lower lobes, and an advanced jaw structure. For much of the Paleozoic, fishes had been represented by

- agnathans (armoured jawless fishes)
- placoderms (armoured jawed fishes)
- acanthodians (spiny bony fishes)

- sarcopterygians (lobe-finned fishes)
- actinopterygians (ray-finned fishes)
- sharks.

Of these, the armoured agnathans, placoderms, and acanthodians did not survive the Paleozoic. The sarcopterygians, which include lung-fishes and coelacanths with fleshy, lobed fins, declined dramatically in the Late Paleozoic but expanded somewhat in the Triassic. Only four genera are alive today – three lung-fishes and one coelacanth. At Wapiti Lake, sarcopterygians are represented by the coelacanth *Whiteia.*

One group of primitive actinopterygians, the palaeonisciformes, was not drastically affected by the terminal Permian extinction. These fishes had large eyes located far forward, thick and shiny scales, and a strongly upturned tail. At Wapiti Lake, the palaeonisciformes are represented by *Pteronisculus, Boreosomus, Birgeria,* and *Bobasatrania.* The only living members of this group are the sturgeon and paddlefish.

In a burst of diversification in the Early Triassic, the palaeonisciformes gave rise to a large number of descendant groups. Some of these groups are represented at Wapiti Lake by *Australosomus, Saurichthys, Albertonia,* and *Watsonulus.* The bowfin *Amia* is the only living member of the group to which the latter two belong. This group is believed to be on the direct line to the teleosts, a group of actinopterygian fishes that began their expansion in the Late Jurassic. With 20,000 species, they are by far the most common fishes in today's oceans, lakes, and rivers.

Lobe-finned Fishes

Latimeria and *Whiteia*

Coelacanths are ancient fishes that differ from the better-known ray-finned fishes in a number of ways, including having fleshy, lobed fins with their own muscles and skeletal system. This group was thought to have been extinct since the end of the Cretaceous (more than sixty-five million years ago). But a few days before Christmas in 1938, a peculiar fish was caught in the nets of a fishing boat in the Indian Ocean, southwest of East London near the mouth of the Chalumna River in South Africa. The fish was brought to the attention of the curator of the East London Natural History Museum, Marjorie Courtenay-Latimer, who realized that it was unlike anything she had seen before. The specimen was very large (150 centimetres long and over fifty kilograms) and steely blue. It had few teeth and was covered with large, hard, bony scales with spines – unlike the scales of typical bony fishes. The tail, also unusual, was symmetrical with an upper and lower lobe, and a third, middle lobe separate from the other two lobes. The fish carried two distinct dorsal fins. Most peculiar, however, was the structure of the paired pectoral and pelvic fins, which were stout, stubby, and partially covered with scales, looking more like legs than typical fins.

Courtenay-Latimer sent a letter and sketch to James L.B. Smith, the only active ichthyologist in South Africa at that time. Both he and Courtenay-Latimer noted similarities between this creature and a lungfish, although they differed in other respects. The more Smith studied the sketch, the more he thought it might be a coelacanth. Because the fish was decaying rapidly, it was decided to make a trophy mount and dispose of the decaying internal organs. Therefore, by the time Smith saw the specimen, seven weeks after it had been discovered, much of the internal anatomy was lost. As well, there were no photographs because the film was spoiled. Fortunately, the taxidermist made detailed notes on his observations as he was preparing the mount.

Smith subsequently wrote a monograph entitled 'A Living Coelacanthid Fish from South Africa,' which was published in February 1940 in the *Transactions of the Royal Society of South Africa.* He named the coelacanth *Latimeria chalumnae* after Courtenay-Latimer and the Chalumna River, where it was caught. Providing a detailed description of the specimen, the monograph made an important contribution to ichthyology.

Although Smith and others at once began an intensive search for more specimens, it was not until December 1952 that a second specimen came to light. This fish, caught off Anjouan Island in the Comoros Archipelago, was also brought to Dr. Smith's attention. The specimen was comparatively well preserved because it had been salted thoroughly and then injected with generous amounts of formalin.

The following year, the French began an intense effort to catch more coelacanths, declaring that only they could continue the search because the Comoros Archipelago was a French territory at that time. This effort resulted in the capture of five more specimens in 1953 and 1954. The next fish to be caught survived for seventeen hours, and thus allowed observation of a living specimen. The French group continued collecting specimens and eventually distributed them to other investigators around the world. At least one specimen has been caught every year since 1953.

Although a great deal of information about the anatomy of *Latimeria* was gained from these specimens, the real breakthrough occurred in 1987 when Dr. Hans Fricke of the Max Planck Institute in Germany managed to get film footage (Fricke 1988). Using a two-person submersible, Fricke became the first human to see a coelacanth swimming freely in its native environment. This research has contributed greatly to our understanding of the behaviour and social interactions of this fish.

When *Latimeria* moves through the water, the forward pectoral fins and rear pelvic fins are synchronized. The front right fin works in tandem with the left rear and vice versa (the same gait as in a trotting horse). In addition, the coelacanth can rotate its flexible forward fins nearly 180 degrees, enabling the fish to 'scull' as it swims. Though some individuals have been seen resting with their fins braced against the sea bottom, none have been seen to walk on the bottom.

The weakly braced snout, toothless jaws, fleshy lips, and restricted gape seem to indicate that this fish feeds primarily by suction. This feeding apparatus fits the concept of a slow-swimming, lurking fish, able to manoeuvre efficiently until prey is within range of its suction force. Coelacanths have large eyes surrounded by numerous well-ossified, bony plates, indicating that these fish were primarily visual feeders. Although they have not been observed while feeding, remains of squid and fish have been found in the gut of *Latimeria.* The biology and paleobiology of *Latimeria* have been detailed by Thomson (1991).

We now know that *Latimeria* can grow up to 180 centimetres long and weigh as much as ninety-five kilograms. It normally lives at a depth of 100 to 500 metres. About 180 specimens have been caught off the coast of Grand Comoro and Anjouan Islands in the Federal Islamic Republic of the Comoros, in the western Indian Ocean between Mozambique and Madagascar. Sightings are quite common in the Comoros. Individuals often congregate in deep caves in groups that include both adults and young. A tagging program has been initiated to determine the habitat frequented by these fish. A population of 300 to 500 individuals is thought to live off the coast of Grande Comoro. *Latimeria chalumnae* is now partially protected by the Convention on International Trade in Endangered Species.

Coelacanth fossils are found around the world in Devonian through Triassic strata. They are moderately abundant and diverse in both marine and freshwater environments. During the Jurassic and Cretaceous, the diversity of coelacanths declined, their geographic range narrowed, and they almost disappeared from the non-marine realm. The world's youngest fossil coelacanth is from the latest Cretaceous, about seventy million years old. It was from nearshore marine deposits in the New Jersey region of the United States. These coelacanths, together with *Macropoma* from the Cretaceous of Europe, appear to be the closest relatives of *Latimeria.*

Whiteia, the coelacanth found at Wapiti Lake (Figure 9.4), had a characteristic body shape and fin placement (its locomotor system) and a characteristic head and mouth shape (its feeding system). Both systems are essentially similar to those of the coelacanth *Latimeria* described above. In general, the locomotor system seems to be well adapted for stalking or lunging at prey, rather than for active pursuit. The fins appear to be

Figure 9.4 **The coelacanth *Whiteia*.** This specimen is 50 cm long.
Source: Collections of the Royal Tyrrell Museum of Palaeontology

superbly structured and distributed around the centre of buoyancy, making them effective for braking, turning, and positional control, but these numerous projecting lobed fins would produce considerable hydrodynamic drag in swimming.

Sarcopterygian fishes, living as well as fossil, are of extraordinary interest to evolutionary biologists because this group included the ancestors of all tetrapods – amphibians, reptiles, birds, and mammals. Coelacanths and lung-fishes are the closest living relatives to the first amphibian that emerged from the water to crawl on to land in the Late Devonian.

Ray-finned Fishes

Albertonia

Albertonia is one of the largest fishes at the Wapiti Lake sites. Its elongated pectoral (front) fins, deep body, and well-developed tail fin suggest that it was a slow, strong swimmer (Figure 9.5). Its small teeth suggest that it fed on small prey or plant material.

When first discovered, *Albertonia* was described as a flying fish because of its greatly elongated front fins. However, its body shape, the placement of its pectoral fins, and the shape of its tail show that it was not capable of gliding like modern flying fishes. Modern flying fishes leave the water and remain airborne for as long as twenty seconds. They can glide for distances of 150 metres or more. Their pectorals are situated high up on

Figure 9.5 **The ray-finned fish *Albertonia*.** This specimen is 35 cm long.
Source: Collections of the Laboratory for Vertebrate Paleontology, University of Alberta

the flanks, the body is torpedo-shaped, and the lower lobe of the tail fin is larger than the upper lobe. It is more likely that the pectorals in *Albertonia* were used for normal swimming. The body form is similar to fishes found in moderate depths where the pectorals are used in manoeuvring and for short bursts of speed.

In addition, the enlarged pectoral fins in *Albertonia* may have had a tactile function. On some fishes, these fins bear taste buds and touch receptors. Several groups of recent fishes use elongated pectoral fins as tactile organs when feeding, extending and fanning them to stir up edible morsels.

Another possibility is that *Albertonia* was sexually dimorphic (that is, the sexes had different forms) and that these enlarged fins served as sexual displays. Recent groups of fishes that display sexual dimorphism in the paired fins include the dragonets, topminnows, snappers, and suckers.

Bobasatrania

Bobasatrania was a laterally compressed, diamond-shaped fish. The largest specimens are nearly a metre long. It had a deeply forked tail fin, long fins on the dorsal and ventral surfaces, and long fan-shaped pectoral fins placed high on the flanks (Figure 9.6). The presence of crushing teeth suggests that these fishes fed on hard-bodied prey such as crayfish or

Figure 9.6 **The ray-finned fish *Bobasatrania*.** This small specimen is 30 cm long. *Source:* Collections of the Laboratory for Vertebrate Paleontology, University of Alberta

shrimp. Although *Bobasatrania* is represented by some of the largest specimens from Wapiti Lake, small specimens of this animal are also known, although they may not represent the same species.

Saurichthys

Several predatory fishes are present in the assemblage from Wapiti Lake. *Saurichthys* has a long, shallow body with the fins placed far back, and elongated jaws with well-developed, pointed teeth. These features indicate that this fish was an active predator. One of the individuals collected had actually died with a small fish in its mouth (Figure 9.7).

Saurichthys was similar to the living long-nosed gar in body and skull shape. Like the gar, *Saurichthys* was probably an ambush predator that struck quickly, used its jaws to grasp and hold its prey, and then swallowed it whole. The specimens from Wapiti Lake are up to a metre long and must have been fearsome predators in the Early Triassic seas.

Australosomus

Australosomus has a slender, elongate body, posteriorly positioned dorsal and anal fins, and a well-developed tail fin (Figure 9.8). The form of its body suggests a potential for rapid bursts of speed. The fish is protected by rows of elongated, interlocking body scales. The almost vertical orientation of the enlarged lateral body scales appears to allow flexibility without sacrificing protection. These scales are reminiscent of the

Figure 9.7 **Head of the ray-finned fish *Saurichthys* fossilized with a small fish in its mouth.** The slab is 30 cm long.
Source: Collections of the Laboratory for Vertebrate Paleontology, University of Alberta

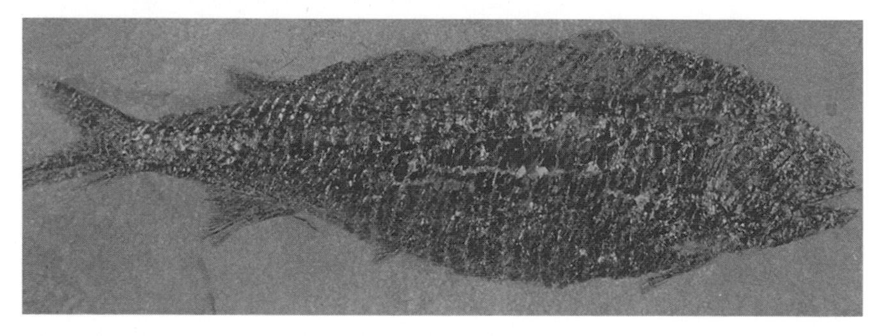

Figure 9.8 **The ray-finned fish *Australosomus*.** The specimen is 15 cm long.
Source: Collections of the Laboratory for Vertebrate Paleontology, University of Alberta

interlocking plates of armour on the suit of a medieval knight. The jaws are long but the teeth are small, indicating that the prey would be relatively small.

Many interesting fish fossils have now been discovered from the sites at Wapiti Lake. New localities give a hint of even more important finds. These specimens have greatly expanded our knowledge of fishes in the seas around Pangaea during the time when dinosaurs and other more familiar animals were just beginning their stay on land. Although many of the fossils from Wapiti have no close living relatives, some, like coelacanths, still survive today. Continued searches in the Lower Triassic rocks of British Columbia should bring to light new and exciting fish specimens that will help to unravel the complex history of vertebrate life on our planet.

References

Fricke, H. 1988. Coelacanths, The Fish that Time Forgot. *National Geographic* 173:824-38

Neuman, A.G. 1992. Lower and Middle Triassic Sulphur Mountain Formation, Wapiti Lake, British Columbia: Summary of Geology and Fauna. *Contributions to Natural Science, Royal British Columbia Museum* No. 16, pp. 1-12

Schaeffer, B., and M. Mangus. 1976. An Early Triassic Fish Assemblage from British Columbia. *Bulletin of the American Museum of Natural History* 156:519-63

Thomson, K.S. 1991. *Living Fossil: The Story of the Coelacanth.* New York: W.W. Norton and Company

Wilson, J.B. 1984. Foraging for Fine Fossils at Fossil Fish Lake. *Canadian Geographic* 104:53-9

10
Ammonoids and Bivalves: Triassic Life at Sea

E. Tim Tozer

During Triassic times, one side of the globe was almost completely covered by the supercontinent Pangaea and the other side by the ancestral Pacific Ocean, Panthalassa. Later in the Mesozoic and Cenozoic, Pangaea broke into fragments along large-scale fractures. The continental fragments gradually drifted apart to form the Americas and the other continents we know today.

Rocks of British Columbia contain a great variety of Triassic fossils. This chapter discusses several of the ammonoids and bivalves that contribute a fascinating fossil record of the form and variety of these extinct mollusks. Their succession in bedded rocks provides a chronology for unravelling and correlating geological events. Also, these fossils indicate the nature of past environments and provide information about continental movement.

A Bit of History

The study of Triassic fossils of British Columbia goes back about 120 years. The explorers who collected the first fossils were three famous Canadian geologists – A.R.C. Selwyn, G.M. Dawson, and R.G. McConnell (Figure 10.1). In 1875 Selwyn traversed the Cordillera from Vancouver to Fort St. John by boat and on horseback. On the Peace River he collected specimens of the bivalve *Monotis subcircularis,* the first Triassic fossils found in British Columbia. Dawson joined the GSC in 1875 and for the next twenty years spent nearly every summer investigating the geology of the Cordillera, despite a considerable physical disability. An important expedition was to the Queen Charlotte Islands in 1878, where he collected Upper Triassic ammonoids. Between 1881 and 1914 McConnell spent every summer doing geological field work in western Canada, mostly in the Cordillera. His most notable expedition was in 1887 and 1888, when he started from Wrangell, Alaska, and traversed the whole of the

Cordillera. He wintered on Great Slave Lake and returned to the Pacific Coast via Yukon River and the Chilkoot Pass. His traverse included a descent of the swift, treacherous Liard River, where he discovered Triassic mollusks. Ottawa paleontologist J.F. Whiteaves identified their finds. Whiteaves described not only the first Triassic fossils from British Columbia but also most of the other macrofossils, both Paleozoic and Mesozoic, collected by his GSC colleagues. Because Whiteaves's report (1889) was widely read, the paleontologic world became aware of the widespread occurrence of fossil-bearing Triassic rocks in British Columbia. The occurrence of Triassic fossils along the Peace and Liard rivers was grounds for suspecting that an extensive belt of Triassic rocks was present in northeastern British Columbia.

In this century many new localities for Triassic fossils have been discovered in British Columbia by government and oil company geologists. Frank McLearn of the Geological Survey of Canada (GSC) was the scientist most concerned with Triassic rocks and fossils during the early part of this century. In 1917 he began a study of the Peace River Triassic. This work, which he continued intermittently until 1938, led to important discoveries of Late Triassic faunas.

Crucial to Triassic studies was the opening of the 2,400-kilometre Alaska Highway, an astonishing feat of construction started and completed by the US Army in 1942. The road made accessible Middle Triassic beds with high concentrations of fossils, and these were studied by McLearn in 1944 (McLearn 1960, 1969). That same year Ed Kindle made a canoe trip down the Fort Nelson River and then up the Liard, discovering the first Early Triassic ammonoids to be found in British Columbia. Now the Triassic record for British Columbia included faunas of all three Triassic series: Lower, Middle, and Upper.

In 1953 McLearn published a summary paper listing the known Triassic fossil occurrences and outlining the geographic extent and time frame of the Triassic in British Columbia. I joined the GSC in 1952, and was McLearn's apprentice until he died in 1964. Work since that time has made the record of Triassic ammonoid faunas in British Columbia one of the most complete records known anywhere in the world (Tozer 1967, 1984, 1994).

Ammonoids

The mollusk phylum (Mollusca) includes cephalopods, bivalves, and gastropods. The most familiar living cephalopods are cuttlefish, squid, octopus, and the pearly nautilus; bivalves are clams; and gastropods are snails and slugs. All three kinds of mollusks are found in Triassic rocks of British Columbia, cephalopods and bivalves being the most common.

Ammonoids are a fascinating extinct group of cephalopods. The

Figure 10.1 **Pioneer scientists who collected and identified Triassic fossils from British Columbia.** (A) Alfred R.C. Selwyn (1824-1902), second Director of the GSC. (B) George M. Dawson (1849-1901), third Director of the GSC. (C) Richard G. McConnell (1857-1942), Deputy Minister of Mines from 1914 to 1921. (D) Joseph F. Whiteaves (1835-1909), paleontologist for the GSC from 1876 until his death in 1909.
Source: From the Archives of the Geological Survey of Canada

ammonoid animal was probably somewhat like the living nautilus, having a large head and tentacles and living in a coiled chambered shell. Ammonoids first appeared in the Devonian, and became extinct at the end of the Cretaceous. They nearly became extinct several times during their 300-million-year history. On each occasion, however, one or two simple types survived, from which a great variety evolved. Large numbers became extinct at the end of the Permian and nearly all were wiped out at the end of the Triassic. Only a single group survived into the Jurassic, and this group evolved into the enormous variety of Jurassic and Cretaceous ammonoids.

Fossils of ammonoids are variable forms of cone-shaped shells. In most ammonoids the cone is coiled in a plane spiral, for example, as in *Anawasatchites* and *Kashmirites* (Figure 10.2). The inside of the cone is divided into chambers by partitions called septa. The animal lived in the last chamber – the body chamber. Several of the forms illustrated in this chapter show the last septum and the body chamber, for example, as in *Discogymnites* (Figure 10.3). The chambered part of the shell, known as the phragmocone, served as a positioning device to keep the animal at neutral buoyancy – most of its chambers would have been filled with gas and a small amount of liquid.

The juncture of the septum and the shell is called the suture. If the shell is peeled away, the nature of the suture becomes evident, as shown in Figure 10.2 in *Anawasatchites*. Such specimens are internal moulds

Figure 10.2 **Lower Triassic ammonoids.** (A) *Anawasatchites tardus*. The last suture line has been highlighted with white paint, showing it to be ceratitic. (B) and (C) *Kashmirites warreni*. This ammonoid is an evolute form with simple ribs, a type very common in the Lower Triassic. Both specimens are from the Toad River, collected by E.D. Kindle in 1944. All are actual size.
Source: Collections of the Geological Survey of Canada

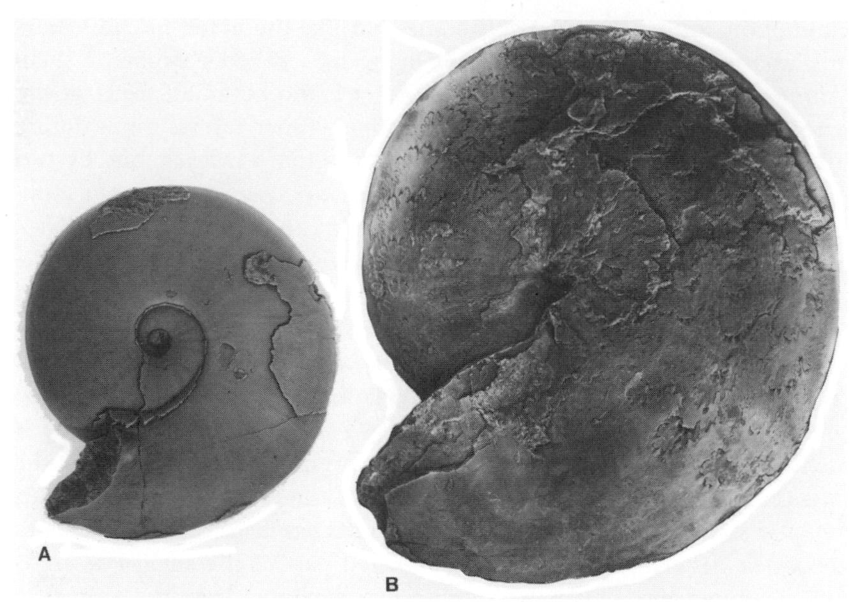

Figure 10.3 **Middle Triassic ammonoids.** (A) *Amphipopanoceras tetsa* from the Chischa River. (X 0.5) (B) *Discogymnites hollandi,* a large, smooth disc-like ammonoid from the Liard River. (X 0.3)
Source: Collections of the Geological Survey of Canada

known as 'steinkerns.' Because most ammonoids have thin shells, surface features such as ribs are often preserved on steinkerns.

The pattern of the suture line is generally distinct for different ammonoid genera. The basic pattern is one of folds. Folds that are convex forward toward the aperture are called saddles. Alternating with saddles are lobes that fold backward. In most Triassic ammonoids the saddles are smooth and rounded and the lobes are toothed. This condition is described as 'ceratitic,' again as shown in *Anawasatchites* (Figure 10.2). In other forms both the saddles and lobes are wrinkled, producing what is known as an ammonitic suture line, as in *Discogymnites* (Figure 10.3).

The shape and the surface ornamentation of Triassic ammonoid shells varies enormously. Some ammonoids, like *Kashmirites,* form an open coil. In others, the later-formed coils cover the earlier ones, either partly, as in *Daxatina,* or completely, as in *Nathorstites* (Figure 10.4). In *Rhabdoceras,* nearly the entire cone is straight, rather than coiled. Other ammonoids, like *Mesohimavatites,* bear nodes, spines, and tubercles. *Malayites* has ribs and a spiral ornament. The outer edge also varies, being smooth or furrowed, or keeled as in *Tropites* (Figure 10.5). In *Paracochloceras,* the shell surface may be smooth or may have simple or branched ribs (Figure 10.6). This ammonoid is coiled in a helix, like a snail.

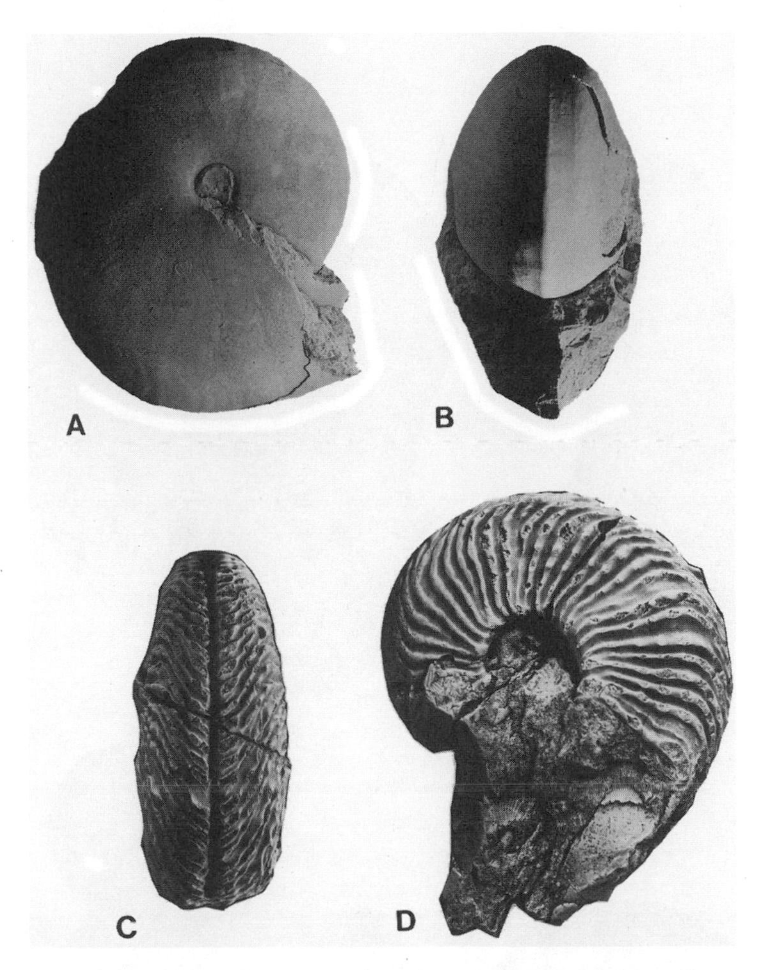

Figure 10.4 **Middle Triassic ammonoids collected by R.G. McConnell on Liard River in 1887 and described by J.F. Whiteaves in 1889.** (A) and (B) *Nathorstites macconnelli.* (C) and (D) *Daxatina canadensis.* All are actual size. *Source:* Collections of the Geological Survey of Canada

Bivalves

Several kinds of bivalves occur in the Triassic rocks of British Columbia. One group consists of very thin-shelled forms (Figure 10.7) such as *Claraia, Daonella, Halobia,* and *Monotis*. These bivalves are sometimes known as paper clams or flat clams. They have no counterpart among bivalves living today.

Many paper clams have a notch in the right valve, which was probably occupied by attachment threads similar to the tuft of filaments by which a living mussel attaches itself to rocks. Paper clams commonly occur in vast numbers, so many that they make up most of the rock. In

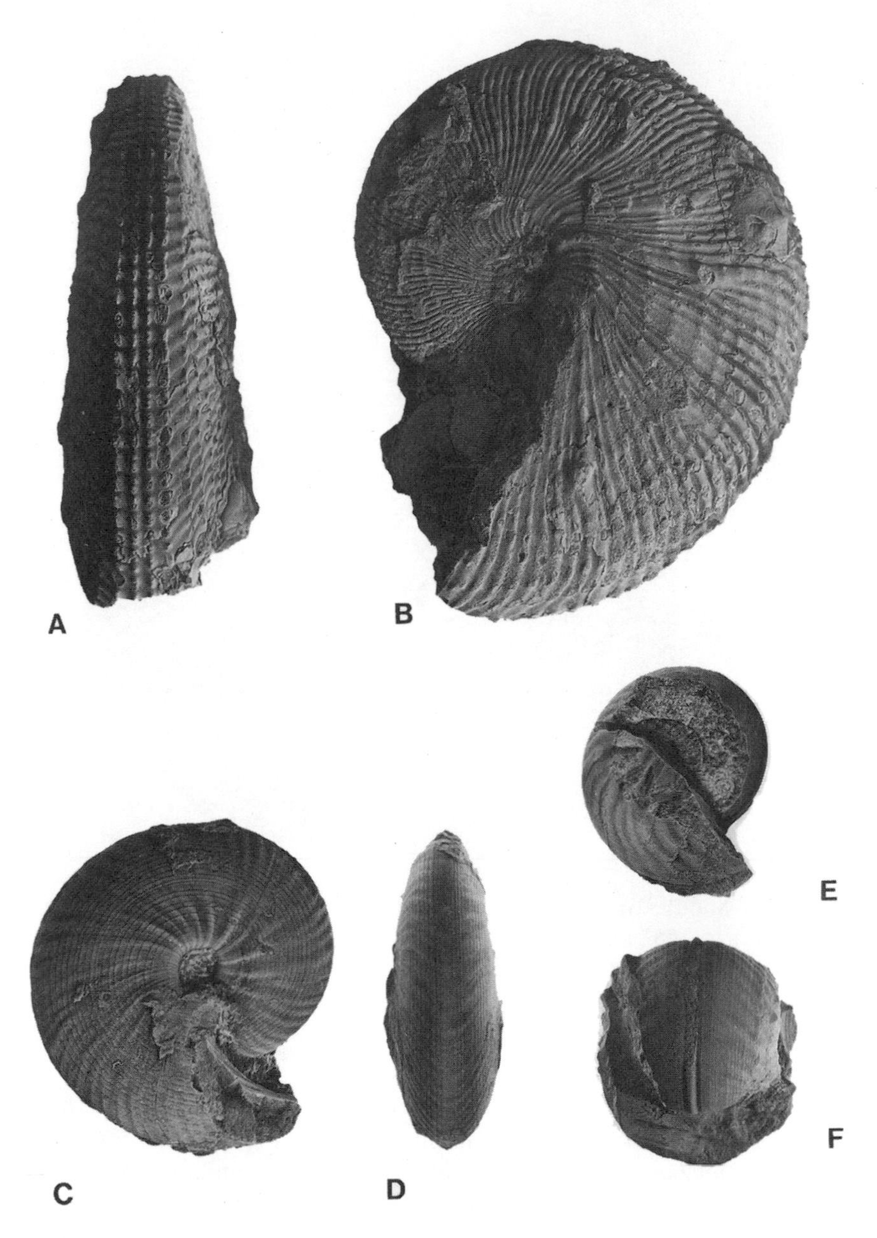

Figure 10.5 **Upper Triassic ammonoids.** (A) and (B) *Mesohimavatites columbianus* from the Sikanni Chief River. (C) and (D) *Malayites dawsoni* from Pardonet Hill. (E) and (F) *Tropites dilleri,* a keeled ammonoid with thick whorls, from Vancouver Island. All are actual size.
Source: Collections of the Geological Survey of Canada

Figure 10.6 **Some of the youngest Triassic ammonoids.** (A) *Choristoceras crickmayi* from Tyaughton Creek. (B) *Paracochloceras canaliculatum* from Vancouver Island. (C) *Rhabdoceras suessi* also from Vancouver Island. (All X 2)
Source: Collections of the Geological Survey of Canada

such strata the only other associated large fossils are ammonoids and ichthyosaur bones. Animals that might have lived on the sea bottom are not present.

Paper clams seem to have been ocean mariners *par excellence.* Some species of *Monotis* achieved an astonishingly broad geographic distribution, occurring in widely separated localities such as Austria, the Himalayas, Nevada, British Columbia, and Alaska. As we shall see, the geographic distribution of *Monotis* species may provide a signal of paleogeographic significance. Because these unusual clams are extinct and without close living relatives, their mode of life is a subject of conjecture. An idea that would account for their wide geographic distribution and their presence in many different kinds of rocks is that they lived attached to masses of floating seaweed, from which they 'rained' down on the sea bottom in millions.

A second kind of Triassic bivalve is the megalodont – literally, 'big tooth.' In form, megalodonts are the complete antithesis of paper clams, being large, up to a metre in length, and having very thick shells. They were first found in the Alps, where they are sometimes so abundant that they form thick beds of limestone. Occurrences in the Mediterranean area suggest that these clams were characteristic inhabitants of low-latitude tropical seas.

The remaining Triassic bivalves may be described as 'ordinary' because

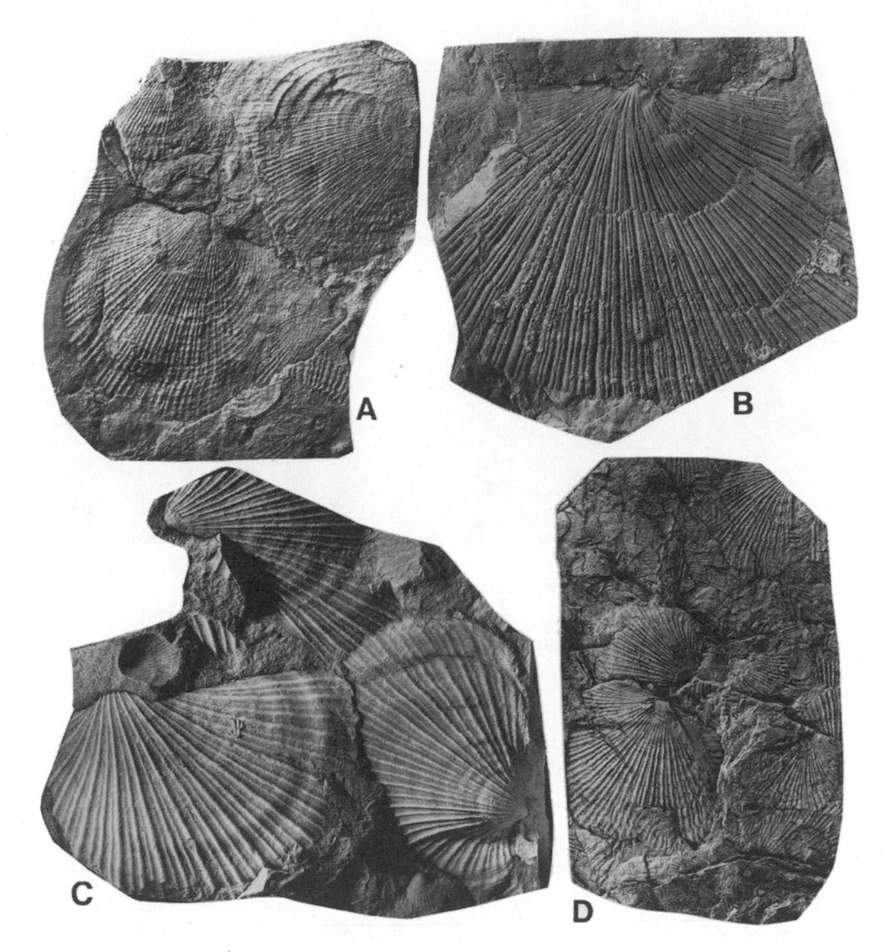

Figure 10.7 **Triassic paper clams.** (A) *Claraia stachei* from the Liard River. (B) *Daonella nitanae* from near Tuchodi Lakes. (C) *Monotis subcircularis* from Pine Pass. (D) *Monotis alaskana* from the Queen Charlotte Islands. All are actual size. *Source:* Collections of the Geological Survey of Canada

most have close counterparts in today's bivalve faunas. They include oysters, scallops, and assorted 'soup clams.'

Telling Time with Triassic Mollusks

The fundamental principle underlying stratigraphic geology is the law of superposition, stating that the higher beds in a sequence of bedded rocks, or strata, are younger than the lower beds. British Columbia is particularly well endowed with sequences that display dramatic differences in the fossil content of successive beds.

The earliest Triassic ammonoids were simple, smooth cones, generally with ceratitic suture lines, as in *Kashmirites* (Figure 10.2). The Early Triassic and Middle Triassic saw the evolution of *Anawasatchites* and other ammonoids with ribs and nodes, as well as some forms like *Discogymnites* with elaborately frilled suture lines (Figure 10.3). In the Late Triassic even more elaborate ornamentation prevailed, as in *Mesohimavatites* (Figure 10.5). By the end of the Triassic some ammonoids like *Rhabdoceras* and *Paracochloceras* were heteromorphs – ammonoids not coiled in a plane spiral (Figure 10.6).

Paper clams also display distinct evolutionary changes (Figure 10.7). *Claraia* was restricted to the Early Triassic, *Daonella* to the Middle, and *Halobia* to the Late Triassic. *Monotis* and its close relatives had a very short time range – the later part of the Late Triassic – and an almost worldwide distribution.

These differences in successive mollusk faunas have been used to partition the Triassic and construct a biochronology. The smallest divisions, 'zones,' are named after a characteristic fossil. The youngest Triassic zone, for example, is the *Choristoceras crickmayi* Zone, generally abbreviated to 'Crickmayi Zone.' At present thirty-seven zonal divisions are recognized in western and Arctic Canada. Next up the scale are 'stages,' Griesbachian to Norian, each consisting of several zones containing related faunas. Larger divisions are 'series,' and are referred to as the Lower, Middle, and Upper Triassic.

Stages derive their names from geographic localities. The Lower Triassic stages are named for places on Ellesmere and Axel Heiberg islands in Arctic Canada (Tozer 1967). The Middle and Upper Triassic stages take their names from places in Austria and northern Italy. This nomenclature is testimony to a particularly useful aspect of biochronology; namely, that stages are commonly recognizable not only within a province or country but throughout the world.

Analyses of the decay products of radioactive minerals show that the Triassic lasted for about forty million years. This means that, on average, each zone lasted for about a million years. Calibration of Triassic biochronological divisions and radiometric ages is an ongoing field of research. At present few zones can be labelled as having a duration expressed in years. Nevertheless, as developed to date, biochronolgy is an indispensable tool for deciphering history during the Triassic. Its study leads to a better understanding of the coming and going of seas, oceans, and lands; and it provides methods for the dating of intervals of folding, faulting, and volcanism.

Messages from Triassic Fossils in the Cordillera

Triassic rocks are most widely distributed on the North American Margin (NAM), where they provide a nearly continuous faunal record. The most extensive NAM outcrops are in the foothills of the Rocky Mountains of northern British Columbia.

Triassic fossiliferous rocks are also known from most of the terranes west of the NAM, particularly Quesnellia, Stikinia, and Wrangellia. The rocks of these terranes probably formed hundreds or thousands of kilometres out in the ancestral Pacific Ocean. Later, they were transported by moving oceanic crust and collided with the North American craton. Therefore they are described as accreted terranes. Those with Triassic rocks probably docked in Jurassic time.

Triassic rocks and faunas of the NAM localities are very different from those of the displaced terranes. The NAM rocks are wholly sedimentary – sandstone, siltstone, shale, and limestone. Faunas of Early, Middle, and Late Triassic age are present in these rocks. Volcanic rocks are conspicuously absent.

Triassic rocks of the accreted terranes differ from those of the NAM in several ways. The rock types and the fossil sequence of an individual terrane are unlike those of adjacent terranes. Of particular significance is the presence of thick masses of limestone in places forming reefs built of corals and sponges. Volcanic rocks are widely present, but their age on one terrane does not correspond with ages on adjacent terranes. Nearly all the mollusks from the accreted terranes are Late Triassic in age, only a few are Middle Triassic, and none is Early Triassic. Nevertheless, biochronological studies using microfossils such as conodonts and radiolarians show that Lower and Middle Triassic rocks are present on the accreted terranes.

The contrast between the NAM and the accreted rocks and among their faunas has considerable paleogeographic significance. Within British Columbia the character of the NAM rocks and faunas suggests that these rocks were deposited in relatively cool water at mid-latitudes. There is a conspicuous absence of limestones containing reef-building organisms and megalodont bivalves, which would indicate deposition in warm water at low latitudes. In fact, evidence of low-latitude deposition in the NAM sequence is not found north of Nevada.

The message from the rocks of the accreted terranes is entirely different. Thick limestones with corals and megalodonts occur in several terranes, notably Stikinia. Limestones of this nature occur as far north as sixty degrees latitude in the displaced terranes.

Another paleoclimatological indicator is the bivalve *Monotis*. Several types have been recognized. In the strict sense, however, *Monotis* is typified by *M. salinaria,* a species originally described from rocks in Austria that are regarded as deposits of the Triassic equatorial belt. In Canada this type of *Monotis* occurs in the accreted terranes Quesnellia and Wrangellia but not in the NAM sequences.

The distribution of warm-water limestones and *Monotis* suggests that in Triassic time the rocks of the accreted terranes formed at low latitudes and at a distance from ancestral North America.

References

McLearn, F.H. 1953. Correlation of the Triassic Formations of Canada. *Bulletin of the Geological Society of America* 64:1,205-27

–. 1960. *Ammonoid Faunas of the Upper Triassic Pardonet Formation, Peace River Foothills, British Columbia.* Geological Survey of Canada, Memoir 311

–. 1969. *Middle Triassic (Anisian) Ammonoids from Northeastern British Columbia and Ellesmere Island.* Geological Survey of Canada, Bulletin 170

Tozer, E.T. 1967. *A Standard for Triassic Time.* Geological Survey of Canada, Bulletin 156

–. 1984. *The Triassic and Its Ammonoids: The Evolution of a Time Scale.* Geological Survey of Canada, Miscellaneous Report 35

–. 1994. *Canadian Triassic Ammonoid Faunas.* Geological Survey of Canada, Bulletin 467

Whiteaves, J.F. 1889. On Some Fossils from the Triassic Rocks of British Columbia. Geological Survey of Canada, *Contributions to Canadian Palaeontology* 1:127-49

11
Ammonoids: Itinerants of the Jurassic

Giselle K. Jakobs

Ammonoids reached their heyday in the Jurassic, a system characterized by their rapid speciation, their ubiquitous presence in sedimentary strata, and their widespread geographic distribution. These features led to the use of ammonoids as precise dating tools for Jurassic sedimentary rocks. Over the last century, paleontologists in Europe have refined the ammonoid zonations that now provide the key to dating many geologic and tectonic events.

Early on, Jurassic paleontologists in North America attempted to force their faunas into European zonations. These attempts inevitably led to failure because certain key European species – and even genera – are absent in North America. In addition, until recently, few attempts were made to collect ammonoids systematically in a stratigraphic sequence. Instead, paleontologists in North America generally worked with isolated collections provided by geologists. In fact, the complexity of the Canadian Cordillera and the inaccessibility of many areas made it exceedingly difficult to collect specimens in any kind of systematic order. Recent studies in the Queen Charlotte Islands, however, have yielded excellent Lower Jurassic ammonoid faunas, which have at last provided a framework for an ammonoid zonation applicable to western North America (Smith et al. 1988).

British Columbia contains extensive exposures of Jurassic sediments. Most are located in Stikinia and Wrangellia, terranes that were accreted to ancient North America. The Lower and Middle Jurassic are the more widely exposed, and these rocks are currently undergoing sedimentologic and biostratigraphic studies. The Upper Jurassic is not as well exposed and thus is poorly understood at this time.

Ammonoids are common throughout the Jurassic, being preserved generally as flattened impressions and less commonly as three-dimensional specimens. BC paleontologists compare their specimens to those from South America, Europe, Siberia, and Arctic Canada. As a result, we know that the ammonoid faunas in British Columbia are more similar to those of South America and the Mediterranean region than to those of closer areas such as Siberia and Arctic Canada. The best examples of Jurassic ammonoids are those from the Lower Jurassic. These fossils are the focus of this chapter.

Jurassic Fossils in British Columbia

Jurassic fossils have been collected throughout British Columbia from the Rocky Mountains in the east to Vancouver Island in the west, and from Manning Park in the south to Atlin Lake in the north. An attempt to document all of these collections would require an entire volume in itself. Some of the best areas are at Harrison Lake, Tyaughton Creek, the Copper River near Smithers, and the Spatsizi Plateau north of Hazelton. The best and most prolific ammonoid faunas of the BC Jurassic are those from the Queen Charlotte Islands.

Some of the earliest geologists to explore the Queen Charlotte Islands were also among the first to collect Jurassic fossils in British Columbia. George M. Dawson collected numerous specimens, some erroneously identified as Cretaceous. These specimens gave the first indication of the extensive and well-preserved ammonoids that occur in this area. When Atholl Sutherland Brown mapped the Queen Charlotte Islands during the 1950s and 1960s, he collected additional material that confirmed the presence of almost the entire Lower and Middle Jurassic succession.

It was not until the late 1970s and 1980s, however, that systematic attempts were made to collect and describe Jurassic fossils from the Queen Charlotte Islands in paleontological rather than geological studies. Hans Frebold, the Jurassic ammonoid expert of the Geological Survey of Canada, identified most of these collections, using his extensive knowledge of northwest European ammonoids.

Key Ammonoid Players in the Lower Jurassic

Because Lower Jurassic ammonoids are best preserved in the Queen Charlotte Islands, this area is the source of most of the important breakthroughs in ammonoid biochronology and paleobiogeography. One of the easiest ammonoids to recognize in the Lower Jurassic is *Amaltheus,* a genus also common in northwest Europe (Figure 11.1). It has a crenulated keel, looking rather like a rope coiled around the outside of the shell. *Amaltheus* has been collected at several localities across northern British Columbia, the Yukon, and Arctic Canada. It occurs only in the middle part of the Lower Jurassic, where it provides direct time correlations among North America, Japan, Siberia, Greenland, and northwest Europe. Because dating rocks is one of the primary uses of fossils, easily recognizable forms are particularly useful tools for a paleontologist. To

Figure 11.1 ***Amaltheus stokesi*, a Boreal ammonoid from the Fernie Group, northeastern British Columbia.** Actual size.
Source: Collections of the Geological Survey of Canada

the astonishment of regional geologists, a small fragment of an *Amaltheus* keel can provide a precise age, whereas relatively complete but undistinguished ammonoids usually provide only a general date.

Another example comes from slightly younger rocks of the uppermost Lower Jurassic. In the 1980s several strange and perplexing ammonoids could not be assigned to an existing species, genus, or even family. Comparisons with European and South American forms were no help. A tiny specimen from the Spatsizi area in north-central British Columbia finally provided the breakthrough. A single spine projecting from one of the nodes proved to be very similar to the genus *Leukadiella* from the Mediterranean region. *Leukadiella* is characterized by nodes and by paired ribs that curve across the flanks (Figure 11.2). A comparison of these North American specimens with European specimens of *Leukadiella* was difficult due to size differences – the European specimens are small and usually less than ten centimetres in diameter, whereas those from North America may reach twenty-five centimetres in diameter. Because the characteristic ornamentation of *Leukadiella* is easily recognizable, however, the specimens from the Hazelton and Harrison Lake areas could be identified confidently even though they were poorly preserved. These ammonoids give precise ages necessary for regional geological studies.

Occasionally paleontologists come to the conclusion that they are deal-

Figure 11.2 ***Leukadiella ionica,*** **a Tethyan ammonoid from the Maude Group, Queen Charlotte Islands.** Actual size.
Source: Collections of the Geological Survey of Canada

ing with a new species or new genus. The uppermost Lower Jurassic in the Queen Charlotte Islands provided such a case. The upper part of the Whiteaves Formation in the Queen Charlotte Islands contains numerous discoidal calcareous concretions that become exposed as weathering erodes the surrounding soft shales. Along one section, the concretions tend to weather out of the shale slope and accumulate along the river bank where constant immersion in water makes them easy to crack open. These concretions have yielded some of the best-preserved Lower Jurassic ammonoids in North America. One of the more common ammonoids collected here is *Phymatoceras hillebrandti,* a newly described species that is related to a South American form (Figure 11.3). This species also appears to be the ancestor of a new genus that occurs in the overlying Phantom Creek Formation of the Queen Charlotte Islands.

The Phantom Creek Formation is a shallow-water sandstone that contains large calcareous concretions greater than one metre long. Some of these concretions are packed with beautifully preserved three-dimensional ammonoids and bivalves. Regular hammers and chisels are not up to the task – it is only with the aid of a sledgehammer that these blocks are broken into manageable pieces. Several genera and species were identified in these concretions but several other forms were not so easily pigeonholed.

Figure 11.3 **Phymatoceras hillebrandti, a common ammonoid from the Whiteaves Formation, Queen Charlotte Islands.** Actual size.
Source: Collections of the Geological Survey of Canada

Comparison of these ammonoids with those previously described suggested that the only other similar forms occurred in the Rocky Mountains and Alaska and that these forms had been equally difficult to correlate with known genera. In the end, a new genus was established – *Yakounia* includes four species, each occurring only in North America (Figure 11.4). Also occurring was *Pleydellia maudensis*, a new species that differs from other species of *Pleydellia* by the prominent umbilical tubercles present at later stages of growth (Figure 11.5).

Paleontologists are intrigued by the similarities and differences in ammonoid faunas that occur on different continents and within areas on one continent. Interest in this subject has led to a separate field of study called paleobiogeography.

Figure 11.4 ***Yakounia freboldi,*** **a common ammonoid from the Maude Group, Queen Charlotte Islands.** Actual size.
Source: Collections of the Geological Survey of Canada

Figure 11.5 ***Pleydellia maudensis,*** **a well-preserved ammonoid from the Phantom Creek Formation, Queen Charlotte Islands.** Actual size.
Source: Collections of the Geological Survey of Canada

Jurassic Paleobiogeography

Biogeography is the study of the geographic distribution of organisms. Biogeographers today are trying to determine why certain species occur only on one continent, or in one ocean, and not another. Are there geographic or climatic barriers that prevent migration?

Organisms that are distributed globally are referred to as pandemic. Endemic organisms, on the other hand, are confined to certain areas, possibly due to geographic or climatic barriers. A modern-day example is the penguin, which is endemic to the southern polar region, or the kangaroo, which is endemic to Australia. An area that contains several endemic organisms is distinguished from other regions and is called a province (Smith 1989).

The distribution of terrestrial and freshwater organisms is influenced by barriers such as climate, mountain ranges, deserts, and oceans. Thus this distribution shows strong endemism. Shallow-water, bottom-living (benthic) marine organisms such as bivalves, gastropods, and echinoids (for example, sea urchins and sand dollars) also tend to show strong endemism due, in part, to their inability to cross deep ocean basins. In contrast, floating (pelagic) and swimming (nektic) marine organisms such as fishes, squid, and whales tend to have a pandemic distribution, although climate can also affect their distribution.

Paleobiogeography is the study of the geographic distribution of ancient organisms. Paleontologists studying the paleobiogeographic distribution of fossils in the Jurassic need to consider several factors. First and foremost is the location of the Jurassic continents. During the Early Jurassic the present continents were grouped together to form one large supercontinent, Pangaea (Figure 11.6). The ancestral Atlantic Ocean was in its initial stages of rifting and the ancestral Pacific Ocean (Panthalassa) was over 10,000 kilometres wide. Europe and Asia were separated from Africa, Arabia, India, and Australia by the wedge-shaped and tropical Tethyan Ocean, of which the present-day Mediterranean is the only remnant. A second factor that must be considered is the climate. The Jurassic was considerably warmer than the present. Paleontological and geological evidence suggests that polar ice caps were non-existent during the Jurassic. The difference between temperatures at the poles and those at the equator was smaller than it is today, resulting in a more equitable global climate.

It is logical to expect Jurassic ammonoids, with their pelagic mode of life, to have a pandemic distribution with little provincialism. As early as the late 1800s, however, paleontologists in Europe documented variations in the distribution of ammonoids that clearly identify distinct realms. The Boreal Realm includes northern high-latitude ammonoids such as *Amaltheus,* whereas the Tethyan Realm encompasses low-latitude

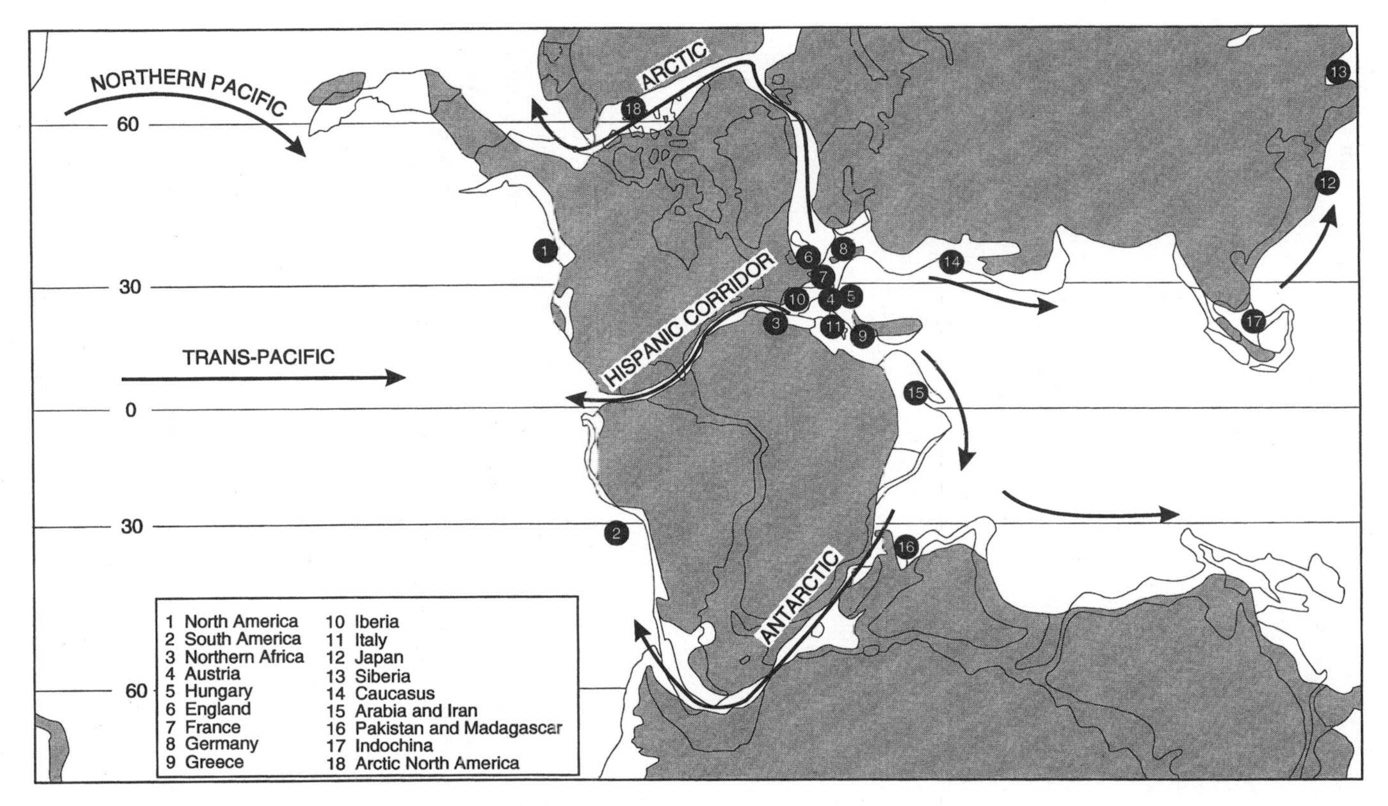

Figure 11.6 **The supercontinent Pangaea during the Early Jurassic.** Possible migration routes of ammonoids from the western part of the Tethyan Ocean to the eastern part of the ancient Pacific Ocean are indicated by arrows.

equatorial forms such as *Leukadiella*. An Austral Realm comprises southern high latitudes, but it was not well developed in the Jurassic. Generally the boundaries between the realms are transitional, their position shifting over time.

In 1967 a revolutionary paper was published on the geographic distribution of Lower Jurassic ammonoids in Europe. D.T. Donovan, a British paleontologist, reported that the majority of genera were pandemic, occurring from Britain in the north to Morocco and Algeria in the south. He had also discovered that, at certain times in the Early Jurassic, a significant proportion of the genera were either northern or southern in their distribution. This distinction was particularly noticeable in the late part of the Early Jurassic when there were relatively few pandemic genera but many Tethyan or Boreal genera. An apt illustration is the distribution of ammonoid families – the Tethyan Hildoceratidae and the Boreal Amaltheidae (Figure 11.7). The southern and northern limits of the hildoceratids and amaltheids are aligned roughly east-west, suggesting that climate probably played a role in these distributions.

At one time, temperature variation across latitudes was a common explanation for the dispersal of ammonoids and other marine animals during the Jurassic. Paleontologists thought that the Jurassic may have been similar to the present, with high-diversity faunas distributed along the equator and low-diversity faunas located at the poles. As mentioned earlier, however, evidence suggests that the Jurassic climate was more equitable than at present. As well, the absence of a south polar Austral Realm during the Jurassic suggests that polar-equatorial temperature differences did not play a large role in the geographic distribution of marine organisms.

Other paleontologists postulated that variations in salt content, or salinity, existed between the northern seas and the low-latitude equatorial belt during the Jurassic. They suggested that the northern seas in Europe, bounded on almost all sides by land masses, may have had slightly lower salinities than normal marine waters. If that were the case, however, ammonoids, brachiopods, bivalves, and echinoids could not have existed there.

Recent studies suggest that one of the key factors controlling diversity in modern oceans is environmental stability. Unstable environments such as nearshore areas and estuaries are subject to rapid changes in temperature and salinity, and tend to have low diversity in animal life. Stable environments, which are generally offshore, experience less change in temperature, salinity, and food supply, and are populated by diverse assemblages of organisms. During the Jurassic, the environment of the Boreal seas was probably unstable due to salinity fluctuations caused by the influx of fresh water off the continents and to a strong seasonal

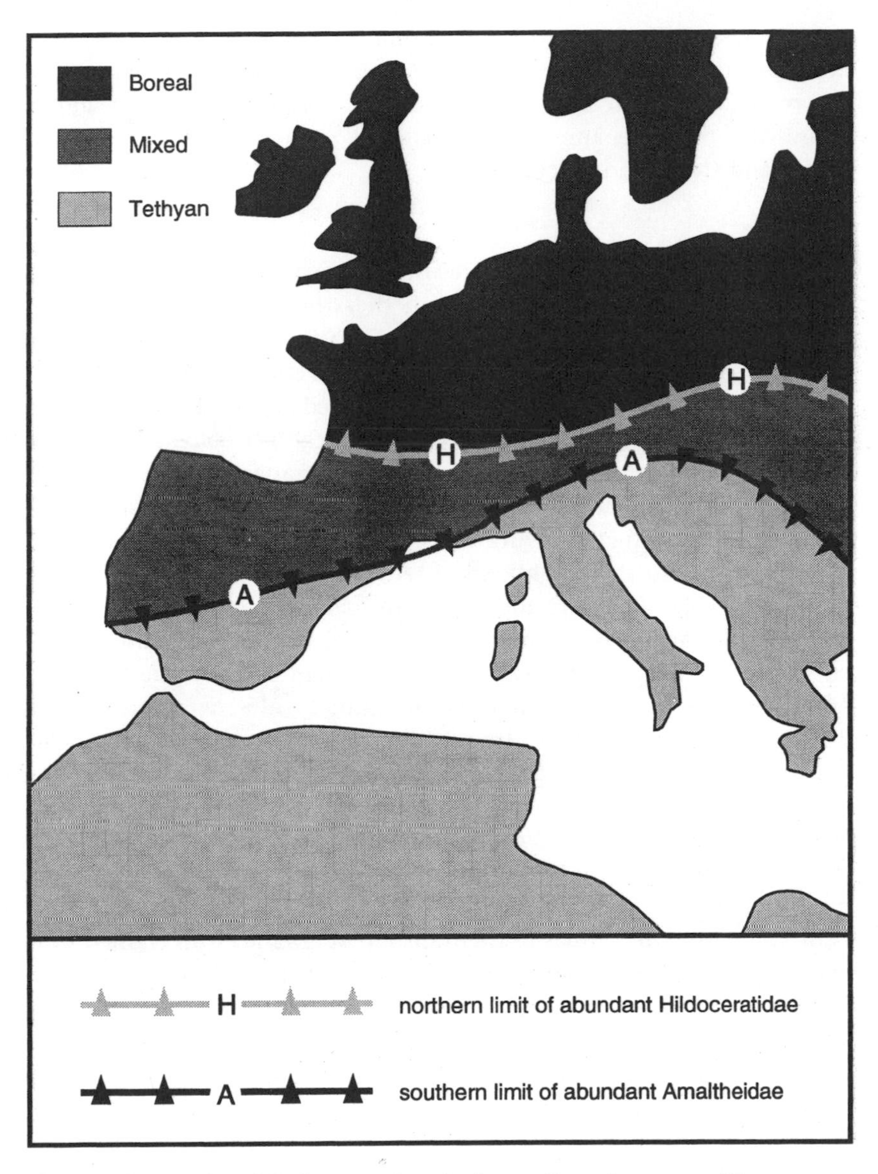

Figure 11.7 **Boreal and Tethyan realms in Lower Jurassic strata of Europe, as defined by abundance patterns of two ammonoid families**
Source: From Donovan (1967)

contrast in temperature and light. These factors were probably intensified by the shallowness of the northern seas. The likely result is that Boreal faunas living in these unstable environments exhibited lower diversity than Tethyan faunas living in stable environments.

Paleobiogeography of British Columbia

BC paleontologists have used Jurassic ammonoid paleobiogeography to determine the extent of movement of the Cordilleran terranes and to define the probable ammonoid migration routes between the Tethys Ocean and the eastern Pacific Ocean.

Terranes

The main focus of recent work on Lower Jurassic ammonoids of British Columbia has been to determine whether the difference in diversity between the Tethyan and Boreal realms in Europe can be recognized in North America (Smith and Tipper 1986). Unlike Europe, British Columbia is a collage of transported terranes, which joined ancestral North America during the Jurassic and Cretaceous (Yorath 1990). Geological evidence suggests that during the Early Jurassic, these terranes were much farther south and at different relative latitudinal positions. Each must be analyzed separately when studying the biogeography of the province, and then compared to the neighbouring terranes and to the craton, that is, the large, stable nucleus of the continent.

The North American craton shows a faunal distribution pattern similar to that of Europe. Low-diversity Boreal faunas occur in the Canadian Arctic and northern Alaska and as far south as Alberta, whereas high-diversity Tethyan faunas are found in Nevada. The terranes Quesnellia and Stikinia have Tethyan faunas in their southern parts and mixed faunas in the northern parts. Wrangellia, a fragmented terrane now located in southern Alaska, the Queen Charlottes, Vancouver Island, and Oregon, shows a similar pattern, with Tethyan faunas in the southern part and mixed faunas to the north (Figure 11.8A).

By comparing the faunas of the transported terranes with those on the craton, paleontologists can reconstruct the probable position of the terranes with respect to the craton and to each other (Smith and Tipper 1986). Thus we know that, during the later part of the Early Jurassic, Wrangellia, Stikinia, and Quesnellia were located at lower latitudes than they are today, approximately at the latitude of Nevada (Figure 11.8B).

Migration Routes

Four migration routes have been proposed to explain the strong similarity between ammonoid faunas of the eastern Pacific and those of the western Tethys during the Early Jurassic (Figure 11.6). An Arctic migration pathway via Great Britain, Spitsbergen, and the North American Arctic is thought to have been open in the Early Jurassic. A southern migration route via eastern Africa and the Antarctic has also been proposed, as has eastward migration along the southern and northern margins of Tethys or dispersal across the ancestral Pacific Ocean. In addition, several paleontologists have postulated the presence of an Hispanic Corridor, a narrow seaway linking the Pacific with western Tethys along the newly opened North Atlantic. Each of these possibilities merits discussion.

Geological and paleontological evidence from northern Europe, Greenland, Spitsbergen, and Arctic Canada suggests that an Arctic migration route persisted throughout the Jurassic as a shallow seaway. This migration route probably helped the spread of Boreal ammonoids such as *Amaltheus* across the Arctic and down the coastline of ancestral North America. This high-latitude route was not feasible for Tethyan ammonoids, however, which were confined to low latitudes.

A southern migration route via eastern Africa and Antarctica is unlikely, since no marine Jurassic rock is known in Africa south of Tanzania. In addition, the Early Jurassic ammonoids of eastern Africa and Arabia are of low diversity and quite different from the high-diversity faunas of southern Europe, North America, and South America. If migration had occurred along this route, we would expect to find a strong similarity among the faunas of these regions. The absence of characteristic Tethyan genera from Africa and Arabia suggests that, if this route was open during the Early Jurassic, it probably was a very shallow seaway of low environmental stability that acted as a barrier to the migration of Tethyan ammonoids.

Eastward migration of Tethyan forms along the southern coastline of Tethys has been suggested, but there is little evidence for this route. Ammonoid collections of late Early Jurassic age from Saudi Arabia, Iran, and Pakistan have very low diversities and lack the distinctive Tethyan elements. Lower Jurassic deposits in Australia and New Zealand have yielded only very few ammonoids, and these are quite different from the high-diversity Tethyan faunas.

Migration eastward via the northern margin of Tethys and the northern Pacific has also been suggested as a means of faunal exchange. If migration had occurred along this route, we would expect to find examples of Tethyan ammonoids in the Caucasus, Indonesia, and Japan. Instead, these areas contain pandemic or Boreal genera but no low-latitude Tethyan genera.

Several paleontologists have suggested that westward migration of marine faunas between the western Tethys and the eastern Pacific occurred along an Hispanic Corridor, a shallow seaway across what are now Central America and the Caribbean (Figure 11.6). Evidence from a variety of fossil groups supports this model, suggesting that this corridor was open from the Early Jurassic on. In addition, numerous Lower Jurassic ammonoid genera such as *Leukadiella* occur only in the western Tethys and the eastern Pacific, suggesting a direct marine connection between

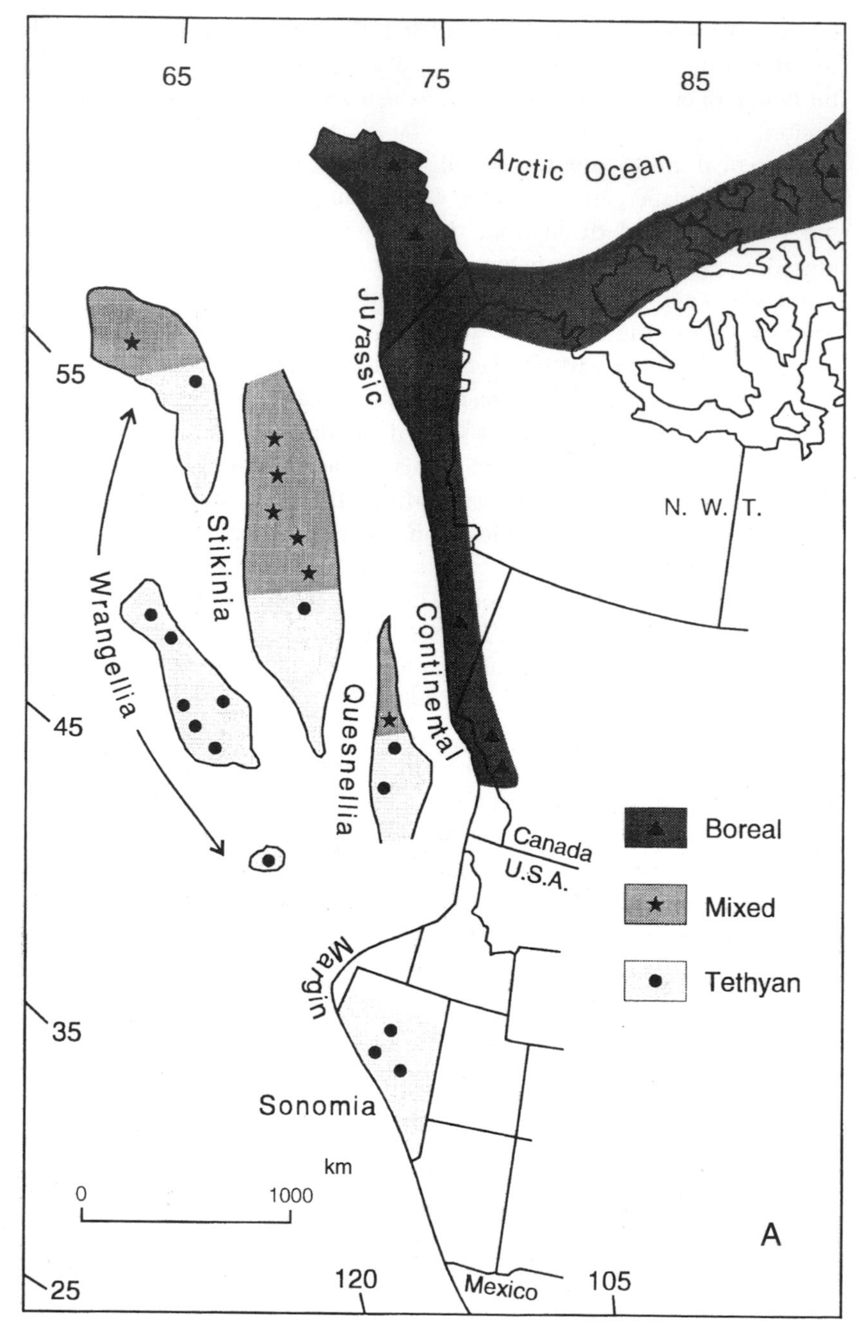

Figure 11.8 **Paleobiogeography and Cordilleran terranes.** (A) Early Jurassic biogeographic realms and transported terranes (shown pulled off the continent).

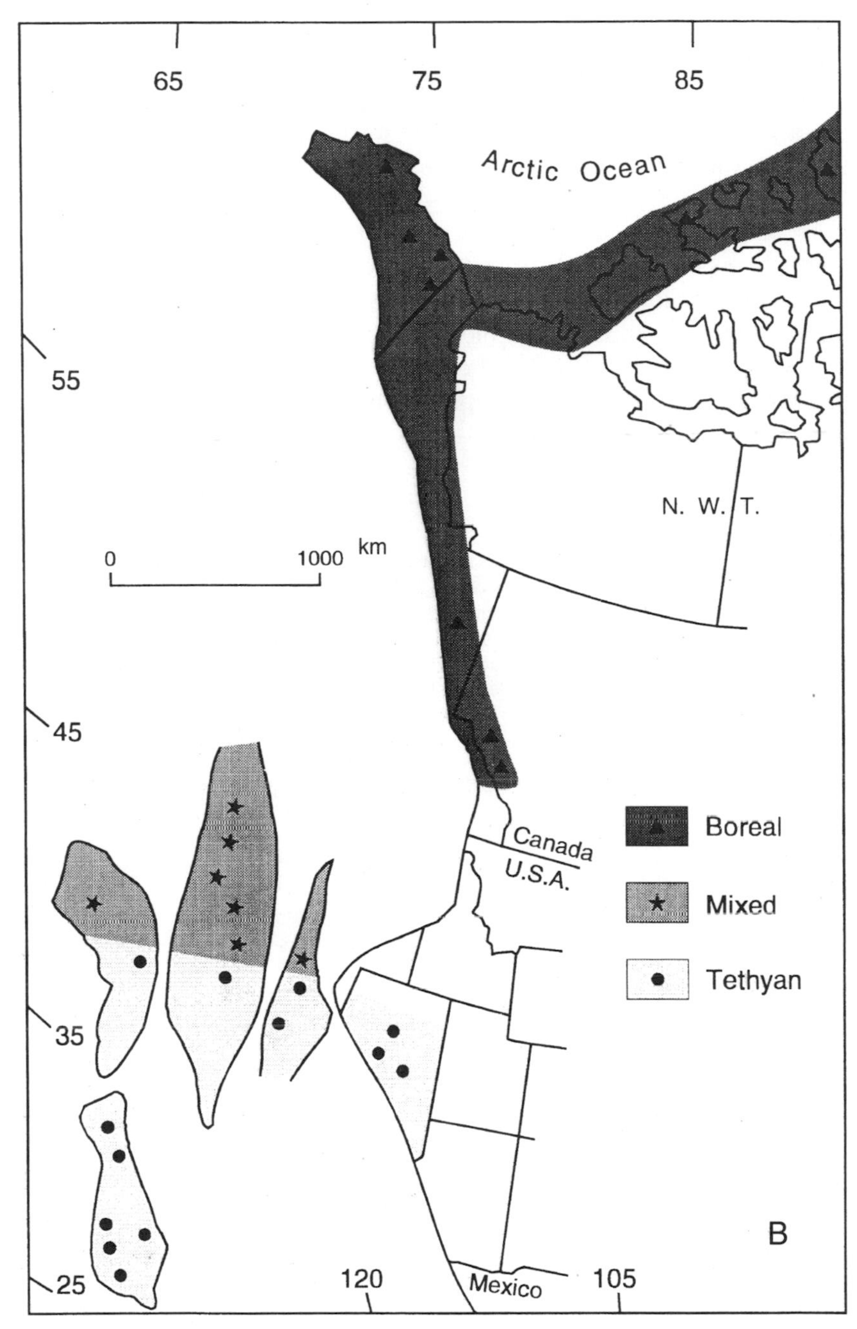

Figure 11.8 [Continued]. (B) Possible location of terranes during the Early Jurassic based on biogeographic assignments. No longitudinal position of the terranes is implied.
Source: From Smith and Tipper (1986)

the two areas. Statistical analyses of the global distribution of Lower Jurassic genera suggest that migration of certain Tethyan forms was directly controlled by sea level. Accordingly, it is probable that a rise in the sea level at that time allowed the migration of low-latitude forms between the eastern Pacific and western Tethys along the Hispanic Corridor.

Lower Jurassic ammonoids from British Columbia are an invaluable dating tool for geologists and paleontologists. These fossils have assisted in unravelling the complicated geologic history of the Canadian Cordillera. In addition, they have improved our understanding of geography during the Jurassic when the ancient supercontinent Pangaea was beginning to fragment.

References

Donovan, D.T. 1967. The Geographical Distribution of Lower Jurassic Ammonites in Europe and Adjacent Areas. In *Aspects of Tethyan Biogeography,* ed. C.G. Adams and D.V. Ager, 111-34. Publ. No. 7, Systematics Association

Smith, P.L. 1989. Paleobiogeography and Plate Tectonics. *Geoscience Canada* 15:261-79

Smith, P.L., and H.W. Tipper. 1986. Plate Tectonics and Paleobiogeography: Early Jurassic (Pliensbachian) Endemism and Diversity. *Palaios* 1:399-412

Smith, P.L., H.W. Tipper, D.G. Taylor, and J. Guex. 1988. An Ammonite Zonation for the Lower Jurassic of Canada and the United States: The Pliensbachian. *Canadian Journal of Earth Sciences* 25:1,503-23

Yorath, C.J. 1990. *Where Terranes Collide.* Victoria: Orca Book Publishers

12
On the Trail of Cretaceous Dinosaurs

Scott D. Sampson and Philip J. Currie

British Columbia is a province rich in fossils. During the Mesozoic Era (which lasted from 250 million years ago to sixty-five million years ago), it was a place where millions of dinosaurs lived. Nevertheless, only a handful of dinosaur bones have been recovered from the province. This fact is surprising considering that the next province, Alberta, is one of the richest areas in the world for dinosaur fossils.

In truth, the record of dinosaur bones from British Columbia is virtually non-existent. In 1979, however, a small display in the offices of Crow's Nest Industries Ltd. in Fernie yielded an unexpected find. In a display of coal pebbles – inclusions found within coal seams – was part of a single toe bone of an ornithopod, or bird-footed, dinosaur. The bone (Figure 12.1) is black and was already water-worn before deposition, so it is not surprising that it was thought to be a stone. Many layers of coal are mined in the Fernie region, and the bone could have come from any one of the uppermost Jurassic to Lower Cretaceous seams. Because the specimen is water-worn and incomplete, it is difficult to identify with certainty. The articulating surface is smoothly concave, suggesting that it is the first phalanx of the finger or toe. It most closely resembles the first phalanx of the fourth (index) finger of *Camptosaurus,* but this identification is tentative. There are also rumours that in 1930 Charles M. Sternberg collected a few bones of an ornithopod along the Pine River of northeastern British Columbia. Yet he did not mention this discovery in his field notes, and the whereabouts of such specimens is unknown. So, at this time the report cannot be confirmed.

Few Bones, Many Footprints

Why are there so few dinosaur bones known from British Columbia? After all, there are extensive Mesozoic terrestrial deposits in British Columbia, and the equivalent beds across the border in Alberta have produced a plethora of dinosaurs. There are several reasons for the disparity. During

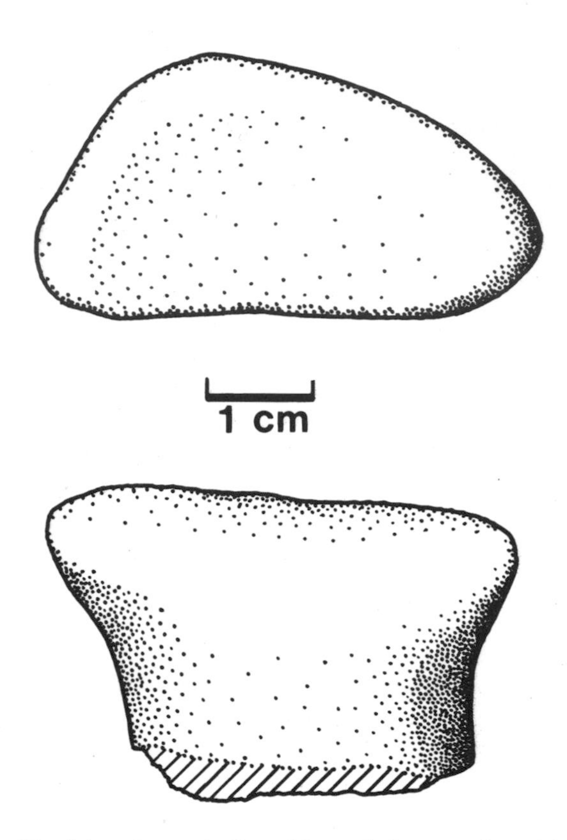

Figure 12.1 **Fossilized toe element of ornithopod dinosaur found in south-eastern British Columbia**

Mesozoic times, much of the interior of North America was covered by an inland sea that at times extended from the Gulf of Mexico to the Arctic Ocean. As mountain ranges were pushed up in what is now British Columbia, wind and water erosion started tearing the mountains down. Most of the eroded sediments were carried by rivers and streams east to the sea, where they were deposited on the coastal plains and deltas of Alberta. Bones and even whole carcasses of dinosaurs were buried in the river-borne sands and muds, greatly increasing their chances of preservation as fossils. Furthermore, sediments were deposited in this region so rapidly that there was little opportunity for the development of long-standing marshes and swamps. Although marshy environments are good for the formation of coal deposits, they create acidic ground water that destroys bones before they can fossilize. Thus conditions were ideal in Alberta for the fossilization of dinosaur bone. In contrast, most of the Interior of British Columbia was being eroded, so there were few opportunities for dinosaur bones to become buried and fossilized.

Much more recently, Mesozoic rocks and fossils were re-exposed in Alberta when younger sediments were scraped off by glaciers. Further erosion by the glacial meltwaters cut huge channels into the rocks and created the badlands of southern Alberta. The modern semi-arid climate is not conducive to heavy plant growth, so the rocks remain exposed. This combination of effects greatly facilitates the discovery and recovery of fossils. In contrast, the heavily vegetated, often mountainous regions of British Columbia make it difficult to find dinosaur fossils, even if they are present.

Nonetheless, although British Columbia has not yet yielded many dinosaur bones, the province has long been known as the best source in Canada for dinosaur footprints and trackways. The kinds of conditions required for the preservation of footprints are usually different than the requirements for preservation of bone. Thus, although the coastal plain deposits in Dinosaur Provincial Park, Alberta, have produced millions of fossilized teeth and bones, only two footprints have been identified and collected to date. Conversely, although low-lying wet regions do not permit the fossilization of bone, the wet muds, slow water currents, and gentle accumulation of sediments offer the ideal conditions for preservation of footprints. It is therefore not surprising that the best footprint sites in the United States, China, Korea, and other parts of the world have produced few, if any, dinosaur bones. Dinosaur footprints offer unique and important opportunities to study the lifestyles of these ancient creatures. The remainder of this chapter focuses on the rich dinosaur footprint resources of British Columbia.

The Nature and Study of Tracks

Fossilized tracks are truly remarkable. They record fleeting moments in the lives of animals that lived millions of years ago. As one paleontologist noted, tracks are about the closest thing we have to 'motion pictures' of dinosaurs. The study of fossilized tracks has become a thriving discipline in recent decades, as evidenced by the number of books dedicated to the topic (for example, Lockley 1991). Unlike bones, shells, or other fossil remains of the animals themselves, footprints in trackways record the activities of animals, and sometimes provide information unavailable from skeletons. Sauropod dinosaurs, for example – the huge *Brontosaurus*-type with long necks and tails – were traditionally thought to be aquatic. It was postulated that their gargantuan bodies were simply too massive to be supported on land. This hypothesis has now been disproved on physiological grounds, but the first and best evidence that sauropods were not limited to an aquatic regime came from tracks. Numerous sauropod trackways demonstrate conclusively that these incredible beasts not only spent time on land but also walked upright with all four legs directly beneath the body and with the tail held aloft instead of dragging along the ground.

Dinosaur tracks have many limitations, however. Paleontologists generally do not attempt to correlate specific tracks with specific dinosaurs. Rather they devise new names, or 'ichnotaxa', for those forms based solely on footprints. The reason for this practice is not surprising – it is difficult to identify the trackmaker with any precision. Feet tend to be rather conservative in dinosaurs, so tracks that appear identical may record the footsteps of quite different animals. To confound matters, the same animal can produce very different kinds of tracks by walking on different surfaces. Deep mud may obscure foot anatomy, while a much harder surface may not record detailed features such as claws. James Farlow, a dinosaur paleontologist and track specialist at Purdue University, has observed ostriches running over a variety of surfaces. He postulates that, depending on the surface, we might even be confusing the clawed tracks of carnivorous dinosaurs with the hoofed tracks of herbivorous forms.

Nonetheless, tracks have much to tell us about dinosaur life and evolution, from walking and running speeds to social behaviour and the turnover of species through time. As it turns out, the numerous dinosaur tracks found in British Columbia are without doubt some of the best and most informative in the world.

Dinosaur Tracks in British Columbia

The earliest record of discovery of dinosaur tracks in North America (and perhaps anywhere) was in 1800, when a large, three-toed footprint was inadvertently ploughed up in Massachusetts by a fellow named Pliny Moody. Dinosaurs were still unknown in 1800, and Moody believed his find to be a track of Noah's raven. The discovery of dinosaur tracks in Canada came more than a century later in 1922, when F.H. McLearn, a Geological Survey of Canada geologist, identified dinosaur tracks along the Peace River Canyon, near Hudson's Hope, British Columbia. At the time, this discovery was the most northerly record of dinosaurs in the world. Following up on this discovery, C.M. Sternberg, one of the greatest collectors of Canadian dinosaurs, ventured to the Peace River site and found over 400 tracks from a variety of dinosaurs (Sternberg 1932). Sternberg named eight new types of dinosaur ichnotaxa, including both bipedal (two-footed) and quadrupedal (four-footed) forms, based on finds from this site alone. Indeed, dinosaur footprints were so common and well preserved in this area that they likely were seen by Alexander Mackenzie when he explored the canyon in 1793.

Unfortunately, we must use the past tense in describing the remarkable Peace River dinosaur tracks. In 1965, with the impending completion of the W.A.C. Bennett Dam, staff of the Royal Ontario Museum visited the construction site to study and cast numerous footprints. More than fifty tracks were observed by that expedition, tracks that are now covered by the dam itself and by Williston Lake, the largest lake in British Columbia. In the 1970s, construction began on the Peace Canyon Dam, and it became clear that another world-class site would soon be under water. Consequently, the Provincial Museum of Alberta (the parent institution of the Royal Tyrrell Museum of Palaeontology) sent four 'salvage operations,' funded largely by BC Hydro, to record data and collect specimens. In total, more than 1,700 footprints were documented prior to the area being flooded (Mossman and Sarjeant 1983). Today, these tracks are inundated by a lake between the W.A.C. Bennett and Peace Canyon dams. Dinosaur tracks can still be found in the area but not nearly in the abundance known previously.

The dinosaur tracks in Peace River Canyon occur in rocks of Early Cretaceous age (about 115 million years ago). At least nine other dinosaur trackway sites have been found in British Columbia (Currie 1989), ranging in time from the Late Jurassic (about 150 million years ago) to the Upper Cretaceous (about ninety million years ago). Most of these localities occur in the eastern and southeastern part of British Columbia, although there are unconfirmed reports of tracks in the extreme northeastern corner of the province along the Liard River. The tracks at these sites vary from a few footprints to lengthy trackways, and they have been discovered in some extraordinary settings.

Some of the best specimens in British Columbia and elsewhere are found on the ceilings of underground coal mines. Coal is composed of organic remains and is often formed in bog-like environments ideal for the preservation of footprints. Large dinosaur tracks have been known to break away from the ceilings of coal mines, posing a hazard to miners. The oldest known tracks in western Canada come from coal mines near Michel, British Columbia. These prints are relatively small (less than thirty-five centimetres long) and represent tracks from bipedal dinosaurs. As with the tracks found along the Peace River, the Michel tracks represent a dinosaur community that inhabited a lowland coal swamp. Some of the original track specimens, recovered from these mines in the 1940s, are preserved in the collections of the Canadian Museum of Nature and the Royal British Columbia Museum, although the majority apparently have disappeared into private collections. Other footprints have now been found near Fernie and Elkford, although these ones are from the sandstone layers between the coal seams. Today, fossilized tracks are found less frequently in mines due to changes in mining technology.

Another major trackway site occurs in Lower Cretaceous deposits along the Narraway River in eastern British Columbia. These dinosaur tracks, which no longer exist, were in a rather different setting, situated on a near-vertical cliff face composed of sandstone (Figure 12.2). The sandstone was covered with ripple marks and likely represented an ancient mudflat. There were at least eight trackways, and more than two hundred footprints, all preserved on a single bedding plane. Although numerous tracks on this cliff face were of small, meat-eating dinosaurs, perhaps the most fascinating tracks are of a large, bipedal dinosaur, probably a carnivore as well. The trackway showed that this animal was walking west when it stopped in its tracks (literally), turned to the right, paused briefly, and then proceeded at a right angle to the original path. One can only imagine what would cause a multi-tonne carnivorous dinosaur to change direction so abruptly.

This site was originally described to us by Don Stott of the Geological Survey of Canada in the 1970s, and was examined in 1981. Photographs were shown to staff of *National Geographic* magazine, who decided to send

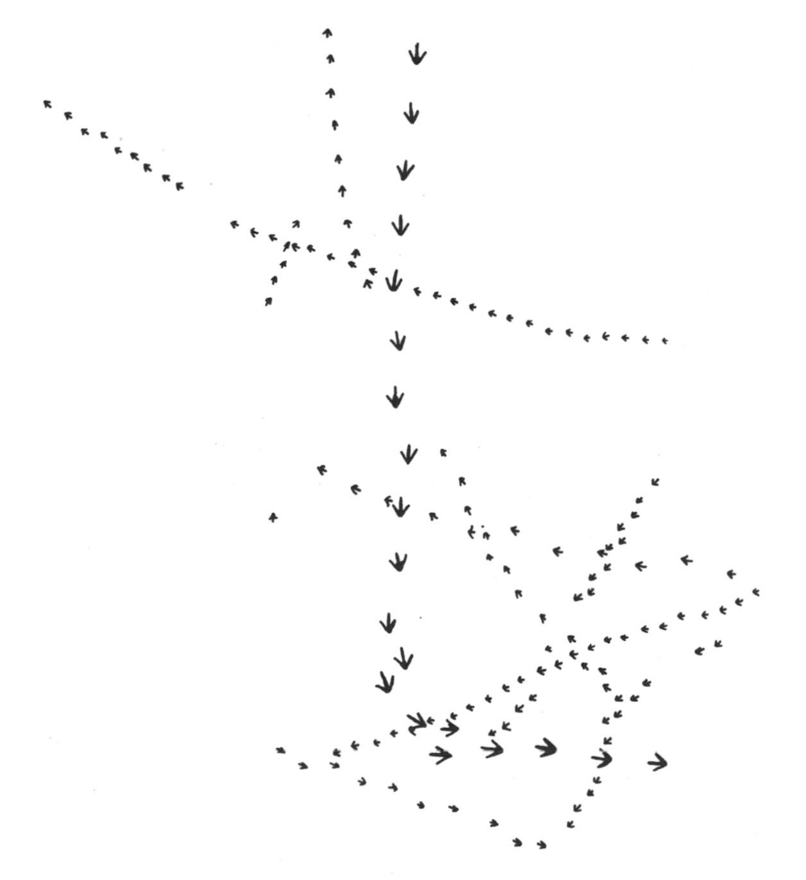

Figure 12.2 **Map of trackway along the Narraway River taken from photographs shot from a helicopter.** Unfortunately, there was no scale on the cliff face, but the largest footprints are about half a metre long.

photographers to take shots suitable for an article on dinosaurs (which subsequently appeared in January 1993). The site is not accessible by road or river, and the only practical way to get to it is by helicopter. After months of planning, a helicopter carried in personnel for the expedition, only to find that the layer with the footprints had slid off the cliff face into the Narraway River. That was an incredible shock, but worse was to come. The backup trackway site for photographs, in Grande Cache, Alberta, collapsed the following night, destroying perhaps another hundred dinosaur tracks. The *National Geographic* crew returned empty-handed, and had to venture as far as Argentina to obtain the equivalent photograph of dinosaur trackways.

Dinosaur Diversity and Distribution

As very few dinosaur bones are known from British Columbia, we must rely on trackways to tell us about the kinds of dinosaurs that inhabited the province (Figure 12.3). The diversity of dinosaurian forms is best documented in the abundant trackways of the Peace River Canyon (Sternberg 1932; Currie and Sarjeant 1979; Mossman and Sarjeant 1983). By far the most common track fossil found to date at this locality is that of a large, bipedal, three-toed animal identified as *Amblydactylus* (Figure 12.4), for which two species have been named. Although originally associated with the primitive ornithopod dinosaur *Iguanodon* (Sternberg 1932), the size and shape of the *Amblydactylus* handprints and footprints indicate that the trackmakers were probably duckbilled dinosaurs, or hadrosaurs (Currie 1983). Well-preserved specimens show impressions made by fleshy webs between the toes. Some footprints show evidence of a large, bulbous pad of soft tissue on the underside of each toe. The tips of all three digits are relatively rounded, indicating the presence of a blunt hoof.

Hadrosaur fossils are extremely common in Upper Cretaceous deposits from around the world. In many regions, these herbivorous ornithopods are the dinosaurs found most frequently, sometimes occurring in massive bone beds containing thousands of individuals. *Amblydactylus* tracks are particularly significant because the oldest hadrosaur bones are of Late Cretaceous age, millions of years younger than the footprints found along the Peace River. Thus the Peace River tracks may represent the earliest record of hadrosaurs, showing that they were alive and well during the Early Cretaceous of British Columbia.

Another dinosaur track commonly found along the Peace River is a bipedal carnivore Sternberg named *Irenesauripus* (Figure 12.3). Like *Amblydactylus,* these tracks are three-toed, but the footprints of *Irenesauripus* are much smaller, with a narrower heel and thinner toes that terminated in sharp claws. Smaller still are the footprints of another biped, *Irenichnites,* represented by several specimens including a slab with five tracks. The

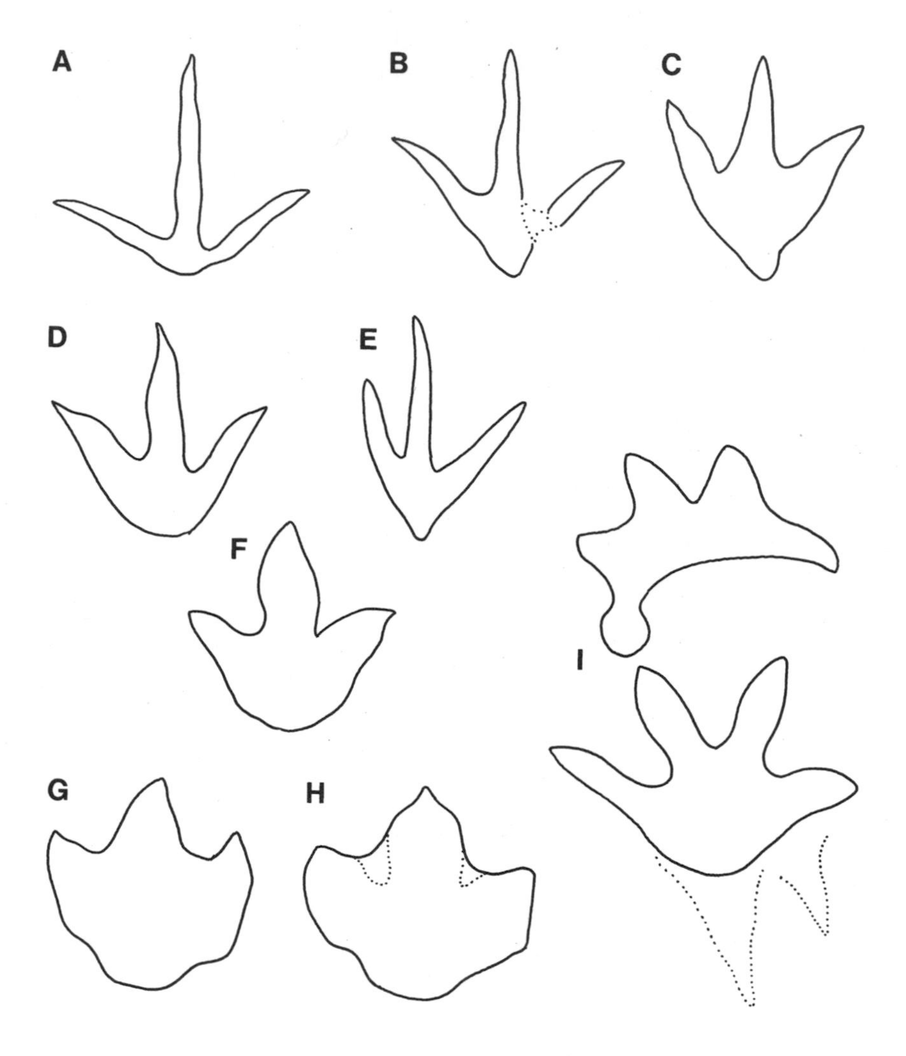

Figure 12.3 **Distinctive types of bird and dinosaur footprints found in British Columbia.** (A) *Aquatilavipes swiboldae,* a bird footprint 4 cm long. (B) *Irenichnites gracilis,* a small theropod track 15 cm long. (C) *Irenesauripus mclearni,* a theropod footprint 33 cm long. (D) *Columbosauripus gracilis,* a possible theropod track 15 cm long. (E) *Irenesauripus acutus,* a large theropod footprint 55 cm long. (F) *Gypsichnites pacensis,* a small ornithopod with a foot 30 cm long (G) *Amblydactylus gethingi,* a 64 cm long track of an iguanodontid or hadrosaur. (H) *Amblydactylus kortmeyeri,* an iguanodontid or hadrosaur footprint 42 cm in length. (I) *Tetrapodosaurus borealis,* left handprint and footprint (the latter is 26 cm long). Dotted lines represent interpolated parts of the prints.

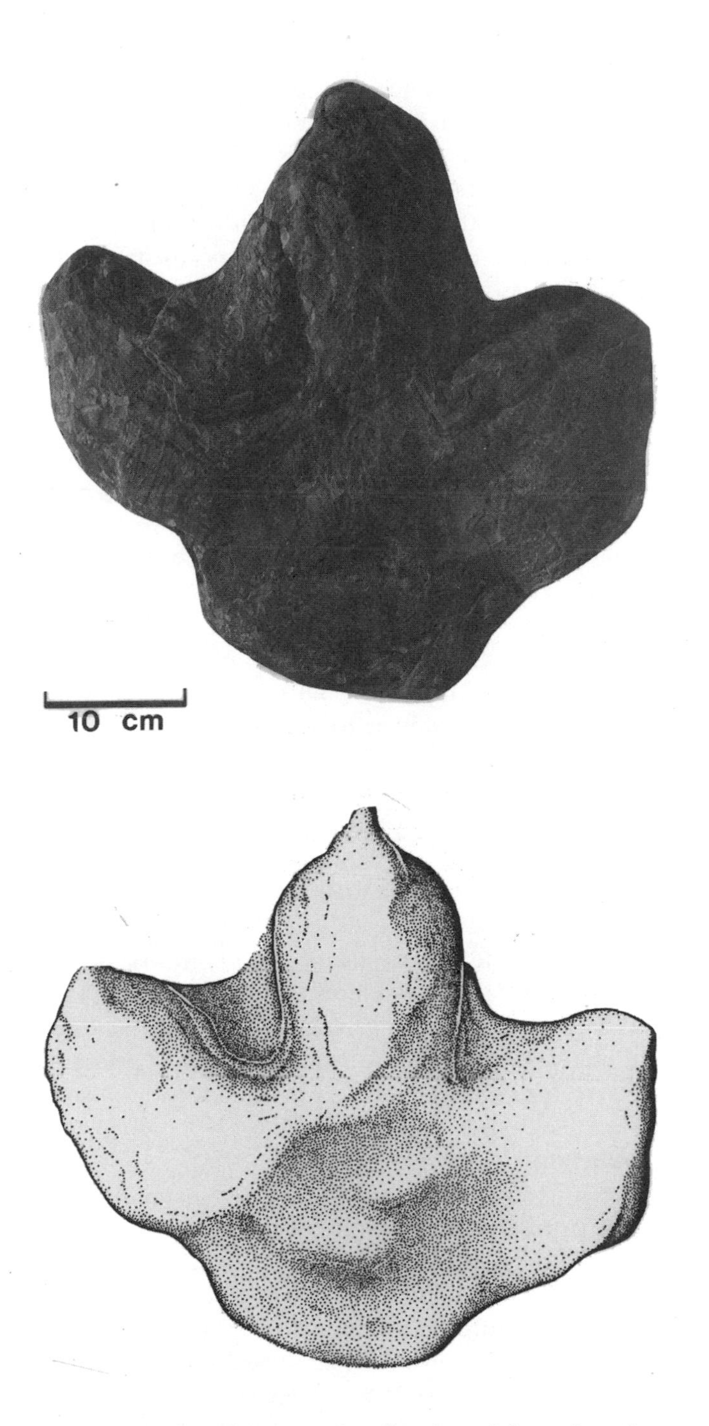

Figure 12.4 **Photograph and interpretive drawing of footprint of *Amblydactylus kortmeyeri* from the Peace River Canyon, British Columbia**
Source: Collections of the Royal Tyrrell Museum of Palaeontology

footprints of *Irenichnites* are of questionable origin, but may represent the tracks of a small carnivorous dinosaur. Another distinctive form found in the Peace River Canyon before it was flooded is *Tetrapodosaurus*. *Tetrapodosaurus* ('four-footed-reptile') was a relatively large quadrupedal dinosaur, as revealed by the trackways, which include imprints of both the fore and hind feet (Figure 12.3). Sternberg postulated that the track-maker in this case was a horned dinosaur, or ceratopsian. The detailed anatomy revealed by the tracks, however, and the age of the track itself – younger than any known ceratopsid dinosaurs – suggest that *Tetrapodosaurus* was an armoured dinosaur, or ankylosaur.

Finally, because owls, ostriches, and indeed all birds are believed to be the direct descendants of dinosaurs (and therefore are dinosaurs in the strict sense), we must consider one last set of tracks. In 1979, bird trackways were found in the Peace River Canyon. At the time, this specimen represented the earliest evidence of bird footprints, although some examples from the Jurassic have been reported recently. Like the tracks of small theropod dinosaurs, the bird footprints are tridactylous (three-toed), and the toe impressions are thin. The bird tracks at the Peace River site and elsewhere are small (less than fifty millimetres), and most were likely made by wading shorebirds similar to modern-day plovers.

Notably absent from the fossil footprint assemblages of British Columbia are sauropod dinosaurs. Sauropods have been found in Jurassic rocks from the United States and Asia, and their tracks are known from Lower Cretaceous deposits in Texas. Yet sauropod fossils, either bones or tracks, are unknown from Canada. A land bridge connected Asia and North America for part of the Cretaceous, and the known distribution of sauropods during the Late Cretaceous included both Asia and the United States. So the complete lack of their bones and tracks in Canada is intriguing. As there is no evidence of a geological barrier preventing access by sauropods to northwestern North America during the Cretaceous, we can only assume that they did not venture into Canada for ecological reasons.

Dinosaur Evolution

Throughout their 160-million-year history, dinosaurs were constantly evolving, making substantial modifications in their body plans. Footprints and trackways can provide a measure of the evolutionary changes taking place in the limbs of dinosaurs. If dinosaur trackways from the Lower Jurassic rocks of the Connecticut River valley are compared with those of the Peace River Canyon, it is evident that the earlier animals moved much more slowly and that they frequently dragged the ends of their tails on the ground. The trackways are wider in the Early Jurassic species, which also tended to have more toes than the Early Cretaceous dinosaurs of British Columbia.

The trackway localities of western Canada cover the last eighty million years of dinosaur history. Cretaceous sites tend to show similar types of faunas, dominated by hadrosaurs and large theropods. On the other hand, the uppermost Jurassic sites of southeastern British Columbia have footprints that are more like those of Nova Scotia and parts of the United States. These tracks generally represent much smaller animals, and there is a significant absence of large ornithopods such as iguanodontids and hadrosaurs. Thus the footprint record of British Columbia helps to document the faunal turnover that seems to have occurred at the beginning of Cretaceous times about 145 million years ago.

Dinosaur Behaviour

In the Peace River Canyon, hadrosaur tracks have been found in most of the rock types represented, including sediments deposited in rivers, lagoons, swamps or marshes, and mudflats on the floodplains. These occurrences show that the hadrosaurs walked (and lived) in all these environments. What is perhaps more significant is that the carnivore tracks show the same kind of distribution, leading one to believe that they would not have been averse to following potential prey into the water.

Trackways provide evidence of animal locomotion. No tail drags are associated with any of the trackways, providing strong evidence that Cretaceous dinosaurs held their tails high above the ground. The footprints also show that dinosaurs were efficient walkers that held their legs directly beneath their bodies, and generally walked with long strides on land. Large hadrosaurs seem to have walked on all four legs, because handprints are usually associated with the footprints. This is not always the case, however, as the smaller individuals seem to have walked mostly on their hind limbs.

Many track sites along the Peace River Canyon were made by dinosaurs while in the quiet waters of ponds and placid rivers. In one case, a hadrosaur was walking in a stream, leaving tracks deeply impressed in the mud. As the water became deeper, its stride decreased, and it appears to have been pushing off the bottom with its toes, because the heel of the foot is only shallowly impressed in each track. At one point, the midline of the trackway shifts more than a metre to the right, and several steps later it shifts to the left again. It is difficult to explain these shifts unless the three-to-four-tonne weight of the hadrosaur was being buoyed up in the water. Other lines of evidence support the viewpoint that many of the tracks were made underwater by swimming animals. It is not surprising that only large footprints are present at some localities, because the legs of younger individuals would not have been long enough to reach the bottom and leave footprints.

The orientation of the Peace River tracks, as well as the form of the trackways, suggests that the ornithopods were somewhat slower and less efficient walkers than the theropods. The ornithopod trackways are 'pigeon-toed,' indicating that the herbivores were rotating their bodies in an arc of about thirty degrees as they walked, swinging their tails from side to side. The theropods, on the other hand, had longer strides, and placed one foot almost directly in front of the other. A formula developed by R. McNeill Alexander uses footprint size and stride length to estimate the speeds of dinosaurs. Applying this formula to the dinosaur trackways in the Peace River Canyon, it is apparent that the hadrosaurs were generally slower-moving than the theropods. On average, hadrosaurs seem to have been walking through the mud at speeds of about six kilometres per hour, compared with theropods at 7.5 kilometres per hour. The highest speed calculated for the ornithopods was only 8.5 kilometres per hour, although it must be kept in mind that all of these animals were walking in mud. One of the Peace River Canyon theropods was calculated to have been running at about fifteen kilometres per hour.

The trackways in the Peace River Canyon provide strong support for the notion that at least some dinosaurs were gregarious. One locality in the canyon preserves the record of nine or ten hadrosaurs *(Amblydactylus)* walking across an ancient flood-plain deposit (Figure 12.5). As they walked, they were actually changing direction. Four of the trackways (C, D, E, and F) follow the same sinuous curves but do not intersect, and seem to have been made by animals that were walking side by side. One interpretation is that these four individuals were walking so close together that when F changed course suddenly, the remaining three animals modified their courses to avoid collision.

Some of the trackways from the Peace River Canyon were made by babies and juveniles. The tracks of one baby *Amblydactylus* are associated

Figure 12.5 **Map of track site in the Peace River Canyon showing multiple trackways attributed to *Amblydactylus gethingi*.** The site shows several animals moving in a single direction and apparently adjusting their paths in unison; the site is therefore good evidence for group behaviour in at least some ornithopod dinosaurs.

with the tracks of another individual of the same size, but no baby footprints were found with juvenile or adult tracks. This is another circumstantial line of evidence supporting the hypothesis that in some species of dinosaurs, the babies did not join the herds of their own species until they were large enough to avoid being stepped on.

More dinosaur tracks will almost certainly be found in British Columbia as more of the province is explored in detail, both above and below ground. Despite the scarcity of dinosaur bones to date, there is tremendous potential for discovering such fossils in British Columbia. While it is true that most Mesozoic terrestrial deposits in the province are well vegetated, there are areas where these rocks are exposed, particularly along river channels. Similarly vegetated areas in Alberta have been explored in recent years, with prolific results; localities discovered near Crowsnest Pass, Edmonton, Hinton, and Grande Prairie are just a few of the newer dinosaur sites. Several rivers draining from British Columbia into Alberta have yielded dinosaur bones along their channels on the Alberta side, while the BC side has never been adequately surveyed for fossils. A number of fossil-rich dinosaur sites have been found in Alaska (also in areas with minimal rock exposures), and it is possible that dinosaurs were migrating between Alaska and Alberta via northeastern British Columbia. Indeed, the probability of finding significant dinosaur localities in this region of the province is considered to be very high. Although the rugged terrain of British Columbia means that the search will be difficult, such efforts are likely to be well rewarded.

References

Currie, P.J. 1989. Dinosaur Footprints of Western Canada. In *Dinosaur Tracks and Traces,* ed. D.D. Gillette and M.G. Lockley. Cambridge: Cambridge University Press

–. 1983. Hadrosaur Trackways from the Lower Cretaceous of Canada. *Acta Palaeontologica Polonica* 28:63-73

Currie, P.J., and W.A.S. Sarjeant. 1979. Lower Cretaceous Footprints from the Peace River Canyon, BC, Canada. *Palaeogeography, Palaeoclimatology, Palaeoecology* 28:103-15

Lockley, M.B. 1991. *Tracking Dinosaurs: A New Look at an Ancient World.* Cambridge: Cambridge University Press

Mossman, D.J., and W.A.S. Sarjeant. 1983. The Footprints of Extinct Animals. *Scientific American* 248(1):74-85

Sternberg, C.M. 1932. Dinosaur Tracks from Peace River, British Columbia, 59-85. National Museum of Canada. *Annual Report, 1930*

13
Ancient Saurians: Cretaceous Reptiles of Vancouver Island

Rolf Ludvigsen

Reptiles now exist under restrained circumstances. No reptile flies and, with the exception of sea turtles, none is truly marine. On land the only large reptiles are crocodiles, monitor lizards, tortoises, and pythons, and those occur only in the tropics. By contrast, during their heyday in the late Mesozoic, saurians – to use the traditional term for extinct reptiles – dominated the air, the water, and all land areas from the tropics to the temperate regions. Pterosaurs soared through the skies; ichthyosaurs, plesiosaurs, elasmosaurs, and mosasaurs plied the seas for prey; and herbivorous and predatory dinosaurs of all sizes lived in a variety of terrestrial settings.

Most of the Cretaceous fossil reptiles known in North America were collected from the broad belt of flat-lying strata that now underlies the Great Plains – from the Gulf of Mexico, north through Kansas and Montana, and into Alberta. These strata were laid down in a shallow sea that bisected the continent during the Cretaceous and in the bordering swampy lowlands. Only a few reptilian fossils have been collected from rocks of this age on the west side of the Rocky Mountains and, until a few years ago, none was known from the Pacific coast of British Columbia.

Since the middle part of the last century, the sedimentary rocks of Late Cretaceous age exposed on eastern Vancouver Island and the Gulf Islands have been studied and sampled by many geologists and paleontologists. These investigations produced thousands of fossils: ammonoids, bivalves, gastropods, crustaceans, and other invertebrates, in addition to a variety of plants. Shark teeth and vertebrae, and fish scales were also recovered from these marine rocks, but reptile bones eluded the collectors.

Starting in the mid-1980s amateur paleontologists began to discover fossil reptile bones in the Upper Cretaceous shales exposed on river banks in the vicinity of Courtenay. Recognizing the scientific importance of these fossils and, perhaps equally as significant, their value in public education, the Courtenay and District Museum secured sufficient funding to excavate a few of the large specimens – notably, an impressive ten-metre-long elasmosaur, which was immediately christened the Puntledge Elasmosaur, and a metre-long row of mosasaur vertebrae.

These fossils, particularly the Puntledge Elasmosaur (invariably misrepresented as a dinosaur), became the focus of a deluge of popular interest. Countless articles appeared in local newspapers, television crews documented the 'dig' on a weekly basis, and an imaginative papier-mâché reconstruction of an elasmosaur was featured in a Canada Day parade. Frequent field trips to the excavation inspired school children to paint colourful dioramas that featured different versions of ferocious extinct monsters. And, when the first fossil bones were put on display, attendance at the Courtenay Museum increased tenfold.

Fossils of four distinct kinds of Cretaceous saurians have now been collected in the Comox Valley: three types of sea-going reptiles (turtle, elasmosaur, and mosasaur) and a single terrestrial reptile (theropod dinosaur). By itself, each of these is a rare and important find, but discovery of this variety of fossil reptiles in such a short time and in such a small area where none had been known before is truly remarkable. The efforts of amateur paleontologists in the Comox Valley have now permitted definition of a new vertebrate fossil region in Cretaceous rocks of the Canadian west coast – one that was entirely unknown a decade ago.

Chelonioid: A Sea Turtle

Turtles first appeared in the Late Triassic as a group of specialized reptiles with a carapace composed of bony interlocking scutes fused to the vertebrae and the ribs. Living chelonioids (sea turtles, including the leatherback turtle) employ 'underwater flight' to migrate great distances across oceans.

In the summer of 1984 amateur paleontologist Richard Boldt of Comox was examining the rubble displaced by recent blasting to create a fish-ladder at Stotan Falls on the Puntledge River, when he spotted some pieces of large bone in the siltstone blocks. Once these bones were prepared by technicians at the Royal Tyrrell Museum of Palaeontology and studied by paleontologist Betsy Nicholls, they proved to belong to the chelonioid *Desmatochelys,* which occurs widely in Cretaceous rocks exposed in the interior of North America. The Stotan Falls material includes a humerus (upper arm bone) thirty centimetres long and a lower jaw. This discovery is the second sea turtle found in Cretaceous rocks on the Pacific coast. To date, it is the only Vancouver Island fossil reptile that has been described formally (Nicholls 1992).

Elasmosaur: A Swan Lizard

The small resort town of Lyme Regis faces the Channel on the Dorset coast in southern England, where it is nestled between cliffs of shales and limestones called the Blue Lias. Packed with ammonoids, clams, sea urchins, and a variety of marine reptiles, these Lower Jurassic strata have attracted English paleontologists and fossil collectors for well over 200 years. As well, this area was the setting for John Fowles's best-selling novel *The French Lieutenant's Woman,* in which a gentleman paleontologist and fossil sea urchins figure prominently.

During the first half of the nineteenth century, Lyme Regis was also the home of Mary Anning – a canny businesswoman and one of the first professional fossil collectors. Her fossil and shell shop was frequented by vacationers seeking unique souvenirs of their stay in Lyme Regis and by collectors and paleontologists who wanted Jurassic display and research specimens of high quality. Anning is reputed to have inspired the alliterative tongue-twister, 'She sells sea shells sitting by the sea shore.'

Anning was certainly not afflicted by the excessive modesty and reticence expected of young unmarried women in early Victorian England. When the King of Saxony visited her shop to buy specimens for his museum in Dresden, she boasted that she was 'well known throughout the whole of Europe.' Over the years, Mary Anning discovered and collected many remarkable skeletons of reptiles in the Blue Lias. When an important specimen was collected, she alerted the paleontologist who would be most interested in it and, more important, who would be most able to pay for it.

In 1811 when she was twelve years old, Mary and her brother excavated the first ichthyosaur ever – a magnificent skull over a metre long. Later, she found the first pterosaur bones in Britain. In 1823 she collected a nearly complete skeleton of a new type of marine reptile – a long-necked animal three metres in length with a small head, a flattened body, and two pairs of large paddles (Figure 13.1). Paleontologists Henry De la Beche and William Conybeare had studied fragmentary fossil remains of such an animal a few years earlier and had christened it *Plesiosaurus* – Greek for 'nearer to lizards' – because it was more similar to a crocodile than to an ichthyosaur. They were delighted when Anning's specimen confirmed their preliminary interpretations.

Since Anning's time, plesiosauroids have been collected from Jurassic and Cretaceous marine rocks of every continent. These reptiles are now divided into two groups – the plesiosaurs with only twenty-eight neck vertebrae and the elasmosaurs with as many as seventy-six.

Numerous complete specimens of plesiosaurs and elasmosaurs have been collected from the central Cretaceous seaway of North America and a few elasmosaurs have been recovered from Cretaceous strata of California. Until a November afternoon in 1988, however, none was known from the Pacific coast of British Columbia.

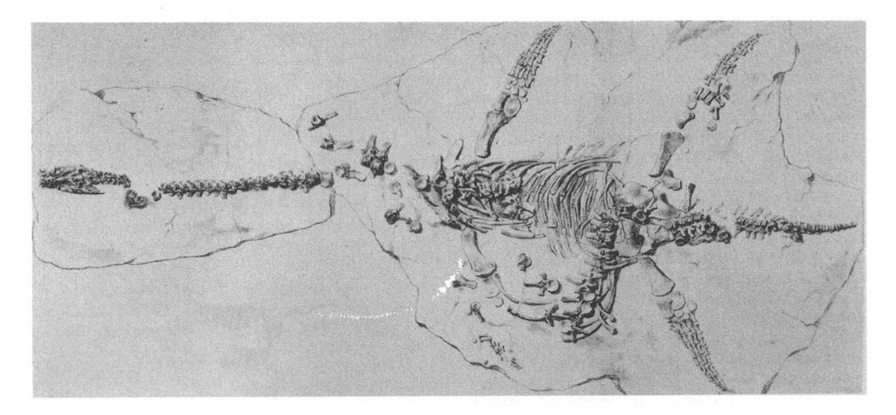

Figure 13.1 **The first complete plesiosauroid, *Plesiosaurus dolichodeirus*.** Collected by Mary Anning from Lower Jurassic rocks near Lyme Regis, England. *Source:* From an 1823 paper by William Conybeare

That afternoon, surveyor and amateur paleontologist Mike Trask and his daughter Heather were searching for fossils in the Upper Cretaceous shales exposed along the Puntledge River just west of Courtenay. Trask had expected to find only the usual ammonoids and clams. So, when they came across a couple of massive barrel-shaped, bony objects protruding from the shale, each the size of a can of beans, Mike knew they had discovered something new and significant. Trask returned over the next few days and uncovered another thirty-eight of these bony cylinders in a shallow excavation. A concretion with teeth and bone fragments protruding from its side provided the promise of a complete jaw and skull.

Because no paleontologist in British Columbia specializes in Mesozoic reptiles, these bones were sent to the Royal Tyrrell Museum of Palaeontology in Alberta. Betsy Nicholls was well qualified to identify them because she had just completed a PhD thesis on Cretaceous marine reptiles of Manitoba. She concluded that the bony cylinders were neck vertebrae of an elasmosaur, the first to be discovered on the Canadian west coast.

An elasmosaur, or swan lizard, was a plesiosauroid with a flattened, streamlined body, two pairs of paddle-like limbs, a long flexible neck, and a small head with long, interlocking teeth used to capture and trap fish. Elasmosaurs grew to lengths of up to twelve metres – half of which was neck (Figure 13.2). A reconstruction would seem familiar to most people because the swan lizard has been accepted as a model for 'typical sea or lake monsters' that are reported regularly from Loch Ness in Scotland and Lake Okanagan in British Columbia. Nevertheless, the available fossil record is unequivocal – elasmosaurs became extinct at the end of the Cretaceous.

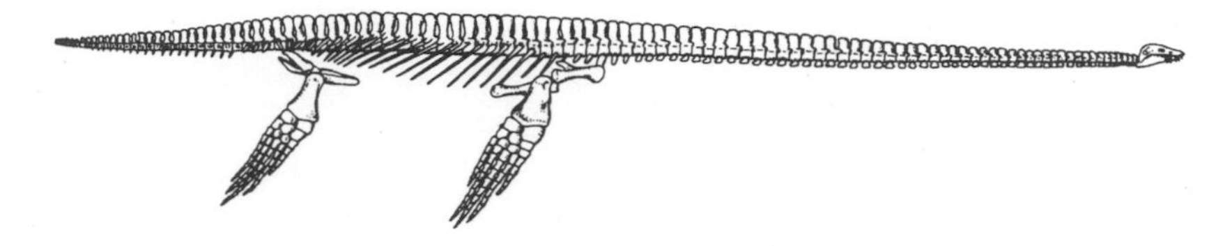

Figure 13.2 **Reconstruction of the Late Cretaceous elasmosaur *Hydrotherosaurus.*** This genus grew to a length of 12 m.
Source: From Carroll (1988)

My involvement with these fossils came a few years later when the Courtenay Museum asked me to supervise the excavation of the body of this elasmosaur skeleton. This task was accomplished over two months in the spring of 1992, thanks to the efforts of a large number of dedicated volunteers from the Comox Valley. When the excavation began, the volunteers were neophytes but, as it proceeded, they became increasingly knowledgeable about paleontological techniques. Hand tools only were used to remove about a hundred cubic metres of shale and tough siltstone in order to expose the fossil-bearing horizon. This horizon contained four oblong, metre-sized concretions encasing massive bones and, in between, additional vertebrae and ribs. After they were assigned identification numbers and their locations accurately mapped, the fossil bones and bone-bearing concretions were carefully chiselled out of the rock and moved to safe storage at the museum.

The time-consuming task of preparation – that is, the meticulous removal of the rock surrounding around each bone – has now been completed, and it is evident that the Puntledge Elasmosaur has some surprising tales to tell. Some of the vertebrae are in excellent condition (Figure 13.3). Unfortunately, the skull was not well preserved. Only the back of the skull survived, but the jaw is well preserved and is unusually large for an elasmosaur. This jaw is V-shaped and half a metre long. Thirty-four long, curved teeth of this snaggle-toothed reptile have now been recovered (Figure 13.4). A formal identification and comparison with other elasmosaurs has not been completed. Therefore, at this time, it is not known to what extent this beast is similar to elasmosaurs of the same age from California. Regardless of its name and its affinity, the cast of the reconstructed skeleton of the Puntledge Elasmosaur makes an impressive display at the Courtenay Museum (Figure 13.5).

The incomplete bony remains of an animal that took its last breath some eighty million years ago can disclose details of its immediate post-mortem fate. We don't know how it died, but we can speculate on what happened to this animal after its death. The humerus of elasmosaurs is a broad, flat triangular bone at the base of the paddle-like limbs. The

Figure 13.3 **Neck vertebra of the Puntledge Elasmosaur.** This vertebra shows the high backswept neural spine and the lateral processes.
Source: Collections of the Courtenay Museum

humerus prepared from the Puntledge Elasmosaur has bite marks on one end. Bite marks by what? A tell-tale shark's tooth embedded in the bone provides the answer. Because a ten-metre-long elasmosaur carcass was a rich storehouse of protein, it is not surprising that it would have been scavenged by the top-level predators of the time.

Mosasaur: A Sea Lizard

One of the more unusual spoils of war claimed by the French republican armies that swept north into Holland in 1795 was a massive fossil skull and jaw a metre and a half long and studded with sharp teeth. It had been kept at a shrine in the town of Maastricht in the valley of the Meuse River. This monster, thought to be an antediluvian crocodile or whale, had been found years before in a local chalk mine. Brought to Paris, it was claimed by Georges Cuvier, Professor of Anatomy at the Musée

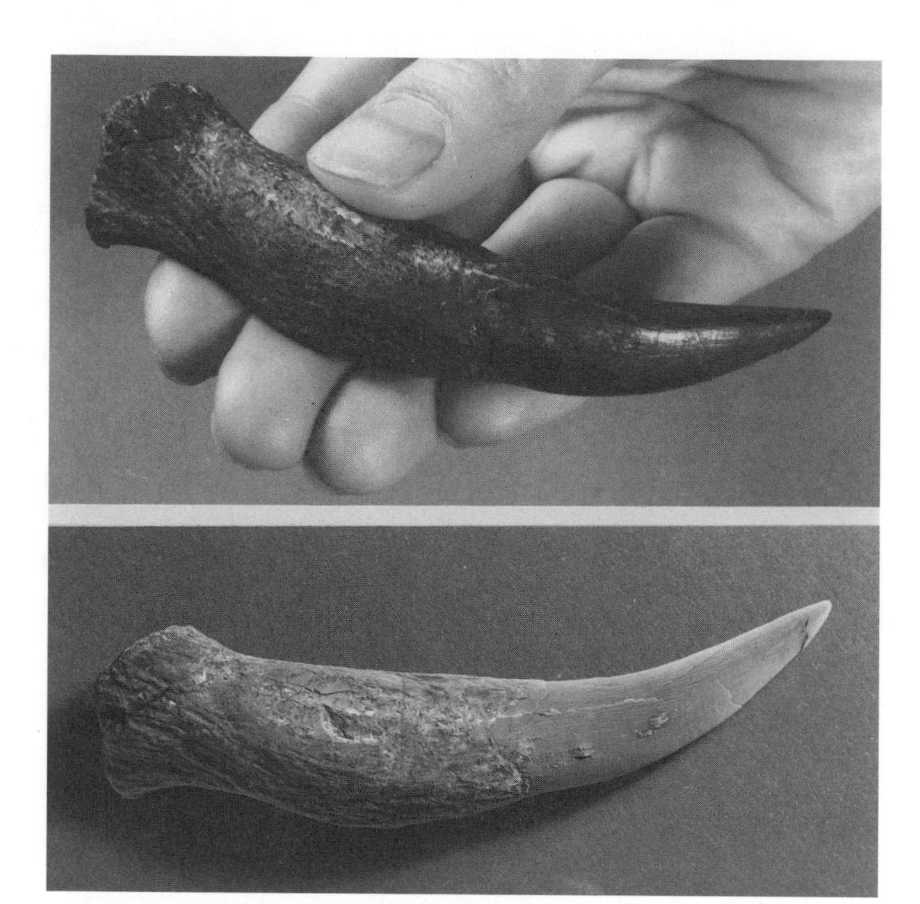

Figure 13.4 **Isolated tooth of the Puntledge Elasmosaur.** The tooth is photographed in its natural state (top) and coated with ammonium chloride to emphasize surface sculpture (bottom).
Source: Collections of the Courtenay Museum

Figure 13.5 **Reconstructed cast of the Puntledge Elasmosaur suspended from ceiling of the Courtenay Museum**

d'Histoire Naturelle, who was soon to become probably the most celebrated scientist in Europe. Cuvier correctly concluded that the Maastricht skull belonged to a gigantic relative of the living monitor lizard. This relic was one of the first fossils of large reptiles collected from Mesozoic strata in northwestern Europe. It was soon to be joined by many other large, perplexing bones of saurians – tangible proof that the distant past was populated by reptilian giants.

Just before the skull reached the museum, Cuvier had demonstrated that some animals had become extinct. At that time, extinction was an almost heretical concept because it implied a waste of the Creator's power. By comparing living elephants with mastodons and mammoths, Cuvier concluded that the fossil elephants had become extinct as a result of a fairly recent catastrophe that abruptly wiped out many of the large animals living on land. To Cuvier, the Maastricht skull, coming from a more distant epoch and apparently from a sea animal, suggested that other extinctions must have happened in the geologic past – the consequence of a series of cataclysms that terminated animal life on land and in the sea. Each catastrophic extinction was followed by divine creation of progressively more modern animals.

The English paleontologist William Conybeare provided the Maastricht skull with a name – *Mosasaurus,* 'the lizard from the Meuse area.' The family Mosasauridae, which comprises twenty genera of sea lizards, is now known from every continent but Antarctica, and always from rocks of Late Cretaceous age.

Mosasaurs were large, slender marine predators with long tails and two pairs of paddle-like limbs (Figure 13.6). These animals used lateral undulations of the body to propel themselves through the water in pursuit of the ammonoids and fishes that were their main prey. On Vancouver Island the fossil record of mosasaurs consists of bones and of holes – the bones are mainly vertebrae and the holes are bite marks on ammonoids.

Mosasaur vertebrae are readily identifiable in the field because they are procoelous, that is, both their front and back sides are convex. Mosasaur vertebrae are now known from two localities on the Puntledge River near

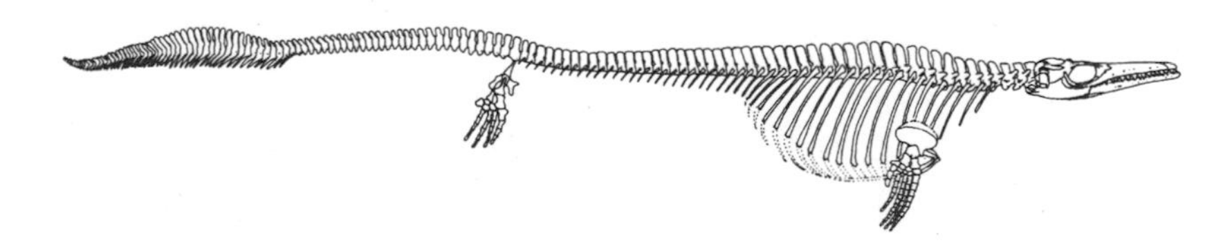

Figure 13.6 **Reconstruction of the Late Cretaceous mosasaur *Plotosaurus*.** This genus grew to a length of 10 m.
Source: From Russell (1967)

downtown Courtenay – one preserving a string of twenty-one vertebrae (one of these is illustrated in Figure 13.7). Such vertebrae cannot be identified more precisely than the family. But some years ago, Qualicum Beach amateur paleontologist Graham Beard and his wife Tina collected a concretion from Upper Cretaceous rocks on Hornby Island that contains an incomplete reptile skull. Mosasaur specialist Dale Russell from the Canadian Museum of Nature identified this skull as belonging to the genus *Tylosaurus* (Ludvigsen and Beard 1994). Beard later collected a femur of a mosasaur from the same locality.

Bite marks are technically trace fossils – fossils that reveal information about the behaviour of ancient animals. Other examples of trace fossils include burrows or borings. A Cretaceous ammonoid from Hornby Island and a nautiloid from Vancouver Island contain rows of puncture holes that appear to have been inflicted by mosasaurs (Ludvigsen and Beard 1994).

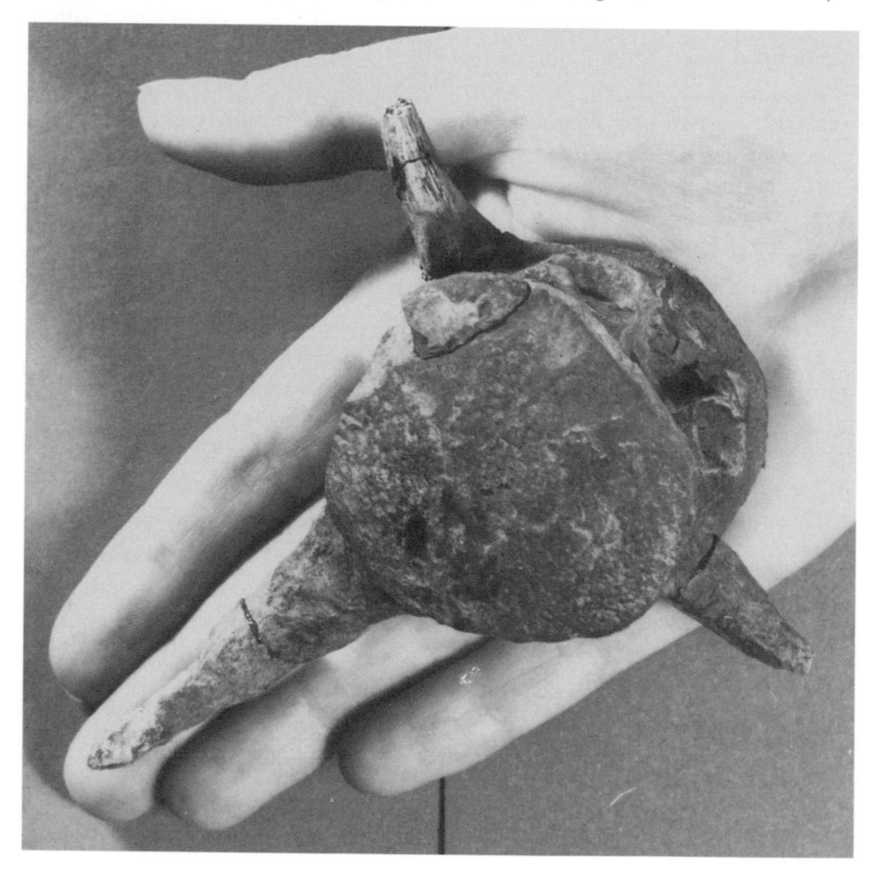

Figure 13.7 **Tail vertebra of unidentified mosasaur from shales exposed on the Puntledge River**
Source: Collections of the Courtenay Museum

Dinosaur: A Theropod

'Much to the disappointment of many children, terrestrial fossil reptiles such as dinosaurs have not been discovered in the rocks of Vancouver Island – yet.' (Ludvigsen and Beard 1994). When these words were written, the fossil evidence of a west coast dinosaur was already in hand but was not recognized. In the summer of 1992 amateur paleontologist Joe Morin of Courtenay and his sons were breaking up pieces of Upper Cretaceous marine shale ripped up by excavators for the natural gas pipeline across the Trent River, when his son Mathew uncovered an unusual tooth. Clearly reptilian, but to what group did it belong? After two years of being passed around among various paleontologists, the small triangular tooth (Figure 13.8) ended up at the Royal Tyrrell Museum of Palaeontology, where Phil Currie, Canada's foremost dinosaur paleontologist, identified it as belonging to a theropod dinosaur — a group that included the archetypal predatory dinosaur *Tyrannosaurus*. This was the first evidence of a dinosaur to be discovered in British Columbia west of the Rocky Mountains.

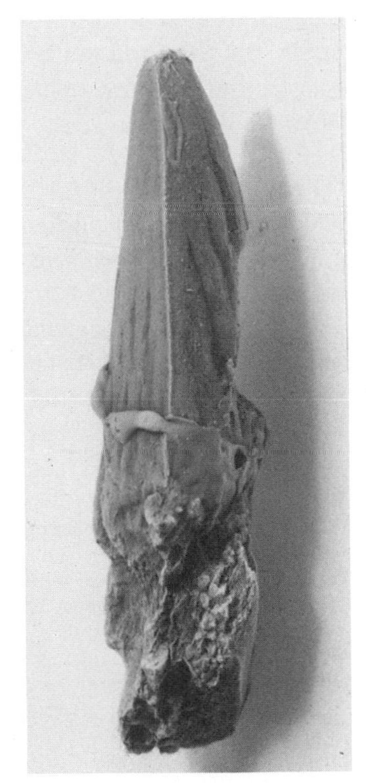

Figure 13.8 **Side and front views of single tooth, one centimetre long, of theropod dinosaur from shales exposed on the Trent River**
Source: Collections of the Courtenay Museum

How did a single tooth of a predatory dinosaur end up in the middle of a deep-marine succession exposed on the Trent River? We'll probably never know, but we can speculate about a small bloated theropod carcass floating on a river down to the sea where it was ripped apart by predatory marine reptiles. The head was ingested and the teeth were excreted to fall to the deep mud bottom.

This small theropod tooth, only a centimetre long, establishes the presence of meat-eating dinosaurs along the shores of the seaway that covered eastern Vancouver Island and the Gulf Islands during the Late Cretaceous. Its discovery greatly expands the known range of dinosaurs in Canada.

In the early Mesozoic, the dinosaurs of the Americas and Asia were markedly different. But, by the Late Cretaceous, the dinosaur faunas of North America were essentially identical to those of central Asia (Russell 1993). The familiar dinosaurs of the Late Cretaceous of Alberta – tyrannosaurs, hadrosaurs, ankylosaurs, and ceratopsids – consist of groups that originated in central Asia and later migrated to North America across a corridor located at the northern end of the Cordillera close to the Bering Strait. The theropod tooth from the Trent River is the only evidence that at least some of these migrating dinosaurs strayed to the far western edge of North America.

'There were giants on the Earth in those days' (Genesis 6:4). The chroniclers who interpret and embellish the various mythologies of different cultures and religions have commonly focused on the remarkable stature and size of ancient beings. If 'those days' include the Cretaceous, the Old Testament writers were remarkably prescient. The Cretaceous giants in the sea and on the land left tantalizing bony fragments of an important fossil record in the sedimentary rocks of Vancouver Island.

References

Carroll, R.L. 1988. *Vertebrate Paleontology and Evolution.* New York: W.H. Freeman and Company

Ludvigsen, R., and G. Beard. 1994. *West Coast Fossils: A Guide to the Ancient Life of Vancouver Island.* Vancouver: Whitecap Books

McGowan, C. 1991. *Dinosaurs, Spitfires, and Sea Dragons.* Cambridge, MA: Harvard University Press

Nicholls, E.L. 1992. Note on the Occurrence of the Marine Turtle *Desmatochelys* (Reptilia: Chelonioidea) from the Upper Cretaceous of Vancouver Island. *Canadian Journal of Earth Sciences* 29:377-80

Russell, D.A. 1967. *Systematics and Morphology of American Mosasaurs (Reptilia, Sauria).* Peabody Museum of Natural History, Bulletin 23

–. 1993. The Role of Central Asia in Dinosaurian Biogeography. *Canadian Journal of Earth Sciences* 30:2,002-12

14
Mollusks: Exotic Shells from Cretaceous Seas

James W. Haggart

No other fossil group has so captured the imagination of fossil collectors in British Columbia as has the Cretaceous Mollusca. The Cretaceous rocks of southeastern Vancouver Island and the adjacent Gulf Islands abound with beautifully preserved fossilized mollusks – mainly ammonoids, bivalves, and gastropods – as well as plentiful fossil plants, arthropods, and the occasional reptile (Ludvigsen and Beard 1994). These fossils represent the remains of plants and animals that once lived in or along shallow seas on the western edge of the continent. Many of the fossil mollusks look just like the shells we pick up on our beaches today, whereas others have strange shapes, unlike anything we see in modern oceans. This diversity of form indicates a once-thriving marine community, strange and exotic compared to the marine life we know today.

The coastal regions of British Columbia constitute the Insular Belt, one of the principal divisions of the Canadian Cordillera (Figure 14.1). Cretaceous mollusk fossils are not restricted to this belt, however. East of the Rocky Mountains, in the northeastern part of the province, Cretaceous rocks form parts of the Foreland Belt and Alberta Plains. These rocks are also rich in mollusk fossils, but types distinct from those found on the Pacific side of the mountains. Cretaceous marine fossils are also found at isolated localities within the mountain region, mainly in the Coast Belt, but typically they are poorly preserved.

What are mollusks? Simply, they are soft-bodied invertebrate animals that possess a hard shell covering their soft tissues, providing them with both support and protection (Bartsch 1968). Today the Mollusca are a diverse and varied group, including the bivalves (clams), gastropods (snails), and cephalopods (squids and octopods) that inhabit both the oceans and the land. The Cretaceous ancestors of these beautiful and complex creatures were no less varied.

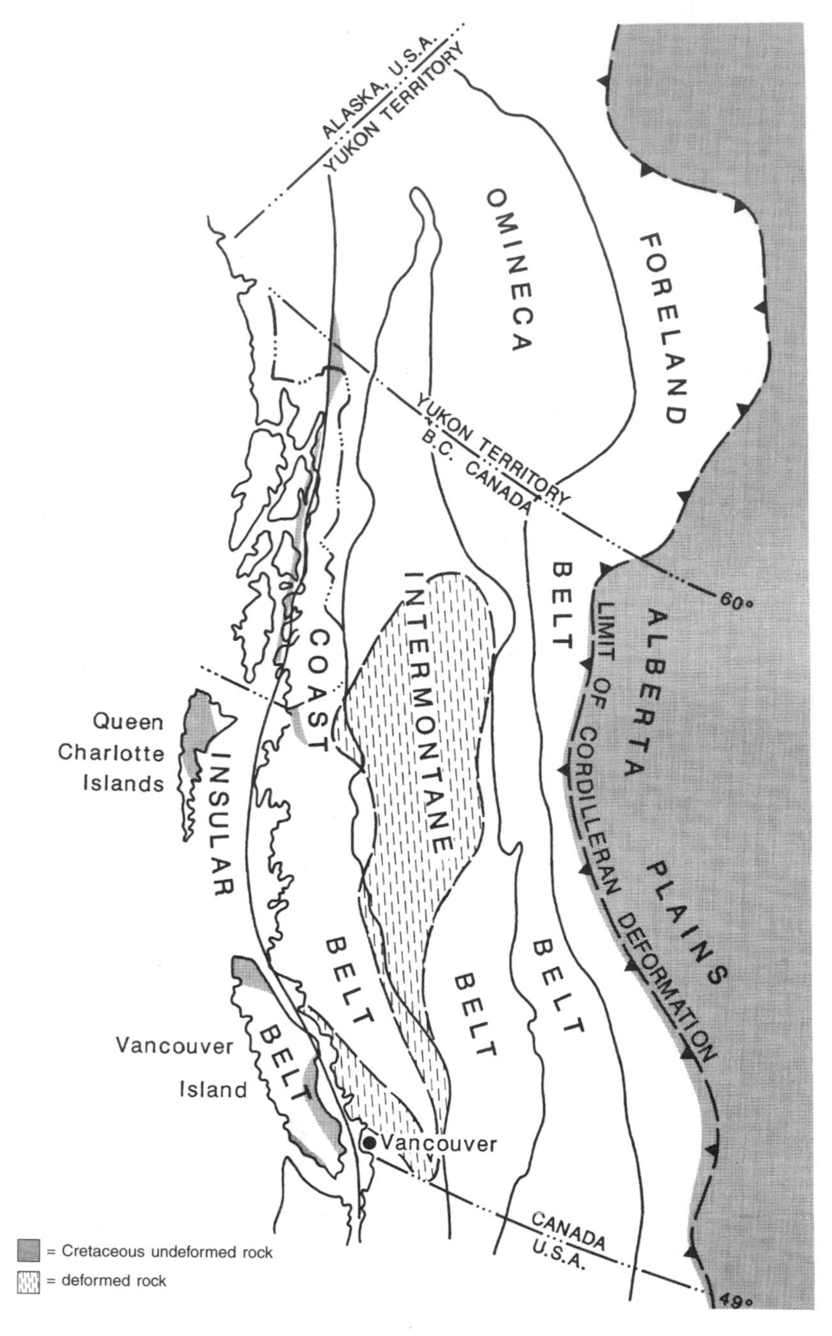

Figure 14.1 **Distribution of fossiliferous Cretaceous sedimentary rock of British Columbia and boundaries of principal geological belts of the Canadian Cordillera**

The Early Finds

One of the earliest descriptions of the Cretaceous mollusks of British Columbia dates from 1857, when F.B. Meek published a report entitled 'Descriptions of new organic remains from the Cretaceous rocks of Vancouver's Island.' Meek worked at the United States National Museum in Washington. As he was pretty much the only authority on Mesozoic fossils in North America at the time, collections from government surveys from both the United States and Canada were often sent to him for study. The fossils submitted to Meek from Vancouver Island had been collected by Dr. J.S. Newberry, a geologist who had participated in several exploratory expeditions to northern California, Oregon, and Washington. Newberry's collections contained scaphopods (tusk shells or dentalia), bivalves, and ammonoids, including *Ammonites Newberryanus,* a species named by Meek and collected from the Cretaceous beds at 'Komooks [Comox], Vancouver's Island.' Meek's species *newberryanus* was subsequently designated the type species of a new genus, *Canadoceras,* named in 1922 for Canada by the English paleontologist L.F. Spath. In the late 1860s and 1870s Meek produced several more reports describing the Cretaceous mollusks from southeastern Vancouver Island.

The systematic explorations of British Columbia's vast regions by officers of the Geological Survey of Canada (GSC) in the late 1800s resulted in discoveries of abundant fossil mollusks throughout the province. No place produced more fossils, though, than the Queen Charlotte Islands. The rich fossil deposits of Skidegate Inlet were first collected in 1872 by the geologist James Richardson of the GSC, and then by George M. Dawson in 1878. A Victoria physician, Dr. C.F. Newcombe, also travelled to the Queen Charlotte Islands about this time and made extensive fossil collections. Richardson's fossils were submitted for identification to the GSC paleontologist J.F. Whiteaves, who was perplexed by what he recognized as the 'mixed' nature of the fossil fauna, which contained elements with both Jurassic and Cretaceous affinities. He attempted to explain this disparity by suggesting that the Skidegate Inlet collections represented an intermediate interval between the Jurassic and Cretaceous systems, an interpretation not readily accepted by his contemporaries. It was not until many years later, principally through the fossil collecting and meticulous mapping of rocks in Skidegate Inlet by F.H. McLearn of the GSC, that it was realized that the fossils had come from several different localities in the inlet. Apparently, all had been thrown into the same sample bag. Both Jurassic and Cretaceous fossils had been mixed in the original collections, resulting in several decades of paleontological confusion. This incident illustrates the importance of recording precise information on locality for all fossil collections.

Although his true love was Triassic ammonoids, Frank McLearn undertook a study of the extensive Cretaceous mollusks from Skidegate Inlet in 1921. This work took many years to complete and ultimately appeared posthumously (McLearn 1972). Like Spath, McLearn named a fossil he collected for Canada, *Kossmaticeras canadense*. McLearn's analysis of ammonoid faunas of the Queen Charlotte Islands subsequently served as the framework for defining a sequence of ammonoid zones of the mid-Cretaceous of the Northeast Pacific.

After the Second World War, Jack Usher was directed by the GSC to undertake a monographic study of the ammonoids of the Nanaimo Group to aid in correlation of the coalfields of eastern Vancouver Island. His fine contribution to ammonoid paleontology (Usher 1952) recognized the strong similarity between the Nanaimo Group faunas and others around the Indo-Pacific region, and introduced these beautiful fossils to the contemporary paleontological community. Usher's work still stands as the basic reference on this diverse fossil group.

The Ammonoids: Ocean Mariners

Cretaceous ammonoids probably are the most commonly collected fossils in British Columbia. For many collectors, especially those based on Vancouver Island or in the Lower Mainland area, ammonoids are likely the first fossils they collect. Many fine and extensive collections of ammonoids have been made from Hornby and Denman islands, from the southern Gulf Islands, and along the many rivers that drain southeast Vancouver Island. These collections are maintained both in public institutions (Royal British Columbia Museum, Geological Survey of Canada, Vancouver Island Paleontological Museum, Courtenay and District Museum) and in private collections. In fact, the collections of some amateurs have provided numerous specimens that proved to be especially valuable to science.

Although ammonoids became extinct at the end of the Cretaceous, a relative survives today. This animal, the pearly *Nautilus* (Figures 14.2B to 14.2D), is a common inhabitant of coral reef environments of the southwest Pacific Ocean. The physiology and ecology of *Nautilus* have been investigated extensively over the past two decades (Ward 1987). As a result, much can now be inferred about the paleobiology of *Nautilus's* distant relatives, the ammonoids.

Ammonoid animals consisted of two principal parts – the soft tissues and the surrounding hard, protective shell (Figure 14.2A). The soft parts included a head with eyes and a highly developed nervous system, a propulsion system much like a squid's, numerous tentacles for holding prey, and a tooth-like beak for breaking and chewing prey (Lehmann 1981). The combination of these characteristics in ammonoids resulted in their position as high-level predators in the food chain.

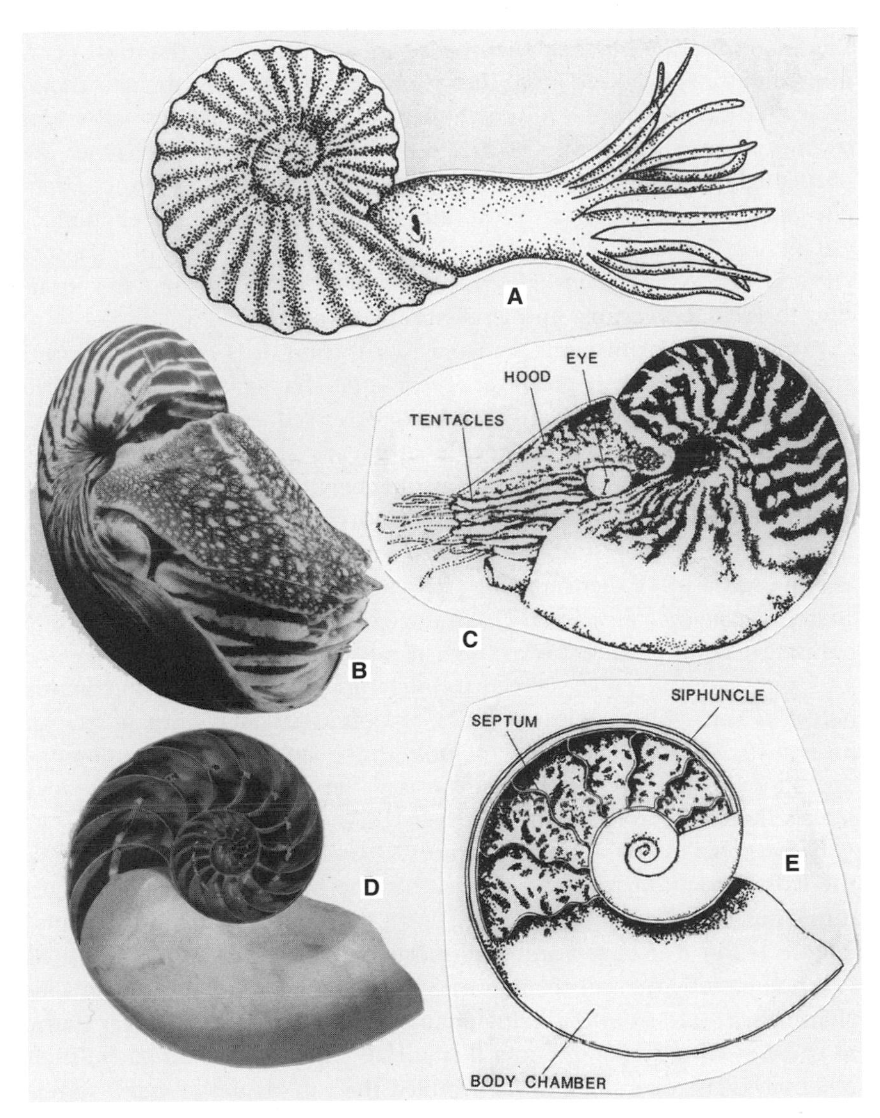

Figure 14.2 **Cretaceous ammonoids and living *Nautilus*.** (A) A living Cretaceous ammonoid, inferred in part from ecology and physiology of *Nautilus*. (B) and (C) Living *Nautilus*. (D) A sectioned *Nautilus* shell showing the chambers and connecting siphuncle. (E) A section through an ammonoid shell, showing the chambers and connecting siphuncle.
Source: (C) and (D) are from Ward (1987).

Fossilized ammonoid shell is what collectors generally find today. The various and diverse shell forms, which can be quite bizarre, are the basis on which paleontologists classify ammonoids. There are two principal types: the ubiquitous planispirals and the less common heteromorphs.

The planispirals, with their shell coiled in one plane, are the most common ammonoids found in British Columbia. Examples include *Canadoceras newberryanum* and *Kossmaticeras canadense,* mentioned earlier, and several others illustrated in Figure 14.3. Some species of planispirals are particularly widespread; for example, *Anagaudryceras sacya* and *Hypophylloceras (Neophylloceras) surya* are known from British Columbia, California, the Far East, Antarctica, Africa, South America, and Europe. Others, such as *Stelckiceras liardense,* have only been found in a few localities in British Columbia and elsewhere in Canada.

Although the planispirals are the easier to find, it is the heteromorph ammonoids, with their great variety of shell shapes, that are the more exciting fossil discoveries. During their extended evolutionary history, heteromorph ammonoids assumed nearly every conceivable shell shape, and the Cretaceous forms represent the height of this morphological experimentation. Indeed, the heteromorph ammonoids of British Columbia come in just about every shape and size of ammonoid anywhere (Figure 14.4): there are long straight shafts *(Baculites);* long recurved shafts *(Polyptychoceras);* open corkscrew coiling *(Pseudhelicoceras);* closed corkscrew coiling *(Eubostrychoceras);* closed conical coils *(Didymoceras);* and tight conical coils succeeded by long, gracefully curving whorls *(Glyptoxoceras).* These various shells are often covered with a strange assortment of ornamentation, including long and short spines, bumps and grooves, ribs and keels. The great variety of heteromorph shapes makes them the favourite of most collectors.

If you were to saw a typical ammonoid shell in half, perpendicular to the axis of rotation, you would find that the interior is partitioned into numerous chambers by curved and intricately fluted walls called 'septa' (Figure 14.2E). The largest and outermost chamber of the ammonoid shell was occupied by the soft tissues of the living animal. Successively smaller chambers inward from the living chamber were occupied at earlier stages in the growth of the ammonoid. Periodically, the ammonoid added shell material to the opening and then pulled the soft tissues forward, secreting a new septum behind them. In this way, the shell slowly grew bigger.

Why didn't the ammonoid simply live in one chamber and slowly enlarge both its soft tissues and shell? What was the purpose of the chambers with their intricate partitions? The answers to these questions are complex and a matter of lively debate. To address them we must look even more closely at the ammonoid shell.

Fortunately, the exquisitely preserved ammonoids of Hornby Island, Denman Island, and Sucia Island in Washington show many of the fine details of the shell structure. Using these and other fossils, Peter Ward (1987) and other ammonoid paleobiologists have theorized about the function of the chambers. It is thought that each chamber within the

Figure 14.3 **Typical Cretaceous planispiral ammonoids.** (A) *Canadoceras newberryanum,* 'Komooks' [Comox], Vancouver Island. (B) *Gaudryceras striatum,* Trent River, Vancouver Island. (C) *Kossmaticeras canadense,* Skidegate Inlet, Queen Charlotte Islands. (D) *Anagaudryceras sacya,* Skidegate Inlet, Queen Charlotte Islands. (E) *Hypophylloceras (Neophylloceras) surya,* Hornby Island. (F) *Stelckiceras liardense,* Liard River, northeastern British Columbia. (All X 0.5)
Source: (A) and (B) are from the collections of the United States National Museum; all other specimens are from the collections of the Geological Survey of Canada.

ammonoid shell was connected to the others by a tube of organic tissue called the 'siphuncle,' often seen in fossils as a thin black tube running along the outer part of each whorl (Figure 14.2E). Using the siphuncle, the ammonoid pumped fluid in and out of the various chambers forming its shell, displacing the gas that normally occupied the chamber. In this way, the buoyancy of the ammonoid could be regulated: water pumped into the chambers made the animal heavier and sent it to the bottom,

Figure 14.4 **Typical Cretaceous heteromorph ammonoids.** (A) *Didymoceras hornbyense,* Hornby Island. (B) *Polyptychoceras vancouverense,* Trent River, Vancouver Island. (C) *Glyptoxoceras subcompressum,* Puntledge River, Vancouver Island. (D) *Pseudhelicoceras* cf. *carlottense,* Cumshewa Inlet, the Queen Charlotte Islands. (E) *Baculites occidentalis,* Hornby Island. (F) *Eubostrychoceras elongatum,* Elkhorn Creek, Vancouver Island. (All X 0.5)
Source: Collections of the Geological Survey of Canada

whereas water pumped out of the chambers made the shell lighter and brought it to the surface. A particular proportion of fluid to gas resulted in neutral buoyancy, allowing the ammonoid to remain at one level within the water column, without exerting energy to keep from sinking or rising.

This chamber-pumping process likely took some time, resulting in significant buoyancy changes over intervals of days rather than minutes. Thus the ammonoid probably did not use the buoyancy system to move up and down through the ocean waters like a submarine; instead, rapid vertical movements were more likely accomplished by active swimming. The ability to change its buoyancy would nonetheless be extremely valuable in the event of an accident to its shell, for example, if a piece of shell was broken off by a predator's attack. If such an attack did not prove fatal, the ammonoid probably found itself either stranded at the surface (if only the living chamber had been broken) or lying on the bottom (if gas chambers had been pierced and flooded). By using the pumping process to compensate for the disruption, the ammonoid might have regained buoyancy and resumed its normal life.

Some Cretaceous ammonoids of British Columbia were surprisingly large for an animal lacking a backbone. Indeed, the Cretaceous ammonoids were the largest ammonoids that ever lived. Ammonoids of up to three-quarters of a metre in diameter have been collected along the Browns, Trent, and Puntledge rivers on Vancouver Island, and at Beresford Bay on the Queen Charlotte Islands. Yet these discoveries pale in comparison to the giant ammonoid shells that have been found in Germany, which are fully two and a half metres across.

The Bivalves: Burrowers and Floaters

Bivalves, or more simply clams, are ubiquitous on British Columbia's beaches and coastlines today. During the Cretaceous, bivalves were probably equally numerous, both along the shorelines and on the continental shelf. In contrast to the free-swimming ammonoids, however, most marine bivalves of the Cretaceous were benthic, that is, they lived on the ocean bottom, either attaching themselves to rocks, shells, or cliffs, or else burrowing into the sediment to avoid predators and to find protection against the vagaries of their environment. Because of their geological importance, several different Cretaceous bivalve groups have received close attention from paleontologists.

Paleontologists classify fossil bivalves on the basis of the morphology of their shells. Many forms abound, often confusingly similar. In complexity of shell form and exotic ornamentation, however, the trigoniids are the winners (Figure 14.5). These three-sided, or trigonal-shaped, clams are commonly found in sandstones from shallow marine areas throughout the Cretaceous of British Columbia. Indeed, the strongly ornamented sculpture of these clams is remarkably similar to that seen on their single living descendant, the genus *Neotrigonia,* which lives on the sandy bottom of shallow marine areas of the tropical southwest Pacific region. The burrowing efficiency of *Neotrigonia* is remarkable – in a matter of minutes it can dig itself into the bottom sand and out of sight. Apparently, the complexly arranged knobs, bumps, and ridges on these bivalves increase their burrowing efficiency. By analogy, we assume that the Cretaceous trigoniid *Pterotrigonia,* which has a similar ornamentation, was also a fast burrower.

A rapid burrowing capability is essential for bottom-dwelling animals living in the rough-and-tumble area just below the tide line and adjacent

Figure 14.5 **Common Cretaceous trigoniid bivalves.** (A) *Pterotrigonia evansana,* Denman Island. (B) *Yaadia lewisagassizi,* Harrison Lake area. (C) *Yaadia tryoniana,* Northwest Bay, Vancouver Island. (D) and (E) *Quoiecchia aliciae,* Harrison Lake area. All are actual size.
Source: Collections of the Geological Survey of Canada

to the beach. Passing storms have a tendency to churn up sands along the coastline, quickly disturbing these animals' home. Trigoniids such as the diminutive *Quoiecchia* and the larger *Yaadia* are found in many areas of the province, including Harrison Lake, the Queen Charlotte Islands, Vancouver Island, and the Gulf Islands – essentially wherever shallow-marine Cretaceous sandstone is known. Thus the presence of trigoniids in Cretaceous rocks is often used as evidence that the environment of deposition was a coastal setting.

In contrast to the complexly ornamented trigoniid bivalves, the buchiids are a rather plain, ordinary-looking group (Figure 14.6). What they lack in exotic shell ornamentation, however, they make up for in geological importance. The buchiids were certainly the most abundant bivalve group in the oceans of British Columbia during the Late Jurassic and Early Cretaceous. Large numbers of these clams are often found at a single locality, sometimes forming beds several metres in thickness (Figure 14.6G). Such occurrences are common at Harrison Lake and in Manning Park. These accumulations are referred to as buchiid 'reefs'; however, there is no evidence that buchiids cemented together, as do the corals of present-day reefs. Instead, these 'reefs' probably formed from thousands of shells accumulating as a result of storm activity or strong currents, much like accumulations of oyster shells today.

Although all buchiid bivalves appear similar at first glance, many subtle differentiations can be recognized. The late paleontologist George Jeletzky studied the buchiids on several small islands off the west coast of Vancouver Island and determined that slightly different forms were found at each level in the rocks. Jeletzky (1965) recognized that these stratigraphically segregated fossils represent different species within a long-lived buchiid evolutionary lineage. Fortunately, buchiids are abundant and widespread in Upper Jurassic and Lower Cretaceous rocks of British Columbia in which other fossils cannot be found. For this reason a geologist is usually pleased to find a plain and ordinary-looking buchiid in a remote area of British Columbia, because such a clam can often provide a precise age for the rocks.

Of all the Cretaceous bivalves, perhaps the most intriguing group is the inoceramids (Figure 14.7). These clams inhabited a wide range of settings in the Cretaceous ocean, from shallow-water nearshore sandy bottoms to deep-marine muds. The inoceramids are a diverse family, but most genera exhibit some aspect of concentric ribbing. Variations include thick ribs and thin ribs, divergent and convergent ribs, double ribs, triple ribs, crenulated ribs, etc. The variations seem almost infinite.

Most inoceramids appear to have lived a stable life on the sea bottom, either attaching themselves to a solid surface or using their wide, thin, flat shells to 'float' on softer sediments. Examples include *Inoceramus vancouverensis, Inoceramus* cf. *balticus,* and *Sphenoceramus schmidti.* Not all inoceramids remained stable, however. Indeed, several paleontologists have suggested that some inoceramids, for example, *Mytiloides labiatus* and *Sphenoceramus? naumanni,* may have lived a pseudoplanktic life, floating around the world's oceans attached to drifting logs or even to seaweed mats. Only such a way of life can account for the numerous finds of inoceramids in rocks that are devoid of all other fossil life. Such sediments are referred to as 'anoxic' because they lacked oxygen when they accumulated and hence were toxic to most life forms. The large numbers of inoceramid fossils sometimes found in anoxic sediments may have

Figure 14.6 **Common Cretaceous buchiid bivalves.** (A) and (B) *Buchia sublaevis,* Tyaughton Creek. (C) and (D) *Buchia inflata,* Tyaughton Creek. (E) and (F) *Buchia crassicollis,* Forward Inlet, Vancouver Island. (A-F actual size) (G) *Buchia uncitoides* 'reef,' Harrison Lake
Source: Collections of the Geological Survey of Canada. (G) is courtesy of T.P. Poulton.

Figure 14.7 **Common Cretaceous inoceramid bivalves.** (A) *Inoceramus vancouverensis,* Sucia Island, Washington State. (B) *Sphenoceramus? naumanni,* Trent River, Vancouver Island. (C) *Inoceramus* cf. *balticus,* Hornby Island. (D) *Sphenoceramus schmidti,* Texada Island. (E) *Mytiloides labiatus,* Skidegate Inlet, Queen Charlotte Islands. (F) *Sphenoceramus orientalis ambiguus,* Puntledge River, Vancouver Island. (G) *Inoceramus* cf. *paraketzovi,* Harrison Lake. (All X 0.5)
Source: Collections of the Geological Survey of Canada

fallen there when the logs or seaweed mats supporting them became waterlogged.

Some inoceramids found in Cretaceous strata of British Columbia are enormous. At Harrison Lake and the Queen Charlotte Islands, specimens of *Inoceramus* cf. *paraketzovi* as long as two-thirds of a metre are commonly found. Individuals can be seen at Vesuvius Bay on Saltspring Island, with thin, flat valves up to one and a half metres in length and only a centimetre or less in thickness. Other inoceramids, for example, those seen at Slatechuck Mountain on the Queen Charlotte Islands, reached similar lengths and had coarse, robust ribs; however, these specimens had somewhat thicker shells, up to several centimetres. In appearance, such clams must have looked very similar to present-day *Tridacna,* or the 'killer clams' of the southwest Pacific. For this reason it has been suggested that, similar to the *Tridacna,* these large inoceramid clams may have harboured symbiotic algae in their soft tissues, enabling them to grow to such large sizes. To date, however, the morphological and geochemical evidence is elusive.

The Gastropods: Solitary Grazers

One Cretaceous mollusk group of British Columbia that has not received extensive study is the gastropods (Figure 14.8). These mollusks are characterized by a simple spiral shell similar to that of some heteromorph ammonoids. Gastropods are found in many different Cretaceous rock types, indicating that they inhabited both shallow-marine and deep-marine habitats. Some Cretaceous gastropods bear strong resemblance to the snails found today in the oceans of British Columbia, for example, the Cretaceous *Gyrodes* and the modern *Polinices.* Compared to the ammonoids and bivalves, however, gastropods were never very abundant in the Cretaceous seas of British Columbia.

Ecologically, most gastropods today are simple animals, and their Cretaceous forebears were likely no different. Constrained by simple body construction and lacking the propulsion system of the ammonoids, gastropods were relegated to a nomadic life at the bottom of the sea. Endlessly grazing on the mud and sand of the Cretaceous sea floor, gastropods ingested the sediment they ploughed through, filtering it through their gut to retain any organic matter. The snail used its siphon to release the waste generated by this filtering process. Often, the only evidence of gastropods that one finds in Cretaceous strata, especially in fine-grained mudstone deposits, is the network of trails and traces that they have left (Figure 14.8A).

By analogy with modern forms, some Cretaceous gastropods such as *Gyrodes* likely were carnivorous, searching out clams and other snails for a meal. The little round boreholes often found in fossil bivalve shells are evidence that such predatory gastropods were able to drill through the protective shells in search of their prey.

Generally the gastropod species found in the Cretaceous strata had a longer geological life than their molluscan cousins, the ammonoids. As well, the shapes of their shells are relatively simple in comparison. These generalizations suggest that gastropods evolved more slowly than ammonoids. In spite of this, some gastropod fossils from British Columbia are highly variable in their shell shapes. The genus *Volutoderma,* with its

Figure 14.8 **Common Cretaceous gastropods.** (A) Gastropod grazing trails in mudstone, northwest coast Graham Island, Queen Charlotte Islands. (B) *Volutoderma averillii,* Sucia Island, Washington State. (C) *Perissitys brevirostris,* Sucia Island. (D) *Serrifusus dakotaensis,* Hornby Island. (E) and (F) *Gyrodes dowelli,* Sidney Island, the Strait of Georgia. All are actual size.
Source: Collections of the Geological Survey of Canada

high spire and robust, strongly knobbed shell, is most often found in rocks indicating shallow-water, turbulent environments such as near the beach. On the other hand, the more delicately ornamented *Serrifusus* is most often found in mudstone and shale typical of a quiet deeper-water environment.

Cretaceous Mollusk Biogeography

Although Cretaceous rocks and mollusks are found in many areas of British Columbia, they are best developed in the Insular Belt, the Foreland Belt, and the prairie regions of the province. Interestingly, British Columbia's present coastal region has been the interface of mountain and sea for millions of years, at least since Cretaceous time. During much of the Cretaceous, an extensive inland sea – the Western Interior Seaway – stretched from the northeastern part of the province across much of central Canada and south toward the Gulf of Mexico. Thrusting skyward between the Interior Seaway and the Pacific Ocean lay an extensive and mountainous highland region, the ancestor of the Canadian Cordillera (Figure 14.9).

The successive rises and falls of sea level during the Cretaceous caused marine waters to migrate back and forth across both the western and eastern margins of the Cordillera, but the area was never fully covered by water. Thus the marine waters of the Western Interior Seaway remained separate from the open ocean of the Pacific, precluding ready migration of marine organisms between the two seas. Due to this faunal isolation, distinct evolutionary successions developed in each region. As a result, Cretaceous mollusk faunas of northeastern British Columbia are more similar to those of the central United States and northern Europe than to those of the coastal region of the province. And the faunas of coastal British Columbia most resemble those of northern California, the Russian Far East, and Japan. These similarities attest to the wide-ranging Pacific Ocean currents and the formidable barrier to trans-oceanic migration presented by the Cretaceous Cordillera.

Biostratigraphy and Biochronology

Using fossils as a tool in interpreting the geological history of a region was one of the earliest scientific applications of fossils. Indeed, the first geological map of England, produced in 1815 by canal builder William Smith, resulted from the use of fossil mollusks to identify sedimentary rocks. Smith used the presence of particular mollusk fossils to correlate rocks in widely separated regions of the English countryside. His correlations and geological maps were the first use of fossils as biostratigraphic tools. 'Biostratigraphy' is the discipline of paleontology concerned with determining the vertical succession of fossils in the bedded rocks of a

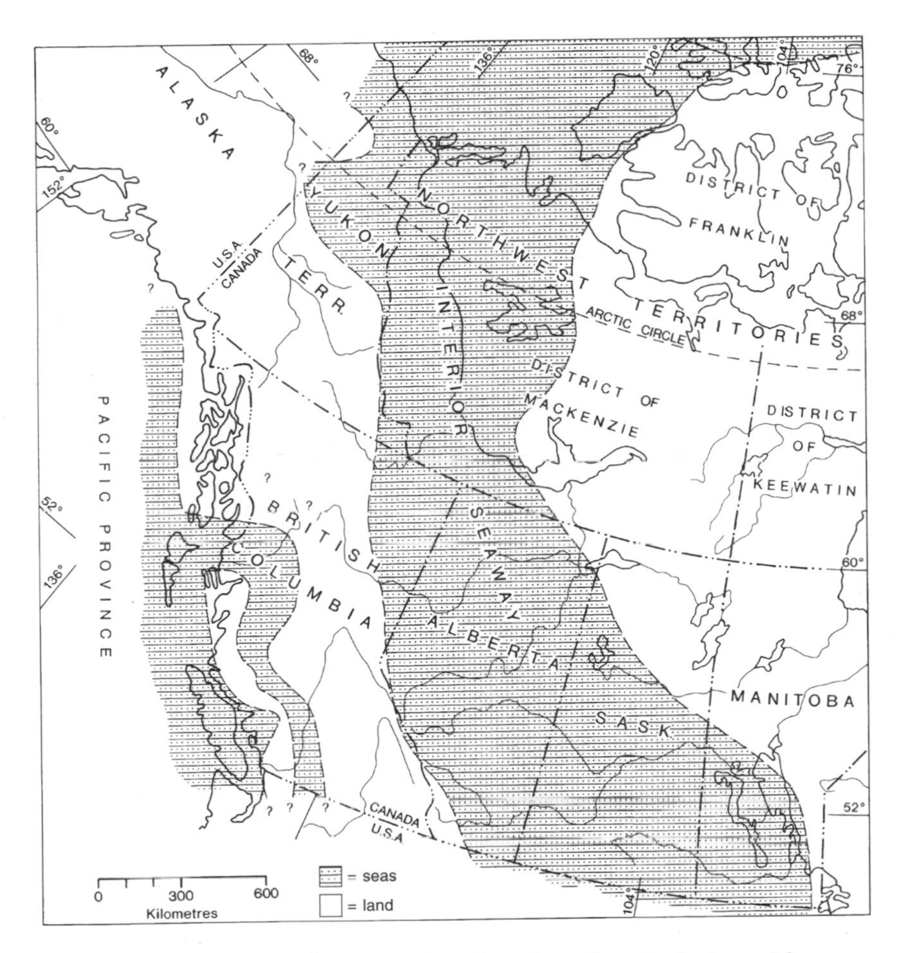

Figure 14.9 **Cretaceous paleogeography of western Canada during mid-Cretaceous time**
Source: Modified from Jeletzky (1970)

region and then using this succession to correlate rocks of similar age from different regions of the earth.

The succession of fossils through time is the basis of the geological time scale. The smallest divisions of this scale are zones characterized by individual fossil species. The most useful fossil species for biostratigraphy are those that were geographically most widespread, that are found in many different environments, and that experienced rapid rates of evolution. A widespread distribution ensures that an organism's fossil remains will be found across widely separated areas of the globe. Equally important, rapid rates of evolution result in individual fossil species with short durations, thus allowing the characterization of short intervals of geologic time. In the Cretaceous, ammonoids are foremost among these quickly evolving and widely distributed fossil organisms. A close second are some bivalve groups such as the buchiids and inoceramids.

The Cretaceous time scale (Figure 14.10) is based principally on the succession of mollusk zones at European localities. George Jeletzky (1970) established sequences of mollusk zones for the BC Cretaceous based on his work on northern Vancouver Island, Manning Park, the foothills of

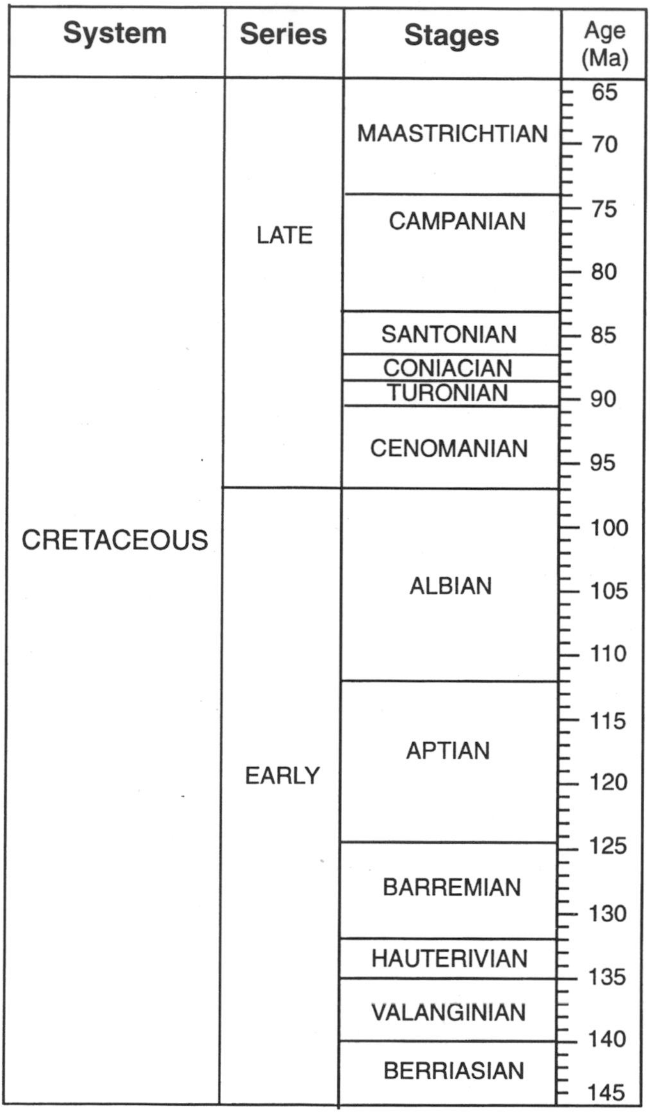

System	Series	Stages	Age (Ma)
CRETACEOUS	LATE	MAASTRICHTIAN	65, 70
		CAMPANIAN	75, 80
		SANTONIAN	85
		CONIACIAN	
		TURONIAN	90
		CENOMANIAN	95
	EARLY	ALBIAN	100, 105, 110
		APTIAN	115, 120
		BARREMIAN	125, 130
		HAUTERIVIAN	135
		VALANGINIAN	140
		BERRIASIAN	145

Figure 14.10 **Cretaceous time scale.** The Cretaceous System is divided into series and stages; each stage is further divided into a succession of zones based on characteristic fossil species and calibrated in years. (Ma = millions of years)

the Canadian Rockies, and the interior plains. By correlating BC zones to the European zones, he was able to apply the Cretaceous stage names to the sedimentary rocks of the Cordillera.

Paleontologists in Canada use the time scale to correlate widely separated and lithologically different Cretaceous rocks and to determine how these strata relate to rocks of similar age in northern regions of North America, Europe, and Asia. In British Columbia, as well as most of the higher latitudes in the northern hemisphere, ammonoids and inoceramids are uncommon in Upper Jurassic and Lower Cretaceous sedimentary strata; consequently, buchiids are used to determine the zonal subdivisions for this interval. In the Early Cretaceous, however, ammonoids became relatively common, so that for the rest of the Cretaceous the time scale is based on the succession of important ammonoid species, supplemented by the inoceramids. By the time of the great extinction at the end of the Cretaceous, buchiids, ammonoids, and inoceramids had all disappeared, and the world's oceans no longer teemed with these enchanting creatures.

The wonderful fossil mollusks of British Columbia's marine Cretaceous rocks reflect a diverse and exotic ocean lifescape. In addition to their intriguing biological and ecological aspects, the fast-evolving ammonoids provide geologists with an invaluable time scale for dating sedimentary rocks. Fossil gastropods and bivalves also supply clues about the ages and environments of the rocks in which they are entombed. Unfortunately, most of these unique animals are long gone from the earth, victims of the extinctions that periodically affected life on our planet. For the patient fossil hunter, however, the record of these animals can still be found – washed up on the beaches of British Columbia's Cretaceous rocks – preserved as life in stone.

Acknowledgments

This chapter is Contribution No. 27794 of the Geological Survey of Canada.

References

Bartsch, P. 1968. *Mollusks.* New York: Dover Publications

Jeletzky, J.A. (George) 1965. *Late Upper Jurassic and Early Lower Cretaceous Fossil Zones of the Canadian Western Cordillera, British Columbia.* Geological Survey of Canada, Bulletin 103

–. 1970. Cretaceous Macrofaunas. In *Biochronology: Standard of Phanerozoic Time,* ed. E.W. Bamber et al. Geological Survey of Canada, Economic Geology Report No. 1, Part B, pp. 649-62

Lehmann, U. 1981. *The Ammonites – Their Life and Their World.* Cambridge: Cambridge University Press

Ludvigsen, R., and G. Beard. 1994. *West Coast Fossils: A Guide to the Ancient Life of Vancouver Island.* Vancouver: Whitecap Books

McLearn, F.H. 1972. *Ammonoids of the Lower Cretaceous Sandstone Member of the Haida Formation, Skidegate Inlet, Queen Charlotte Islands, Western British Columbia.* Geological Survey of Canada, Bulletin 188
Usher, J.L. 1952. *Ammonite Faunas of the Upper Cretaceous Rocks of Vancouver Island, British Columbia.* Geological Survey of Canada, Bulletin 21
Ward, P.D. 1987. *The Natural History of Nautilus.* London: Allen and Unwin

15
Plant Life during the Great Cretaceous Transformation

James F. Basinger and Elisabeth McIver

The Cretaceous is perhaps the most significant period in the evolution of modern land vegetation, since it was the time when the flowering plants (angiosperms) made their first appearance. Today angiosperms make up the vast bulk of the world's vegetation. Indeed, it is difficult to imagine earth without them. Yet, the floras of the Early Cretaceous, as preserved in the coal fields of eastern British Columbia, lack flowers, grasses, reeds and rushes, herbs, palms, and all of the broad-leaved trees and shrubs so familiar to us. Instead, we find profusions of ferns, cycads, conifers, seed ferns, and ginkgos.

The Early Cretaceous marked the end of the great age of cycads and conifers that began about 250 million years ago. The dinosaurs were at their peak then, and the towering forests must have looked quite primeval. By the Late Cretaceous, as seen in the coal fields of east-central Vancouver Island, the transformation was complete and the vegetation was dominated by broad-leaved angiosperms.

The climate of British Columbia during the Cretaceous was much warmer than at present, and the province lacked altogether the dryness of the modern Interior. The mountain ranges of the Interior were present, but the Coast Range had not yet formed and the Rockies were only beginning to rise. The land was clothed in vegetation much more lush than is possible at comparable latitudes in modern times. It is unlikely that permanent ice existed anywhere, except perhaps on the highest mountaintops, and even Antarctica was mild and fully forested.

Ironically, the great forests of British Columbia today are again dominated by giant conifers, thanks to a peculiar coincidence of temperature, rainfall, and postglacial plant migration, and to the relatively recent conifer diversification, especially among the members of the pine family. By standing in the towering old-growth forests of the coastal ranges we can capture some of the feeling of the ancient Early Cretaceous landscape. Missing, thankfully, are the lurking carnivorous dinosaurs.

Cretaceous Plant Localities in British Columbia

The most abundant occurrences of Late Cretaceous fossil plants are in the riverside exposures of coal-bearing rocks of the Nanaimo Group, in the Nanaimo and Courtenay areas on Vancouver Island (Figure 15.1). Although coal mining in these areas ceased years ago, many of the old mine dumps still yield plant remains. Exposure causes the rocks to crumble, but it is usually possible to locate some more resistant blocks.

The coal-mining districts of the Dawson Creek and Tumbler Ridge areas are more difficult to explore for fossil plants. The territory is rugged and access is difficult. Also, active mining precludes access to mine dumps and open-cut exposures. Nevertheless, it is worthwhile to explore the

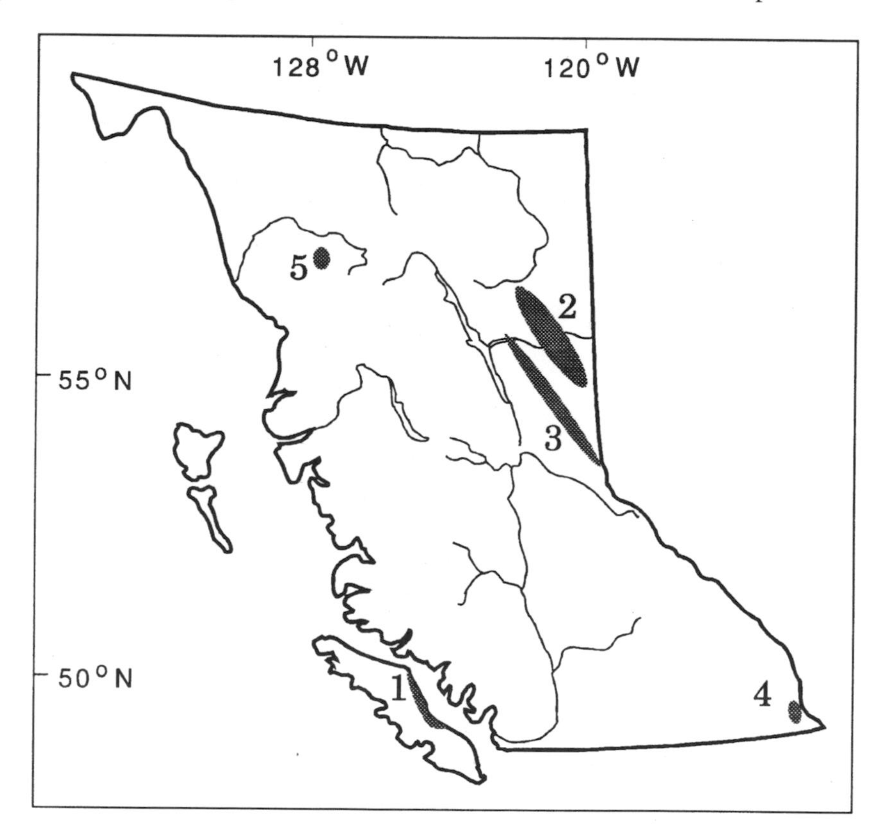

Figure 15.1 **Cretaceous plant localities.** (1) Upper Cretaceous Nanaimo Group, principally Nanaimo and Courtenay areas, 80 million years old. (2) Upper Cretaceous Dunvegan Formation, Dawson Creek and Fort St. John area, 100 million years old. (3) Lower Cretaceous Gething and Gates formations, Tumbler Ridge and Hudson's Hope area, 110 million years old. (4) Lower Cretaceous Kootenay Formation and Blairmore Group, Fernie and Crowsnest Pass area, 110 million years old. (5) Lower Cretaceous Bowser Lake Group, Groundhog Coalfield, Klappan River, 140 million years old.

canyons and riverbanks of the Peace River and its tributaries, where coal-bearing rocks of the Gething, Gates, and Dunvegan formations of Early to mid-Cretaceous age are exposed. Unfortunately, many of the best exposures are now underwater as a result of dam construction in the region.

The coal-bearing rocks of the Blairmore Group in the mountainous Crowsnest Pass area of southeastern British Columbia are essentially equivalent in age to the Gething and Gates rocks, and are similarly difficult to explore for fossil plants. The long history of open-cut mining in the area has, however, exposed vast amounts of material. Rocks of the underlying Kootenay Formation, which straddles the boundary between the Jurassic and Cretaceous, are also exposed in Crowsnest Pass and are known to yield fossil plants.

A review of the fossil flora of the Gates Formation is nearing completion at the University of Saskatchewan. By contrast, the Blairmore and Gething floras have not been seriously studied since Walter Bell's work in the 1950s, so that these formations remain largely unknown paleobotanically. Nevertheless, it is possible to speak of the Blairmore, Gething, and Gates fossil floras in the same breath, because they are so similar. Even the flora of the older Kootenay Formation is, for our purposes, essentially indistinguishable from these three floras.

On the other hand, the Dunvegan Formation, of earliest Late Cretaceous age, is of particular interest to paleobotanists because it preserves the remains of some of the earliest angiosperm-dominated forests. Dunvegan fossils are poorly understood at this time, and represent an untapped scientific resource.

Sharon MacLeod of the University of Calgary has studied the practically inaccessible exposures of the Upper Jurassic and Lower Cretaceous coal-bearing Bowser Lake Group in the Spatsizi wilderness area of northwestern British Columbia. As a coal geologist at the newly developed Groundhog coal field, MacLeod had access to many fossils from the Klappan open-cut mines. She has produced one of the best reviews of the Lower Cretaceous floras of BC (MacLeod 1991).

Walter Bell, the paleobotanist for the Geological Survey of Canada and its director in his later years, was responsible for bringing together a hundred years of reports and collections of Cretaceous fossil plants amassed during the exploration of the Cordillera (Bell 1956, 1957, 1965a, 1965b). Many of the early collections of fossils were made by George Mercer Dawson, son of Canada's pre-eminent paleobotanist Sir J. William Dawson and Director of the Geological Survey of Canada at the turn of the century. Although G.M. Dawson, for whom Dawson Creek and Dawson City are named, suffered from birth from a severe deformity of the spine that would have discouraged most of us, he led field parties into some of Canada's most formidable wilderness areas between the 1870s and 1890s.

The science has moved on since Bell's contribution to the field, so that his descriptions of plants are in need of updating; however, his writings remain the most comprehensive that are available on Cretaceous floras of British Columbia.

Ferns

Ferns can be traced back more than 350 million years. Today they are the second most diverse group of land plants, after the flowering plants. People living in British Columbia have some appreciation for the importance of ferns as understorey forest herbs. To truly understand ferns, however, we must see them in their glory in the tropical rain forests, where we find the closest parallels to the role they play in the Cretaceous floras of British Columbia.

Ferns most commonly occur in the forest understorey, and most are very well adapted to moist, low-light conditions. This habitat has been exploited by ferns since the Carboniferous. Because ferns lack the secondary wood that strengthens the stems of seed plants, few have been capable of achieving large size, although 'tree ferns' have found other ways of stiffening their stems to stand many metres tall. It is impossible to imagine the ancient landscape of British Columbia, either the forests of giant conifers of the Early Cretaceous or the mixed forests of conifers and flowering plants of the Late Cretaceous, without a carpet of lush growth of ferns.

Many ferns have found their way into fossil deposits. Plants of the understorey, especially ferns, whose leaves tend not to fall off but to rot in place, are less commonly transported to sites of fossilization than are leaves of canopy trees. The sheer abundance of some types of fern leaves in many Cretaceous deposits indicates that tree ferns were a significant part of the Cretaceous forest.

Although some types of ferns produce simple leaves, that is, undivided leaves with a single petiole (leaf stalk) and blade, most ferns produce branched (compound) leaves. Such leaves generally have a stout rachis (the stalk of the leaf) that gives rise to pinnae (lateral 'branches' of the leaf) to produce the flattened, feather-like structure of the leaf. Each pinna then bears two rows of small leaflets called pinnules. Some ferns have even more complex leaves, with each pinna subdivided into secondary pinnas. Most fern leaves are quite large, some types being a few metres long, but it is unlikely that a fossil specimen constitutes more than a small fragment of the entire leaf.

Coniopteris is one of several generic names applied to numerous species of ferns whose precise relationships to modern ferns are not clear (Figure 15.2). Many families of ferns can be traced back to the Paleozoic, but modern ferns continue to be an actively evolving group. This character-

Figure 15.2 **Frond of the fern *Coniopteris*.** Lower Cretaceous Gates Formation, Tumbler Ridge. Actual size.
Source: Collections of the Department of Geological Sciences, University of Saskatchewan

istic, plus the fact that many groups of ferns produce similar foliage, makes identification of fossil fern foliage difficult. Furthermore, fern leaves are commonly very large, and the pinnules of different parts of the same leaf may appear quite dissimilar. Although fragmentary fossils of ferns may not look alike, they may well be different parts of the same large leaf.

A casual glance through Bell's *Lower Cretaceous Floras of Western Canada* (1956) reveals a long list of fern names and numerous vaguely similar photographs, all of which tend to blur in the mind. Serious effort is required to identify fossil fern foliage to genus. Fortunately, it is not necessary to achieve this level of identification in order to grasp the significance of ferns in the interpretation of past climatic and ecological conditions.

Redwoods

Conifer remains are abundant in Cretaceous rocks. Most resemble an assortment of modern conifers, but to link a fossil twig with any particular living genus of conifers is difficult because the preservation is usually poor and the remains rarely include the diagnostic seed cones necessary for precise identification. In addition, evolutionary changes and extinction during the Cretaceous and Cenozoic make direct comparisons difficult.

Elatides exemplifies some of the difficulties encountered. *Elatides* is a name commonly applied to plants with needle-leaved foliage typical of that of the redwood family, the Taxodiaceae (Figure 15.3). This foliage appears to have been highly variable, even on a single tree, and many different genera of the family are known to have produced similar foliage. Cones found attached to this type of foliage are similar to those of the redwood. They are about as large as the last joint of a finger and composed of numerous scales with enlarged, shield-shaped tips.

The leaves of *Elatides* appear stiff and typically are broadly attached to the twig. Similar types of foliage may be assigned to the genus *Elatocladus,* which is a handy repository for a wide variety of redwood-like foliage (Figure 15.4). *Elatocladus* is considered an 'unnatural' genus because it includes many plants that are similar but may not be closely related. Such a genus name is used for convenience when there is insufficient information about members of the genus to allow them to be segregated.

Athrotaxites is another common type of foliage belonging to the redwood family. This foliage, with tiny scale-like leaves, looks superficially

Figure 15.3 **Two common types of Lower Cretaceous conifers, *Elatides curvifolia* and *Athrotaxites berryi*.** *Elatides curvifolia* has the leafy twigs; *Athrotaxites berryi* has the twigs with the minute, scale leaves. Gates Formation, Tumbler Ridge. (X 0.7)
Source: Collections of the Department of Geological Sciences, University of Saskatchewan

Figure 15.4 **The conifer *Elatocladus brevifolia,* with attached cones.** Gates Formation, Tumbler Ridge. (X 0.7)
Source: Collections of the Department of Geological Sciences, University of Saskatchewan

like cedar twigs. A closer look reveals that the leaves are arranged in a helix, not in whorls or opposite pairs like the cedars.

During the Cretaceous, Taxodiaceae dominated many forest ecosystems. The pine family, Pinaceae, was much less important, at least in the moist lowland sites where most fossils are formed. Possibly the pine family was important in the cooler, higher elevations, but such settings leave little in the way of a fossil record. In many habitats the Taxodiaceae was displaced by the flowering plants, which appeared first in the Early Cretaceous and became very common in the Late Cretaceous. During the Cenozoic, the Taxodiaceae became increasingly restricted as a result of global climatic cooling. The Pinaceae, apparently better adapted to cooler conditions, was able to take advantage of the cooler environments, and now covers most of British Columbia.

Cycads

As with the ferns, the cycads proliferated in the Cretaceous forests of British Columbia. At present, cycads are on the verge of extinction, with only about a dozen genera scattered about the New and Old World tropics. Living cycads are characterized by their squat, generally unbranched trunks and large feather-like leaves. They bear a superficial resemblance to palms, and sometimes are referred to as 'sago palms,' but they are not related to true palms, which are flowering plants.

During the Cretaceous, they declined from prominence to obscurity as the flowering plants expanded. Although cycads are very common in Lower Cretaceous rocks of British Columbia, they form a far less significant component of the Upper Cretaceous flora of Vancouver Island than do the flowering plants. In Cenozoic rocks of British Columbia, and indeed anywhere in the world, they are almost unknown.

Early Cretaceous cycads display a wide variety of leaf forms, from simple leaves to large pinnate leaves. The general impression is of tough, stiff leaves, a texture also typical of living cycads. Such coarse vegetation would have discouraged foraging by any but the most determined herbivorous dinosaurs. From the little we know of the growth habit of Cretaceous cycads, most of them probably had the same type of stocky, unbranched trunks typical of living cycads. They may have been understorey plants but may also have formed dense thickets in more open or drier spots.

Ptilophyllum is an example of the more common Lower Cretaceous cycads. Leaflets of *Ptilophyllum* typically have a few to several parallel veins running the length of each leaflet. Many different species occur in the fossil record, including the two types illustrated in Figures 15.5 and 15.6.

Nilssonia represents a highly unusual group of cycads that bore simple undivided leaves, as well as compound leaves. Leaves may be as long as a finger or an outstretched hand. The simple types of leaves, which are

Figure 15.5 **Large, feather-like compound leaves of the cycad *Ptilophyllum* sp.** Gates Formation, Tumbler Ridge. (X 0.5)
Source: Collections of the Department of Geological Sciences, University of Saskatchewan

Figure 15.6 **Simple, strap-shaped leaves of the cycad *Nilssonia canadensis*.** Associated with these are pinnate leaves belonging to another species of *Ptilophyllum*. Gates Formation, Tumbler Ridge. Actual size.
Source: Collections of the Department of Geological Sciences, University of Saskatchewan

common in the fossil record of British Columbia, were tapered at each end and attached to the stem by a short, stout leaf stalk. The mid-rib is thick and prominent, and the secondary veins come off almost at right angles and run parallel out to the margin. *Nilssonia* is one of the few cycads that survived into the Late Cretaceous.

One aspect of cycad evolution not apparent from fossil leaves is that two great groups of cycads existed in the Cretaceous. One, the true cycads, survives to the present day; the other, referred to as the 'cycadeoids,' became extinct in the Cretaceous. Both groups apparently had a common ancestry among the ancient Paleozoic seed ferns. People familiar with living cycads know that these plants are either male or female. The female plants produce some of the largest reproductive structures in the plant kingdom, with some seed cones weighing more than twenty kilograms. The cycadeoids, however, produced bisexual cones that superficially resembled large flowers. A close look at these peculiar reproductive structures reveals that they were unlike flowers, and unlike the cones of the true cycads as well.

The cycads and cycadeoids, nevertheless, produced almost indistinguishable types of foliage. As well, they appear to have occupied similar habitats, as they are commonly found mixed at the same locality. Microscopic examination of the pores of the leaves is usually necessary in order to differentiate the foliage of the two groups. The cycads and cycadeoids provide an exceptional example of parallel evolution, in which two related groups of plants have developed many similar features not found in their common ancestors, probably as a result of exploitation of similar environments. Even after evolving independently for 200 million years, members of the two groups are easily mistaken for one another.

Most artistic reconstructions of life in the Cretaceous include the usual array of dinosaurs, pterosaurs, early birds, and tiny mammals living among forests of conifers and cycads. Although these reconstructions usually do justice to the great importance of cycads, they seldom portray the extensive proliferation of forms achieved by the group during the Early Cretaceous – the Age of Cycads and Conifers.

Seed Ferns

Sagenopteris represents a group of seed plants that were common in British Columbia in the Early Cretaceous, occupying important roles as shrubs and trees in the lush vegetation of that time. Because leaves of this type have been linked with the seed-producing structures referred to as *Caytonia,* they are considered part of the family Caytoniaceae. As seed plants, they differed considerably from other gymnosperms (plants that bear exposed seeds) of the Mesozoic. Unlike the conifers they bore large, compound leaves, and unlike the cycads and cycadeoids their leaves were palmately compound, with the leaflets radiating from the end of the petiole like the fingers on a hand. In many ways, the leaves of the Caytoniaceae and related groups superficially resemble some ferns. Although no close relationship exists between seed plants and any fern, they are commonly called 'seed ferns.'

Foliage of *Sagenopteris* typically occurs as isolated leaflets that have been shed from the petiole before fossilization. In some species, the leaflets appear to have been sessile, that is, they had no stalk but were attached directly to the end of the petiole. As a result, each leaflet has a tapered base so that they all fit together with no overlap. Other species appear to have borne leaflets on the ends of secondary petioles so that there was no problem with overlap. These types of leaflets are more circular and usually have a long petiole attached. *Sagenopteris williamsii* is an example of this type of leaf (Figure 15.7).

Sagenopteris leaflets typically range from as long as a finger to twice that long. The compound leaves are the size of a palm of a hand to as large as an outstretched hand. The leaflets have a midrib and usually have a rounded margin, although many leaflets may be found with a wavy or lobed margin. The veins of the blade repeatedly fuse and divide to form a highly distinctive network. This feature distinguishes *Sagenopteris* from most other fossil plants.

Figure 15.7 **Leaf of the seed fern *Sagenopteris williamsii*.** This leaf has a lobed margin rather than the more common rounded margin. Gates Formation, Tumbler Ridge. Actual size.
Source: Collections of the Department of Geological Sciences, University of Saskatchewan

Reproductive organs of these plants are rare finds – inconspicuous narrow stalks with small seed packets. Because finding these remains is very important scientifically, however, collectors should watch for them wherever *Sagenopteris* is encountered. It occurs throughout the Lower Cretaceous coal fields of the Crowsnest Pass and northeastern British Columbia.

The Caytoniaceae is one of many families of seed ferns that flourished in the Mesozoic. These groups have their origins in the Paleozoic seed ferns, which are preserved in abundance in Carboniferous rocks of the Maritime provinces. Because the fossil record of Mesozoic seed ferns is incomplete, little is known about their evolution. The Caytoniaceae was one of the last of the group but it, too, became extinct, disappearing from the fossil record shortly after it left its imprints in Lower Cretaceous rocks of British Columbia.

The story of the Caytoniaceae may not end with the disappearance of plants like *Sagenopteris*. Mesozoic seed ferns are now viewed by most paleobotanists as the source for the flowering plants. Although no seed fern known appears to be a suitable ancestor, the group as a whole possesses many of the characteristics predicted for the ancestral stock of the flowering plants. There is little overlap between the last of the seed ferns and the first of the flowering plants. Thus it appears that the swift ascension of flowering plants in the mid-Cretaceous eclipsed the very groups that gave rise to them.

Maidenhair Tree

From a once-diverse and venerable group, only a single species – the maidenhair tree, *Ginkgo biloba* – has survived to the present day. The ancestry of this group can be traced back to the Paleozoic, nearly 300 million years ago.

What is remarkable about *Ginkgo* is that it has changed so little in over 100 million years. The fossils commonly found in Lower Cretaceous rocks, as illustrated by *Ginkgo pluripartita* (Figure 15.8), cannot be distinguished from leaves of living *Ginkgo biloba* unless one scrutinizes minute details of the veins and the leaf surface. Leaves of this type are fan-shaped, and are typically cleft one or more times. The central cleft is the deepest, so that the leaf appears to be made up of two or more wedge-shaped segments. The enormous variability in shape includes leaves that are scarcely cleft, if at all, to leaves that are highly dissected. If the leaves are well preserved, one can see the open dichotomous venation; that is, the vein in the petiole divides into two at the base of the leaf, and these veins then repeatedly dichotomize (split into two equally) throughout the leaf without rejoining. The effect is unlike the network formed in the leaves of *Sagenopteris*.

Figure 15.8 **Leaves of the maidenhair tree, *Ginkgo pluripartita*.** The dissected, fan-shaped leaves are typical of many extinct Cretaceous and Cenozoic ginkgos as well as the single living species, *Ginkgo biloba*. Gates Formation, Tumbler Ridge. Actual size.
Source: Collections of the Department of Geological Sciences, University of Saskatchewan

Ginkgo biloba, the only living species of the entire group, represents only one lineage of ginkgos. Many other species existed in the Cretaceous. One type of leaf, of the genus *Baiera,* is so highly dissected that it appears to lack a leaf lamina altogether and to consist of only the veins. This leaf may have been somewhat feathery in life, and must have contributed to a splendid-looking tree.

Flowering Plants

Although the modern forests of British Columbia are dominated by conifers, this situation is atypical among the world's forests. Even in British Columbia, however, the vast majority of species are flowering plants (angiosperms). On a global scale, flowering plants have, by all criteria, overwhelmed the earth's surface as has no other single plant group throughout time.

Two examples of the many flowering plants found in Upper Cretaceous rocks of the Nanaimo Group of east-central Vancouver Island are *Cupanites* and *Zizyphus* (Figure 15.9). Ludvigsen and Beard (1994) discuss and illustrate several more. These rocks of the Nanaimo Group are roughly eighty million years old, and until recently they were an important source of coal. Old mine dumps may still yield fossils, but the rocks fracture

Figure 15.9 **Pinnately compound leaves of the angiosperm *Cupanites crenularis.*** The five leaflets at the top of the photo are attached to a common petiole. The broad leaf below is classified as *Zizyphus,* although it is highly unlikely that this plant has any relationship to living species of this genus. Upper Cretaceous Comox Formation, Vancouver Island. (X 0.7)
Source: Collections of the Geological Survey of Canada

badly on weathering, so that this once easily accessible source is now disappearing.

This section only touches on the diversity of flowering plants in the Late Cretaceous. Although palms were present in the Cretaceous, the bulk of the record of angiosperms is represented by broad-leaved trees and shrubs. Flowers are delicate and rarely preserved, and fruits and seeds are seldom identifiable once compressed and turned to coal. Leaves, however, were shed in abundance, and are generally readily recognized by their distinctive shapes and venation.

Leaves of most angiosperm trees and shrubs have a broad blade with a conspicuous primary vein extending from the petiole toward the tip. Secondary veins arise from the primary ones and run toward the margins. Other veins of decreasing prominence (tertiary, quaternary, etc.) form a complex, interconnected network that is quite unlike the more homogeneous venation of the ferns, cycads, and seed ferns. This distinctive pattern of a net-veined leaf is something we unconsciously associate with angiosperms.

Flowering plants first appear in the fossil record in rocks of Early Cretaceous age. Pre-Cretaceous angiosperms have been reported, but these reports failed to withstand close examination. Paleobotanists generally agree that the flowering plants evolved during the Early Cretaceous from a group of Mesozoic seed ferns that is yet unknown.

The increasing abundance of angiosperms is evident in the fossil deposits of British Columbia. The Lower Cretaceous coal-bearing Blairmore, Gates, and Gething deposits in the eastern part of the province, for example, contain few angiosperm leaves. Diversity and abundance of angiosperms rises dramatically from the Early to the Late Cretaceous. The Dunvegan Formation of the Dawson Creek area (near the base of the Upper Cretaceous) preserves a variety of angiosperms. The Late Cretaceous floras of the Nanaimo Group are, of course, dominated by angiosperms.

At one time paleobotanists believed that most Cretaceous remains of flowering plants represent modern families and genera. A quick review of Walter Bell's reports of Cretaceous floras shows that he recognized numerous modern genera, such as oak, poplar, alder, fig, cinnamon, breadfruit, maple, and ash. We now know that in most cases these fossils represent extinct genera and even families of flowering plants. Paleobotanists have difficulty in determining the relationship between the flowering plants of the Cretaceous and those living now. Nevertheless, it is clear that the Late Cretaceous forests are the first that would truly be familiar to us, were we able to take a stroll through them.

Modern vegetation emerged during the Cretaceous. As a result, the fossil floras of this period are a fascinating blend of the strange and the familiar. British Columbia is a rich source of fossil evidence for this evolution. And here, more than anywhere else on earth, it is almost possible to step back millions of years in time by entering the magnificent, primeval coniferous forests of present-day coastal British Columbia.

References

Bell, W.A. 1956. *Lower Cretaceous Floras of Western Canada.* Geological Survey of Canada, Memoir 285

–. 1957. *Flora of the Upper Cretaceous Nanaimo Group of Vancouver Island, British Columbia.* Geological Survey of Canada, Memoir 293

–. 1965a. *Illustrations of Canadian Fossils: Lower Cretaceous Floras of Western Canada.* Geological Survey of Canada, Paper 65-5

–. 1965b. *Illustrations of Canadian Fossils: Upper Cretaceous and Paleocene Plants of Western Canada.* Geological Survey of Canada, Paper 65-35

Ludvigsen, R., and G. Beard. 1994. *West Coast Fossils: A Guide to the Ancient Life of Vancouver Island.* Vancouver: Whitecap Books

MacLeod, S.E. 1991. Late Jurassic to Early Cretaceous (Tithonian to pre-Albian) Plant Macrofossils, Northern Bowser Basin, British Columbia, Canada. *Review of Palaeobotany and Palynology* 70:9-45

16
Paleogene Mammals on Land and at Sea

Lee McKenzie McAnally

The shoreline of southwestern Vancouver Island is composed of sandstone outcrops of the Sooke and Hesquiat formations. These rocks lie exposed on the beaches and form cliffs above the high-tide line. From the window of the Canadian Coast Guard helicopter on which I am riding, these cliffs form a spectacular panorama along the coast. Their top is fringed by forests and sporadically veiled by waterfalls; their base is pounded ceaselessly by waves that have etched caves into the cliff faces (Figure 16.1). These are truly some of the most unspoiled and ruggedly beautiful landscapes in Canada. As I gaze down on these outcrops, I cannot help but wonder how many fossils lie undiscovered in this largely unexplored expanse of rock.

Figure 16.1 **Paleogene rocks of Hesquiat Formation eroded into seaside cliffs along southwestern Vancouver Island coast**

The flight from Victoria is not long enough to fully appreciate the scenery. All too soon my destination is in view – a lightstation perched on a rock high above the sea. The contrast of the red and white buildings against the green lawns has a jewel-like quality from the air. As the pilot circles and approaches the landing platform, I see my hosts Janet and Jerry Etzkorn and their children Jake and Justine waiting to greet me.

The Etzkorns are no ordinary lighthouse keepers. They have a passion for the natural history and paleontology of the Pacific margin of Vancouver Island. The fact that I am here to investigate and collect fossils of ancient marine mammals is due in no small measure to their thorough scrutiny of the rocks and fossils of this area.

Fossils of vertebrates are not abundant here, so nearly every find has the potential for scientific importance. The fossils are solidly embedded in very resistant sandstone and are difficult to remove without experience and adequate equipment. These particular fossils also lie within the boundaries of Pacific Rim National Park, and stiff fines are levied against anyone attempting to collect them without a permit. The Etzkorns, aware of all these considerations, wrote to Dr. Christopher Yorath, a geologist with the Geological Survey of Canada. Chris passed their letters on to me and that, in a nutshell, is why I am here, collecting permit and geologist's hammer in hand.

The Paleogene Rocks

The sandstones now exposed along these beaches were deposited twenty-five to thirty million years ago, at the end of the Paleogene – the oldest of the three Cenozoic systems. The Paleogene, in turn, comprises three series: Paleocene, Eocene, and Oligocene. Over the past four decades, the rocks and fossils of the Hesquiat and Sooke formations have been studied and interpreted by several geologists. Recent work suggests that the older Hesquiat Formation represents a deep-water environment and the overlying Sooke Formation a shallow-water environment. Wave action was sufficient to sort shells according to size but not strong enough to damage them. Thus lovely coquina-like beds are preserved. The Hesquiat and Sooke formations represent what geologists call a shallowing-upward sequence; in other words, the water was becoming shallower over time. Geologists apply this term, even though the rocks of the Sooke Formation cannot be observed directly on top of rocks of the Hesquiat Formation. The main exposures of the Sooke Formation are exposed farther south, between Sooke and Port Renfrew (Muller et al. 1981).

Marine Mammals

Mammals have adapted to many environments, from the equator to the poles and from the mountains to the sea. Some have taken to the air;

others lead an entirely subterranean existence. But surely one of the most remarkable trends in mammalian evolution has been the move to the marine realm. Three grades of marine adaptations are found among mammals today. Whales are warm-blooded, air-breathing mammals that do a remarkable job of mimicking fishes. They lack hind limbs and cannot survive out of water. Being mammals, they give birth to live offspring and nurse their young. Seals and walruses are capable of spending a great deal of time on shore but are slow and cumbersome there. The third type of marine adaptation is found in sea otters, creatures that appear equally at home in or out of water.

A Whale

To say that fossils of marine mammals are not abundant in the rocks of southwestern Vancouver Island is an understatement. The only mammal fossils from the Hesquiat Formation that have been formally described belong to the small whale *Chonocetus sookensis*. *Chonocetus* means literally 'little whale.' In 1968, Loris Russell, now retired from the Royal Ontario Museum, illustrated part of a skull and six vertebrae from what appeared to be an archaeocete, a primitive whale. Recent work by Larry Barnes, a paleontologist at the Natural History Museum of Los Angeles County, indicates that *Chonocetus* was actually a toothed mysticete – a baleen whale with teeth.

Whales evolved more than fifty million years ago in the Tethys Ocean, a large equatorial sea of which the Mediterranean is the only remnant. Tethys (named for the Greek goddess who was the wife of Oceanus) was warm and probably was a veritable storehouse of fishes and invertebrates of all kinds. The predatory marine reptiles of the preceding Cretaceous were extinct, leaving these well-stocked seas wide open for exploitation by marine mammals.

Philip Gingerich, Director of the Museum of Paleontology at the University of Michigan in Ann Arbor, is probably the world's foremost expert on fossil whales. His field seasons are spent in Egypt and Pakistan, arid regions that are now landlocked but which once bordered the Tethys Ocean. The discoveries made by Gingerich and his colleagues have resulted in a fascinating story of the evolution of whales. *Pakicetus* was described as 'perfectly intermediate, a missing link between earlier land mammals and later full-fledged whales' (Gingerich 1994). This ancient whale was about the size of a dog and probably fished along the shores of the Tethys Ocean. A nearly complete skeleton of another ancient whale has demonstrated that early whales did indeed have functioning hind limbs. This fossil has been christened *Ambulocetus natans* – the walking whale that swam. The distribution of whales had expanded dramatically by the end of the Oligocene, when they were essentially cosmopolitan.

They had not developed the sonar capability of modern whales, even though they were fully aquatic by that time.

Just as whales evolved from four-legged land mammals that lived near the sea, so also did seals, sea lions, walruses, and manatees, along with a number of other marine mammals that are now extinct. The ancestor of early whales was a small ungulate, an animal that had hoofs. Whales are closely allied with even-toed ungulates such as pigs, camels, cattle, and hippopotamuses. The hippo connection is the closest.

A Desmostylid

Only one other fossil mammal from Vancouver Island has been described in a scientific journal. Several teeth from the desmostylid *Cornwallius sookensis* were collected from the Sooke Formation near Otter Point, on the southernmost tip of Vancouver Island. The first tooth was collected in 1916 by a Miss Egerton of Victoria. She reported her find to the Provincial Museum, and the fossil was added to the collections. Lawrence Lambe, a paleontologist with the Geological Survey of Canada, and C.F. Newcombe, director of the museum at that time, determined that the tooth belonged to a desmostylid – a large and unusual four-footed amphibious mammal that fed on sea grasses and algae. Even though fossil desmostylids have only been collected from marine sediments, they were clearly capable of movement on land, suggesting that they lived in much the same way as seals and walruses live today. Desmostylids are considered a sister group of the Proboscidea (elephants), since the two groups shared a common ancestor (Ray et al. 1994).

The discovery of this fossil tooth caused a bit of a stir and more than a little speculation about the geologic history of the Victoria area. In September 1916, the *Victoria Colonist* published several stories by Lambe and Newcombe on the discovery of the desmostylid, along with an editorial about Miss Egerton's find. The editor took advantage of the discovery to speculate on the timing of Pleistocene glaciation and about a lost human civilization. 'He [the desmostylid] lived ... before the Ice Age, when this island had a tropical climate.' With reasonable accuracy the editor estimated that the ice age ended about 7,000 or 8,000 years ago, but he went on to add that 'suggestions have been made that some great cataclysm occurred at that time which extinguished an ancient civilization in America ... Very many Indian legends appear to relate to a period before the Great Winter. The discovery of the remains of the demostylus [*sic*] may stimulate the investigation into the geological history of British Columbia' (*Victoria Colonist,* 16 September 1916, editorial).

In spite of evidence given by a reputable paleontologist, the editor did not seem to consider the possibility that other life existed prior to the appearance of humans! Or, he was unable to fathom ancient life outside a human framework. We now know that desmostylids became extinct before the end of the Miocene, five million years ago – long before the advent of human civilization.

Following its brief moment in the limelight, the desmostylid tooth was put away in the museum collections. It was soon joined by another one collected by Rev. Robert Connell of Victoria. These teeth were described in several scientific publications during the 1920s, but in 1928 the two teeth disappeared from the museum collections. Museum records indicate that the fossils were stolen on 24 May of that year. The thief left no clues. Then, just as mysteriously, the fossils reappeared. They had been carefully wrapped in a *Vancouver Province* newspaper dated 22 August 1932, placed in a Player's cigarette tin, and left on the moulding by a fossil case in the museum. This second 'discovery' of the desmostylid fossils, this time by museum staff, took place on 17 September 1932. No one knows who took the fossils, why they were stolen, or what prompted their return almost four and a half years later. The desmostylids are now safe within the Royal British Columbia Museum collections and plaster casts of the teeth still reside in that old cigarette tin (Figure 16.2).

Ira Cornwall (the namesake of *Cornwallius*), another Victoria resident, collected about a dozen other marine-mammal fossils from the Sooke

Figure 16.2 **Cast of tooth of *Cornwallius sookensis* collected by Miss Egerton.** The tooth is shown here with a pill box containing a fragment of another fossil tooth and the cigarette tin left behind by an unknown fossil thief.
Source: Collections of the Royal British Columbia Museum

Formation. Following a fire in his home at William's Head in the early 1920s, Cornwall decided that he should not keep fossils in a private collection. Thus he donated all his fossils to the National Museum of Natural History in Washington, DC, where they remain to this day. Among these fossils, which consist of desmostylid and whale remains, are two desmostylid femurs, part of a whale mandible, several vertebrae, fragments of ribs, and two scapulae. This single collection of marine-mammal fossils is undoubtedly the largest of its kind from Vancouver Island (Hay 1924).

A Beach Bear

The most recent discovery of a fossil mammal from southwestern Vancouver Island is a partial lower jaw, including one molar tooth, of a small marine carnivore (Figure 16.3). Peter Bock of Nanaimo collected this fossil from the Sooke Formation exposed along the beach near Muir Creek. This little animal, now extinct, was similar to an otter in size and habit but more closely aligned with the ursids, the group that includes modern bears. The fossil jaw is not unlike the jaws of the slightly younger *Kolponomos clallamensis* and *Kolponomos newportensis,* from Washington and along the coast of Oregon (Tedford et al. 1994). It differs enough, however, that it will be assigned to a new genus. This animal lived along the shores, probably in and out of the water, dining on mollusks pried from the rocks. Paleontologist Larry Barnes affectionately refers to *Kolponomos* as a 'beach bear.'

Terrestrial Mammals

The record of terrestrial Cenozoic mammals in British Columbia is no more complete but just as intriguing as that of the marine realm. This record is really little more than a series of vignettes – three minuscule fossil collections from localities scattered across the province that provide brief glimpses of the past and just enough information to spark curiosity.

A Tillodont

In 1935 Loris Russell described the first fossil mammal in British Columbia. The material came from a coal mine along the bank of the Similkameen River near Princeton. It consists of two teeth from a strange animal known as a tillodont and named *Trogosus minor.* Tillodonts are rare fossils of unknown affinity. No complete skeleton has been collected anywhere in the world. By piecing together the fragmentary material, however, paleontologists have deduced that tillodonts were quite large – about the size of a modern brown bear. The evident wear of the teeth suggests an unusually abrasive diet.

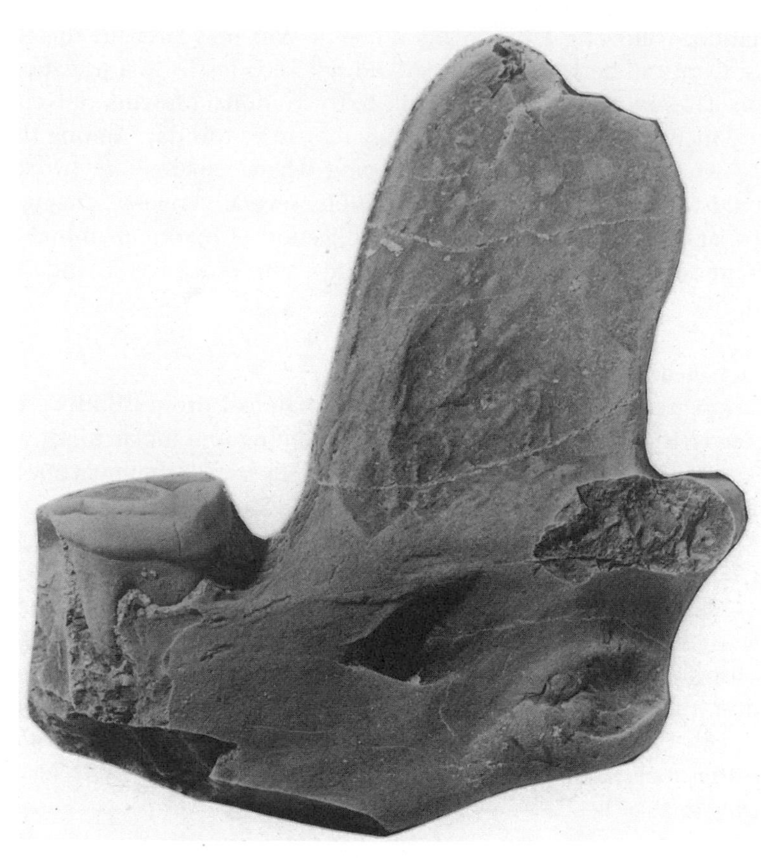

Figure 16.3 **Lower jaw and second molar of small marine carnivore from Sooke Formation.** (X 2)
Source: Collections of the Vancouver Island Paleontological Museum

A Titanothere

Our next window into the past again comes as several isolated teeth – two lower molars and a possible premolar of a titanothere from Quesnel (Figure 16.4). Although these Quesnel specimens have not been formally named or described, a great deal more is known about titanotheres than tillodonts. These huge herbivores roamed the interior of North America in great herds during the Eocene. In those twenty or so million years, they evolved from small, dog-sized animals to giants with a shoulder height of more than two metres. Titanotheres are characterized by a prominent hump on the back and a pair of blade-like horns on the end of the snout. Their brawn was not matched by wit, apparently, as the largest titanothere had a brain not much bigger than a human fist.

Fossil titanothere bones feature in Sioux legend as the remains of a thunder horse – a huge beast that descended from the heavens to hunt

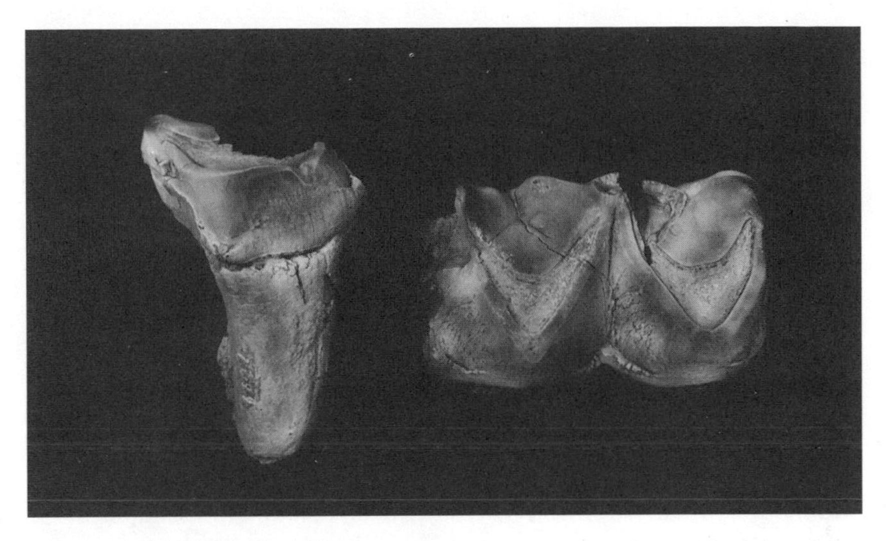

Figure 16.4 **Titanothere teeth from Quesnel.** A partial molar on the left and a nearly complete molar on the right. (X 0.5)
Source: Collections of the Department of Vertebrate Palaeontology, Royal Ontario Museum

bison. And, a century and half ago, a titanothere was the first vertebrate fossil from western North America to be described in a scientific journal. Titanotheres originated on this continent and expanded into eastern Asia via the Bering land bridge. Perhaps our Quesnel fossil records that migration.

The Kishenehn Mammals

The third land-mammal fossil site is located in southeastern British Columbia along the banks of the Flathead River. It was here that Loris Russell collected a small mammalian fauna from the Kishenehn Formation of Early Oligocene age. Seven species representing five orders were described by Russell in 1954.

Three rodent species are known from this locality. *Pseudocylindrodon silvaticus* belongs to the earliest family of rodents, the family which probably gave rise to modern dormice and squirrels. *Desmatolagus,* an early rabbit, is also represented. Rabbits are an excellent example of that old adage, 'If it isn't broken, don't fix it.' By Oligocene time they had evolved their characteristic hopping gait and high-crowned teeth, and they changed little over the next thirty million years. The early ruminant *Leptotragulus* was a rabbit-sized artiodactyl (a deer-like animal) that probably found food and protection in the dense vegetation along streams and waterways.

The Kishenehn fauna also includes a marsupial, *Peratherium,* which represents the Didelphidae, the oldest family of marsupials. Marsupials constitute a very successful group of mammals, one that has existed since the Late Cretaceous. The living opossum *Didelphis* belongs to the same family. The last, and perhaps most intriguing, animal in the fauna is *Thylacaelurus,* which Loris Russell thought was a cat-like marsupial but which turns out to be a plagiomenid. Plagiomenids are considered to belong to the order Dermoptera (McKenna 1990). Living dermopterans are called 'flying lemurs,' even though they do not fly and are not true lemurs.

The descriptions of fossil mammals from the Kishenehn locality were made on the basis of isolated teeth or partial jaws containing a few teeth. Based on the kinds of animals present, Russell believed that a woodlands-type habitat was represented.

Russell believed that the various Cenozoic mammal sites scattered around the province would eventually permit 'detailed correlations in the manner so successful in the great plains and basins of the western United States' (Russell 1935). So far, this has not been possible. Instead, all terrestrial fossils described in this chapter could be held in the palm of a hand. The localities identified to date are isolated in intermontane valleys that are not easily accessible. As well, they do not yield the quantity of fossil material needed to justify field expeditions. Thus it seems that the discovery of additional fossil mammals will require luck more than anything else.

The weather on the west coast of Vancouver Island is unpredictable, especially in the fall. The sound of the fog horn has been reverberating off the rocks for two days while I searched for fossils. It is time for me to go home and, while I enjoy one last meal with the Etzkorn family, the call comes over the radio: the weather has cleared a little and the Coast Guard helicopter is on its way. Several fossil bones, including a complete radius (a bone in the forearm) from a yet-to-be-determined animal, are securely wrapped in my pack. This project is still in its infancy. There are many fossils to collect, more research to be done, and new information to be gained. Over the years these rocks have reluctantly yielded a small but tantalizing collection of fossils of marine mammals. Every new discovery is exciting, and each has the potential to alter our understanding of the evolution and geographic distribution of early marine mammals.

Acknowledgments

Many people contributed to this chapter, particularly Chris Barnes, my academic supervisor, Clayton Ray of the Smithsonian Institution, and the Etzkorn family. This chapter is dedicated to Loris Russell of the Royal Ontario Museum, whose work on the Cenozoic mammals of western Canada spans six decades.

References

Gingerich, P.D. 1994. The Whales of Tethys. *Natural History* 103 (4):86-8

Hay, O.P. 1924. Notes on the Osteology and Dentition of the Genera *Desmostylus* and *Cornwallius*. *Proceedings of the United States National Museum* 65:1-8

McKenna, M.C. 1990. Plagiomenids (Mammalia: ?Dermoptera) from the Oligocene of Oregon, Montana, and South Dakota, and Middle Eocene of Northwestern Wyoming. Geological Society of America Special Paper 243:211-34

Muller, J.E., B.E.D. Cameron, and K.E. Northcote. 1981. *Geology and Mineral Deposits of Nootka Sound Map Area, Vancouver Island, BC*. Geological Survey of Canada, Paper 80-16

Ray, C.E., D.P. Domning, and M.C. McKenna. 1994. A New Specimen of *Behemotops proteus* (Order Desmostylia) from the Marine Oligocene of Washington. In *Contributions in Marine Mammal Paleontology Honoring Frank C. Whitmore, Jr.*, ed. A. Berta and T.A. Deméré. Special issue of *Proceedings of the San Diego Society of Natural History* 29:205-22

Russell, L.S. 1935. A Middle Eocene Mammal from British Columbia. *American Journal of Science* 29:54-5

–. 1954. Mammalian Fauna of the Kishenehn Formation, Southeastern British Columbia. *Annual Report of the National Museum of Canada for 1952-1953*. Bulletin No. 132, pp. 92-111

–. 1968. A New Cetacean from the Oligocene Sooke Formation of Vancouver Island, British Columbia. *Canadian Journal of Earth Science* 5:929-33

Tedford, R.T., L.G. Barnes, and C.E. Ray. 1994. The Early Miocene Littoral Ursoid Carnivoran *Kolponomos*: Systematics and Mode of Life. In *Contributions in Marine Mammal Paleontology Honoring Frank C. Whitmore, Jr.*, ed. A. Berta and T.A. Deméré. Special issue of *Proceedings of the San Diego Society of Natural History* 29:11-32

17
Fishes from Eocene Lakes of the Interior

Mark V.H. Wilson

During the middle part of the Eocene, about fifty million years ago, countless fishes, insects, and plants were buried in sediment on the bottoms of the small and large lakes that were plentiful in western North America. One of the most interesting areas where beds of several smaller Eocene lakes are preserved is the Interior of British Columbia. These lake beds preserve an assemblage of freshwater fishes and other organisms that rivals those found in any other part of the world. The fossils are extremely well preserved and they provide an outstanding amount of information about the biology of the organisms and the conditions in the lakes. Although they have been known for over a century, the fossil fishes of British Columbia were not studied intensively until eighty-one years after their discovery. Fortunately, some of the early finds were preserved in museums where they provided essential clues pointing the way for more recent investigators.

Discoveries by Pioneer Geologists and Amateurs

The first recorded discovery of a fossil fish of Cenozoic age in British Columbia was made near Princeton in 1888 by George M. Dawson, while carrying out geological surveys near the US border. These early explorations were designed to discover mineral resources such as coal and gold. Dawson, himself the son of a well-known geologist, made important contributions to the study of geology through his many explorations, maps, and reports on the geology of Alberta, British Columbia, and the Yukon. He was also a student of Native customs and, like many other geological pioneers, a field biologist. The fossil fish that Dawson found was sent to the famous American vertebrate paleontologist E.D. Cope, who recognized it as a new species of *Amyzon* (a genus in the 'sucker' family Catostomidae). Cope had previously named this genus from fossils found in Colorado and Nevada. In a publication appearing in 1893, Cope named the new species *Amyzon brevipinne,* because its dorsal fin was shorter than that of other *Amyzon* species. In 1890, two years after his first find, Dawson again collected some fossil fishes, this time from a site near Kamloops Lake.

During the 1890s the Cariboo region of British Columbia was a hotbed of gold exploration and mining activity. In 1895, J.B. Hobson, owner of a placer gold mine near Horsefly, came into possession of several fossil fishes that were sent to the National Museum in Ottawa in 1898. These fishes were subsequently studied by Canada's first vertebrate paleontologist, Lawrence M. Lambe, who identified them as species of *Amyzon.*

Lambe himself made a collecting trip to British Columbia in 1906, no doubt attracted by the fossils that had found their way to the national collections through both Dawson's work and Hobson's generosity. Travel in those days was by train or riverboat, where available, or otherwise by wagon, horseback, or foot. Lambe visited the Princeton area, confirming Dawson's discovery by finding some fish fragments. Then he visited a coal mine east of Merritt at Quilchena, where he was the first to report finding fish skeletons and scales. He also found fish fossils near Kamloops Lake. On the same journey Lambe visited Hobson's mine in the Horsefly area and collected fish fossils. Unfortunately, some of the rock outcrops that Lambe examined near Horsefly can no longer be located, perhaps because they were exposed by mining activities and have since been reburied.

Another pioneer worthy of mention is Charles Camsell, who studied the geology of the Tulameen area in 1909 and 1910, finding pieces of fossil fish that are still in the National Museum collections. Camsell later became Deputy Minister of Mines of Canada. Additional fossil fishes from the Kamloops area were found in 1912 by one B. Rose and subsequently named *Eohiodon rosei* after the collector. The first fossil fish discovered in the Smithers area was a fragment found by E.J. Lees in 1936 and deposited in the National Museum of Canada collections.

In time, vertebrate paleontologist Loris S. Russell came to know all of these specimens – those collected in these early years by the Survey, those donated to the National Museum in Ottawa, and those collected from the Horsefly area by geologist R.B. Campbell's field parties and by staff and students of the University of British Columbia. Russell, a geological and paleontological pioneer, was in turn Director of the National Museum and Chief Biologist of the Royal Ontario Museum. Knowing about the earlier discoveries and having himself done field work in British Columbia, Russell suggested study of the fishes to me when I was beginning graduate studies. Thus the work of those early scientists and amateurs eventually led to my special interest in fossil fishes and provided clues about places I was likely to find them when I began field work in 1969, some eighty-one years after Dawson's first discovery.

The Eocene Lakes

During the Eocene the Interior of British Columbia and Washington was the scene of numerous volcanic eruptions, which spewed lava over large areas, dammed rivers, and built up extensive plateaus. The eruptions in some regions were accompanied by faulting that created rift valleys. In some places large depressions were formed where sediments accumulated, promoting the formation of coal swamps that were subsequently drowned by lakes and then covered by lava.

Sediment deposited on the floors of the many lakes entombed the skeletons of fishes, the bodies of insects, and parts of plants that grew on the shores of the lakes and rivers. Fossil-bearing rock types include shales, siltstones, and even laminated diatomites (rocks made from the microscopic, silica-rich skeletons of algae called diatoms), one of the oldest occurrences of freshwater diatoms in the world. These sediments are often found in close association with coal seams or with volcanic rocks. In some areas the solidified lava now forms cliffs that cap the modern-day valley walls. Often lake beds can be found near the base of such lava cliffs, and many of these beds contain fish fossils. Unfortunately, in a large number of cases the rocks under the lavas have been altered, perhaps by the hot lava, and many details of the fossils have been destroyed. In other places, lake beds with good specimens are exposed in river cuts on valley bottoms. The best specimens, however, have been found in delicately laminated, fine-grained rocks such as shales and diatomites, representing sediments that were deposited under quiet conditions in the deeper waters of some of the lakes. Well-preserved examples can be found today in the Smithers, Horsefly, Kamloops, and Princeton areas of British Columbia and in the Republic area of Washington.

The Fishes

Apart from the volcanic activity and the more southerly appearance of some of the vegetation, to a casual observer an Eocene lake might have resembled any lakeside scene of today. The fishes would also at first glance have appeared familiar. On closer inspection, however, it would have become apparent that all the species and nearly all of the genera of these Eocene fishes are now extinct, and that several of their families no longer survive in Pacific drainages of British Columbia.

The only Eocene fish belonging to a still-living genus is *Amia hesperia,* a species of the bowfin family Amiidae. A fifty-five-centimetre-long skeleton of *Amia hesperia* has been found near Princeton. Although skeletons of this fish are rare, its scales are found in many deposits, particularly those near coal seams. The only member of the bowfin family alive today, *Amia calva,* is an inhabitant of southeastern North America. Bowfins have a single, long dorsal fin that they undulate to move slowly through the water. Like the modern bowfin, the Eocene species probably fed mostly on other fishes, supplemented by crayfish and frogs. It probably preferred weedy bays of warm lakes and rivers, and may have gulped air at the surface to supplement its oxygen supply.

The family Hiodontidae is another group of fishes that, at present, is not found west of the Rocky Mountains. This family has only two living species, the mooneye and the goldeye, both in the genus *Hiodon.* The Eocene hiodontids are assigned to a different but closely related genus, *Eohiodon.* Specimens of *E. rosei* are among the most commonly found Eocene fish, having been collected in the Princeton, Tulameen, Quilchena, Falkland, Cache Creek, Kamloops, and Horsefly areas. A second species, *E. woodruffi,* was first recognized from specimens collected near Republic, Washington, and was named after the Republic youth Robert Woodruff, who found the best specimens. This species can be identified in collections from the Horsefly area, including specimens found as long ago as 1906 by Lawrence Lambe. Hiodontids eat mostly insects, the roof and floor of the mouth having prominent sharp teeth. Mature males develop thickened anal fin rays that can be recognized in fossils (Figure 17.1).

The sucker family Catostomidae is now a prominent and diverse part of the North American fish fauna. It competes with the closely related minnow family that dominates many freshwater habitats. During the Eocene, however, the minnow family had apparently not yet arrived in North America, leaving suckers without close competitors for a time. Two species of the extinct sucker genus *Amyzon* have been described from British Columbia: *A. brevipinne,* usually the smaller of the two, has a

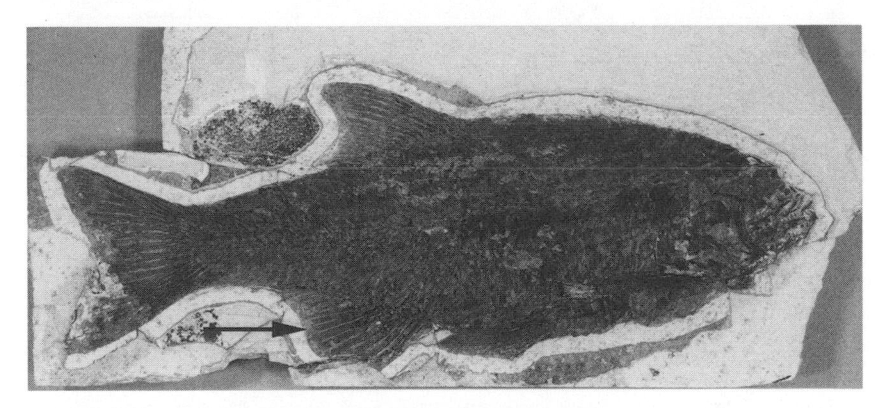

Figure 17.1 **Male specimen of the hiodontid (mooneye relative) species *Eohiodon woodruffi* from the Horsefly area.** This relatively large individual is 18 cm in length. The anal fin with its thickened rays and convex outline is indicated by the arrow.
Source: Collections of the Laboratory for Vertebrate Paleontology, University of Alberta

shorter dorsal fin and is known from the Princeton, Tulameen, and Quilchena deposits; *A. aggregatum,* usually the larger, has a longer dorsal fin and occurs in fossil beds near Republic, Kamloops, and Horsefly (Figure 17.2). Specimens of *A. aggregatum* discovered at one locality were thirty-five centimetres long or more.

Perhaps the most important of the Eocene fishes is *Eosalmo driftwoodensis,* which was first recognized through specimens from the Driftwood Creek fossil beds near Smithers. It has since been found at other localities, including ones in the Princeton and Republic areas. As the oldest fossil member of the family Salmonidae (salmon, trout, char, grayling, and whitefish), *Eosalmo* represents a nearly perfect missing link (intermediate form) between the subfamily Salmoninae (salmon, trout, and char) and the subfamily Thymallinae (grayling), supporting the theory that salmon evolved from an ancestor with grayling-like features (Wilson and Williams 1992). *Eosalmo* also shows that the first salmon-like features to evolve were modifications of the body, tail, and scales, because these are present in *Eosalmo.* Other skull features, such as the salmon's much longer jaw, are not present, so these characteristics probably evolved later.

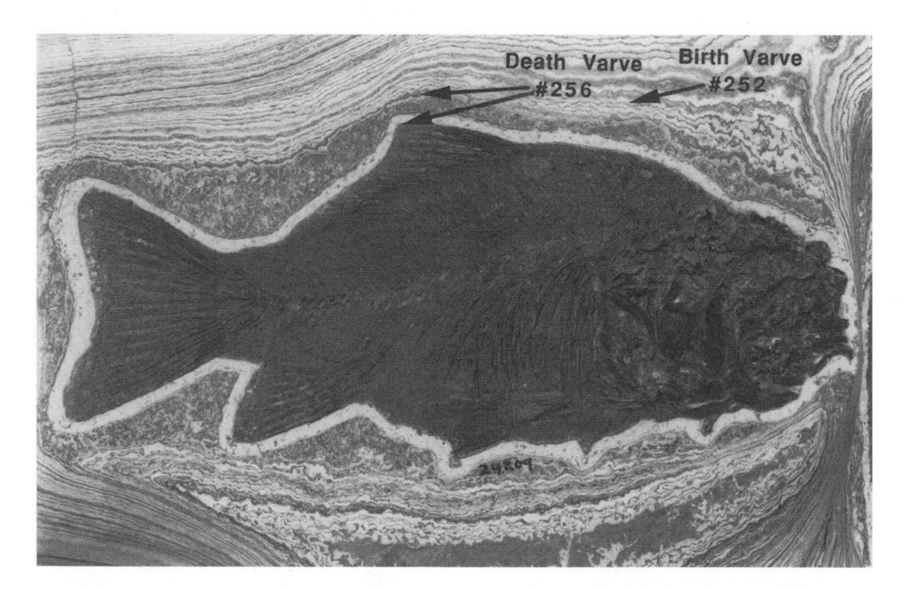

Figure 17.2 **Specimen of the catostomid (sucker) species *Amyzon aggregatum* from the Horsefly area.** The fish is relatively deep bodied for its size (length 22.7 cm) and is probably a mature female specimen. Labels and arrows indicate the dark (winter) lamina of the 256th varve ('death varve') and the light lamina of the 252nd varve ('birth varve') for this fish.
Source: Collections of the Laboratory for Vertebrate Paleontology, University of Alberta

Eosalmo probably lived in lakes and rivers draining to the Pacific. Today, the Pacific drainages are the home of the most advanced sea-running salmon species, which migrate to the sea at a young age and return to rivers and lakes only to spawn. It appears, however, that *Eosalmo* individuals remained in the lake most of their lives, rather than migrating to the sea (Wilson and Williams 1992), because a nearly complete age range of *Eosalmo* specimens, including probable juveniles (Figure 17.3), occurs in the Driftwood Creek beds. This evidence supports the theory of Stearley (1992) that sea migration was not a feature of salmon and trout ancestors but evolved relatively recently.

The trout-perch family Percopsidae has few living species, none of which occurs in British Columbia west of the Rockies. During the Eocene, however, distant relatives now classified in the family Libotoniidae were fairly common inhabitants of bottom waters in the shallow parts of lakes. Libotoniids were small fish only about four centimetres long, with weak fin spines and flattened heads. *Libotonius blakeburnensis* was named after René Liboton of Oliver, who gave specimens found in the Princeton area to the Royal Ontario Museum, and for Blakeburn Mine near Tulameen, where the type specimen was collected. A second species, *L. pearsoni,* occurs in the Republic area of Washington.

Figure 17.3 **Specimen of the salmonid (trout) species *Eosalmo driftwoodensis* from the Smithers area.** This salmonid is the smallest specimen of *Eosalmo* yet known (15 cm in length, as compared to about 30 cm for full-size individuals) and supports the idea that *Eosalmo* did not migrate to the sea.
Source: Collections of the Laboratory for Vertebrate Paleontology, University of Alberta

Among the most common and most attractive fishes in the world-famous Green River fossil beds of Wyoming are those belonging to the genus *Priscacara*. Although they resemble modern freshwater sunfish, they differ in many details, and their closest living relatives are still unknown. *P. aquilonia* is known only from the Eocene rocks of British Columbia; small specimens have been found near Horsefly (Figure 17.4) but so far are unknown in other places. Perhaps, like sunfish, they fed on insects, small clams, and snails, using strong tooth plates in their throat to crush their food. Their dorsal and anal fin spines might have been erected to avoid being swallowed by predators.

There are some surprising absences from these inhabitants of the Eocene lakes of British Columbia. Although bowfins occurred, neither gars nor sturgeons have been found, which is surprising since they are common in fossil deposits on the other side of the continental divide. Herring relatives are the most abundant fishes in the Eocene lake beds of Wyoming, but they too have not been discovered in British Columbia's rocks. Catfish might also be expected on similar grounds. Pike and smelt relatives occur in slightly older (Paleocene) deposits of rivers and shallow

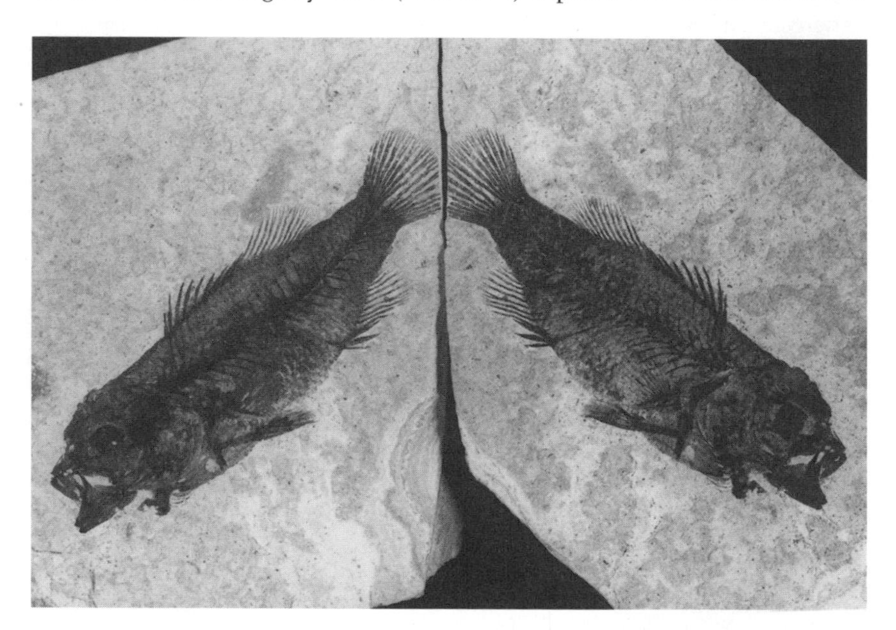

Figure 17.4 **Specimen of the priscacarid (sunfish-like) species *Priscacara aquilonia* from the Horsefly area.** All known specimens of this species are small, this one measuring only 3.7 cm in length. The mirror image results from the fact that the rock split parallel to the bedding planes of the varve layers, with the right and left sides of the fish adhering to opposite pieces.
Source: Collections of the Laboratory for Vertebrate Paleontology, University of Alberta

lakes of Alberta, but so far have not been found in the Eocene rocks of British Columbia.

On the other hand, it would be surprising to find certain other fish groups, since currently they are not known from any Eocene or older rocks of North America (Cavender 1986; Wilson and Williams 1992). Minnows (Cyprinidae), for example, have not been found anywhere in North America in rocks older than Oligocene. It is quite possible that they did not arrive on the continent until after the Eocene. Perches and darters (Percidae), sticklebacks (Gasterosteidae), and sculpins (Cottidae) also are not expected in the Eocene rocks of British Columbia because they are relatively advanced groups that have no definite Eocene fossil record anywhere in the world.

Fossils Reveal Distance from Shore

A single outcrop of an ancient lake bed preserves either shallow (near-shore) or deep (off-shore) fauna and flora but usually not both (Wilson 1988). Likewise, each rock sample records only a tiny part of the floor of the lake. Typically a single outcrop contains only a few species of fish, perhaps only those frequenting a little bay or seeking refuge in a deep part of the lake during the winter. A paleontologist wishing to assemble a more complete picture of the fauna must therefore sample many outcrops, noting similarities and differences.

Usually the rocks that result from sediment deposited in shallow water near the shore are darker and have less perfect laminations than those deposited off-shore in deep water. In addition, characteristic fossils are found in each. Small, juvenile fishes seem more common in near-shore sediment, for example. Fossils of land-dwelling organisms such as insects and plants differ as well, being more common in near-shore rocks and varying according to how easily they could be carried from shore by winds and by flotation. The state of preservation also differs, since turbulence, decay, and scavengers are more prevalent in the warmer, shallower waters near shore.

In near-shore deposits (Figure 17.5) the most abundant fishes are trout-perches *(Libotonius)* and juvenile suckers *(Amyzon),* along with predators such as bowfins *(Amia).* Bundles of pine needles and leafy twigs of the deciduous conifer *Metasequoia,* along with March flies, beetles, and bugs, are also common in these rocks. Both the fishes and the insects are more likely to occur with body parts disarticulated by waves and scavengers.

In the deeper, quieter waters off-shore (Figure 17.6), cool temperatures slowed decay, and oxygen-poor, poisonous water often excluded scavengers. Therefore, off-shore fossils are often complete. In addition, most are larger. The most abundant fishes are hiodontids *(Eohiodon),* adult suckers *(Amyzon),* and trout *(Eosalmo).* Relatively few plants and insects are

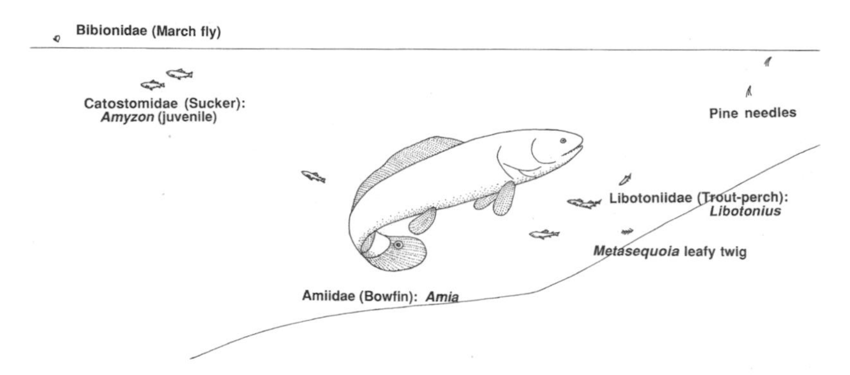

Figure 17.5 **Scene typical of shallow, near-shore environment in an Eocene lake**

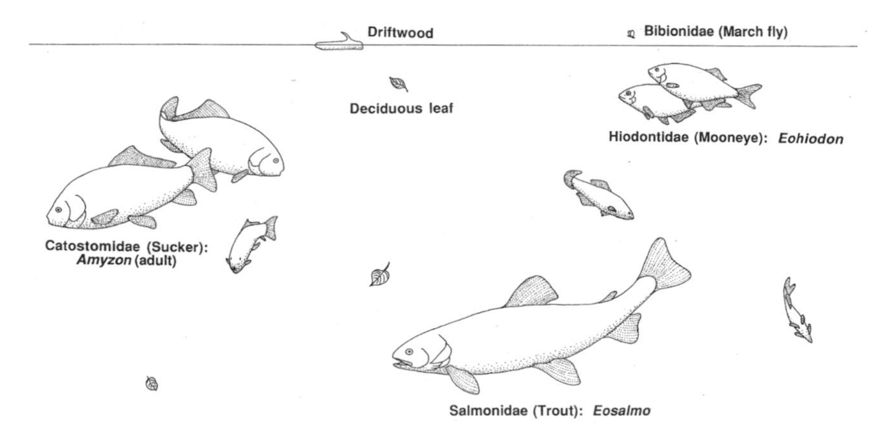

Figure 17.6 **Scene typical of deep, off-shore environment in an Eocene lake**

found with these fish, but those found are well preserved. They include small, easily blown deciduous leaves, pieces of wood that could float far from shore, surface-dwelling water striders, and strong fliers such as wasps and flies.

Signs of Predators and Scavengers

In some deposits many of the fossil bones are partly dissolved, possibly having passed through the gut of a predator such as a bird. At one site near Quilchena, most of the fossil bones appear to have been corroded by stomach acid (Wilson 1987). In modern lakes, the number of fish eaten by birds such as loons and grebes is surprisingly large, owing to their high metabolic rate, large size, and need to feed their young. In fact, a pair of loons, living and raising their young on the lake only part of the year, can consume a total mass of small fish comparable to that consumed by the resident fish-eating predators such as trout.

Many birds regurgitate partly digested remains of their prey as 'pellets' rather than passing them through their gut. These pellets often settled to the bottom as loosely associated clusters of bones that represent the remains of several prey. At other sites, blobs of amorphous brown material (mostly apatite, the mineral found in bone) are common, occasionally containing recognizable fish bones and scales. These materials are also interpreted as gut contents but, in view of their greater degree of digestion, they probably represent fossilized feces (coprolites).

Isolated scales are extremely common in some fossil beds. Undoubtedly many represent dead fish whose scales fell from their rotting carcasses. But fish can survive the loss of scales, for example, when the scales are dislodged during an unsuccessful attack by a predator. The replacement scales that a fish grows are known as regenerated scales, and can be recognized by their large central area devoid of normal ornament. Many regenerated scales are found as fossils, proving that the fish survived an attack.

Signs of scavenger activity are not always obvious. Currents and waves tend to displace bones from a skeleton in one or two preferred directions and often displace the delicate fin rays more than the heavy bones. Scavengers, both vertebrate and invertebrate, tend to displace bones in random directions and concentrate on attacking the fish's head, leaving body and fin rays intact (Elder and Smith 1988). The presence of scavengers implies that the water over the carcass was able to support animal life at the time of scavenging; conversely, lack of scavenging implies either rapid burial or anoxic, poisonous water.

Growth and Sexual Differences

Freshwater fishes typically grow rapidly during the summer and slowly during the winter, giving rise to growth rings on some bones and in their otoliths, or inner-ear stones. By counting growth rings or noting the fish's size, it is possible to estimate the age of the fish when it died. For many kinds of fishes, growth is rapid during the first years of life, slowing thereafter. In many groups, males and females acquire one or more distinctive features when they reach sexual maturity.

At the localities in the Horsefly area, the fish mostly seem to have died during the winter months, as they are preserved in the winter laminae of the annual layers (Wilson 1984). In these deposits, fossils of the sucker *Amyzon aggregatum* have body lengths clustered around four, eight, twelve, seventeen, and twenty-two centimetres. These lengths likely represent the average sizes reached by fish in their first, second, third, fourth, and fifth winters, respectively, and suggest that the individuals grew about four centimetres annually. There is also evidence that some of the fish, presumably the females, became significantly deeper-bodied during their third and fourth years of life. The fish illustrated in Figure 17.2 has a length of 22.7 centimetres and is deep-bodied, suggesting that it was a female that died during its fifth winter (that is, at four years of age).

Another common sexual modification is the development of thickened rays in the anal fin of males. Such dimorphism (difference in form between the sexes) occurs today in males of Hiodontidae (mooneyes) and Osmeridae (smelts). Modified anal fins can also be detected in both fossil species of *Eohiodon*. The example illustrated in Figure 17.1 has an anal fin with thickened anal rays and a distinctively convex margin. Therefore, this fish is likely a mature male. In living species of *Hiodon,* this change occurs in fairly large fish, at an age of three or more years and a length of more than twenty centimetres; however, in both species of *Eohiodon* the modified anal fins are seen in much smaller individuals, as small as ten centimetres in length for *E. woodruffi* and five centimetres for *E. rosei.* Evidently the life history of the Eocene hiodontids differed significantly from that of their Holocene relatives (Wilson and Williams 1992). Either the anal fin dimorphism occurred long before sexual maturity, or sexual maturity was reached in much smaller or younger fish.

The Passage of Time

Sediment deposited on the floor of a deep lake often records the passage of time in great detail. This record is clearest when layers of sediment accumulate steadily, without the disturbance of water movements or animals. An excellent example is the Eocene lake at Horsefly, which left a detailed rock record. In fact, the sediments of the Horsefly Eocene lake record the passage of the seasons within each year, being composed of 'varves,' that is, pairs of laminae that represent alternating summer and winter deposition.

Varves can be composed of clastic sediment such as mud and silt, or of the remains of organisms. The varves at Horsefly consist of slightly thicker, light-coloured, summer laminae composed of the tiny siliceous skeletons of diatoms (algae), alternating with thinner, dark-coloured, winter laminae composed mostly of clay particles. The annual nature of varves is confirmed by the differing fossil content of the summer and winter laminae, the former containing most of the leaves, insects, and fossilized gut contents (pellets and feces) and the latter containing most of the fish (Wilson 1984).

The seasonal cycle is envisaged as follows: each spring and summer, insects flew, were blown, or drifted into the lake, where they drowned and sank to the bottom to be buried by the ooze of the tiny diatom skeletons that resulted from the algal blooms of spring and summer; each summer, also, pellets or feces from predators such as aquatic birds sank to the bottom and were buried; each summer and fall, leaves fell or were blown into the lake and sank; each winter, fishes died, perhaps through failure to survive the winter or through asphyxiation when lake waters overturned, mixing oxygen-poor bottom water with the surface water. This cycle was repeated many times, laying down hundreds or thousands of varves in succession.

Occasionally, storms, earthquakes, or slumps caused dense clouds of sediment-laden water to flow rapidly over the lake floor, depositing layers of sediment called 'turbidites.' Usually, however, surface and deep waters did not mix. Water near the bottom was still and cold, often lacked oxygen, and could have been poisonous, thus excluding living organisms and preserving sediment layers and dead animals and plants in an undisturbed state.

The distinct annual laminations make it possible to determine the varve in which a given fossil is buried. One thirty-three-centimetre-thick sequence of varves and turbidites, for example, records 715 years of deposition. About 240 varves in the sequence yielded one or more of the fishes – 167 varves contained one fish and the remaining varves contained more, one containing nine fish, for a total of 358 fish for the entire sequence.

The fish illustrated in Figure 17.2, one from this sequence, shows how it is possible to estimate the relative year of 'birth,' or hatching. This fish died in the 256th varve, counted from the bottom of the sequence of 715; because the fish is thought to have been four years old (termed '4+,' as it died in its fifth winter but before reaching the age of five), it likely hatched during the spring season represented by the diatom-rich, light-coloured lamina four varves below it, that is, the 252nd varve. In the same way, year of death and year of hatching were estimated for the hundreds of other fish found in the same layers. By careful study of the morphology of each fish, its conditions of preservation, and the characteristics of the varve deposited during its year of hatching, paleontologists at the University of Alberta determined the relationship between changes in the fish population and changes in this lake over the course of seven centuries, fifty million years ago.

Much of the work reported in this chapter depends largely on the early contributions of pioneer geologists and paleontologists such as Dawson and Lambe, and the generosity of interested amateurs such as Hobson and Liboton. Had these individuals not shared their finds with others and, especially, had their discoveries not been preserved in museums, their efforts would not have contributed to scientific knowledge. In fact, many important fossils languish in obscurity, and most of them will never be scientifically studied. Another example of a fossil that escaped such a fate is the fish featured in Figure 17.3, the youngest and smallest known specimen of *Eosalmo*. Like many other important fossils, it belonged for years to a private collector; it only came under study when its owner responded to a newspaper advertisement. Many other potentially important finds remain unknown because they have not been reported or deposited in museum collections. Do you have a skeleton in your closet of possible importance to paleontology?

References

Cavender, T.M. 1986. Review of the Fossil History of North American Freshwater Fishes. In *The Zoogeography of North American Freshwater Fishes,* ed. C.H. Hocutt and E.O. Wiley, 699-724. New York: John Wiley and Sons

Elder, R.L., and G.R. Smith. 1988. Fish Taphonomy and Environmental Inference in Paleolimnology. *Palaeogeography, Palaeoclimatology, Palaeoecology* 62:577-92

Stearley, R.F. 1992. Historical Ecology of Salmonine Fishes. In *Systematics, Historical Ecology, and North American Freshwater Fishes,* ed. R.L. Mayden, 622-58. Stanford: Stanford University Press

Wilson, M.V.H. 1984. Year Classes and Sexual Dimorphism in the Eocene Catostomid Fish *Amyzon aggregatum. Journal of Vertebrate Paleontology* 3:137-42

–. 1987. Predation as a Source of Fish Fossils in Eocene lake Sediments. In Paleo-Lacustrine Theme Issue, ed. A. Cohen. *Palaios* 2:497-504

–. 1988. Reconstruction of Ancient Lake Environments Using Both Autochthonous and Allochthonous Fossils. In *Paleolimnology: Aspects of Freshwater Paleoecology and Biogeography,* ed. J. Gray. Amsterdam: Elsevier. Reprinted from *Palaeogeography, Palaeoclimatology, Palaeoecology* 62:609-23

Wilson, M.V.H., and R.R.G. Williams. 1992. Phylogenetic, Biogeographic, and Ecological Significance of Early Fossil Records of North American Freshwater Teleostean Fishes. In *Systematics, Historical Ecology, and North American Freshwater Fishes,* ed. R.L. Mayden, 224-44. Stanford: Stanford University Press

18
Insects near Eocene Lakes of the Interior

Mark V.H. Wilson

Standing on the shore of a modern lake one sees many insects flying, floating, or being blown away from the shore. The same must have been true on the shores of the numerous lakes that existed during the Eocene (about fifty million years ago) in the Interior of British Columbia and Washington. The rocks that formed from sediments of the ancient lake floor have yielded many fossil insects. Since at least the beginning of the Paleogene Era, the numbers of insect species and of individuals have far exceeded those of any other group of land-dwelling animals. Fossil insects are so numerous and so diverse, and the number of insect paleontologists so small, that only a tiny fraction of the story has been deciphered. The little we do know suggests that insect fossils have much to tell us – about changes in the composition of insect faunas over time, about the environments in which the insects lived and were buried, about the evolution of interdependence of insects and plants, and about rates of change within and among species.

Insect Diversity

As with many fossil fishes and plants, Eocene fossil insects have been studied for more than a century. Many of the pioneer geologists who collected Eocene fishes and plants also sent fossil insects to museums and to experts for study. As a result, early studies on fossil insects of British Columbia were published by Samuel Scudder and Anton Handlirsch. As well, geologist Tony Rice described fossil insects, including March flies (Bibionidae) and sawflies (Tenthredinidae). Paleontologists in the United States, particularly Standley Lewis and Wesley Wehr, have studied insects in closely related fossil beds near Republic, Washington (Lewis 1992; Wehr in press). My own research on BC fossil insects, a spin-off of my fish studies, has focused on their usefulness as indicators of past environments (Wilson 1988a).

These studies combine to show that Eocene insects were very similar to those of today, although probably not so similar that they belonged to the same species. Many kinds of Eocene insects are related to types that are rare or missing in the modern BC fauna, either because they became extinct throughout North America or because they are adapted to climates much warmer than those now found in British Columbia. A few rare examples of spiders have also been discovered in Eocene deposits of British Columbia and Washington.

Many of the most primitive living orders of insects are represented among the fossils; these include mayflies (Ephemeroptera), dragonflies (Odonata), cockroaches (Blattodea), grasshoppers (Orthoptera), and lacewings (Neuroptera). At least one group of primitive termites (Isoptera), now extinct in North America, has been found. Some of the fossil grasshoppers and other insects were much larger than their living relatives in British Columbia (Figure 18.1), suggesting that warmer climates prevailed then. A large proportion of the insect fossils are Hemiptera (bugs), including water striders (Gerridae), froghoppers or spittlebugs (Cercopoidea), aphids (Aphididae), and even stink bugs (Pentatomidae). Many of these were also large (Figure 18.2). As might be expected for a group adapted to live on the water surface, water striders are extremely common at some fossil sites (Figure 18.2E).

Fossils of adult caddisflies (Trichoptera) and alderflies (Megaloptera) are rare, but the cases built by young water-dwelling caddisflies are sometimes very common. Butterflies and moths are extremely rare, as yet represented by only a few specimens, which include a measuring-worm moth (Geometridae) recently found near Republic.

As in the modern fauna, the fly order is well represented both in quantity and in variety. Surprisingly, the family represented by more fossil specimens than any other is the March fly family (Bibionidae), which contributes more than half the specimens in many collections of Eocene insects of the Interior (Figure 18.3). Modern March flies are so named because their adults emerge in large numbers in the early spring. The family is well represented as fossils (Figure 18.3) but today is more characteristic of tropical regions. Other fossil flies include the crane flies (Tipulidae), fungus gnats (Mycetophilidae), wood gnats (Anisopodidae), and flower flies (Syrphidae). The fossil flower fly even has a striped abdomen like many of its modern relatives.

The wasps and bees (Hymenoptera) are well represented by primitive families, including sawflies (Tenthredinoidea) (Figure 18.1E) and parasitic wasps (Ichneumonidae and Braconidae) (Figure 18.1F). Among the groups usually considered more advanced are rarer examples of ants (Formicidae), paper wasps (Vespidae), and leaf-cutter bees (Megachilidae).

The beetle order (Coleoptera) is today one of the most diverse groups of organisms, but many Eocene fossil beds are without beetles (Wilson 1982). This unexpected finding must result from the particular environments

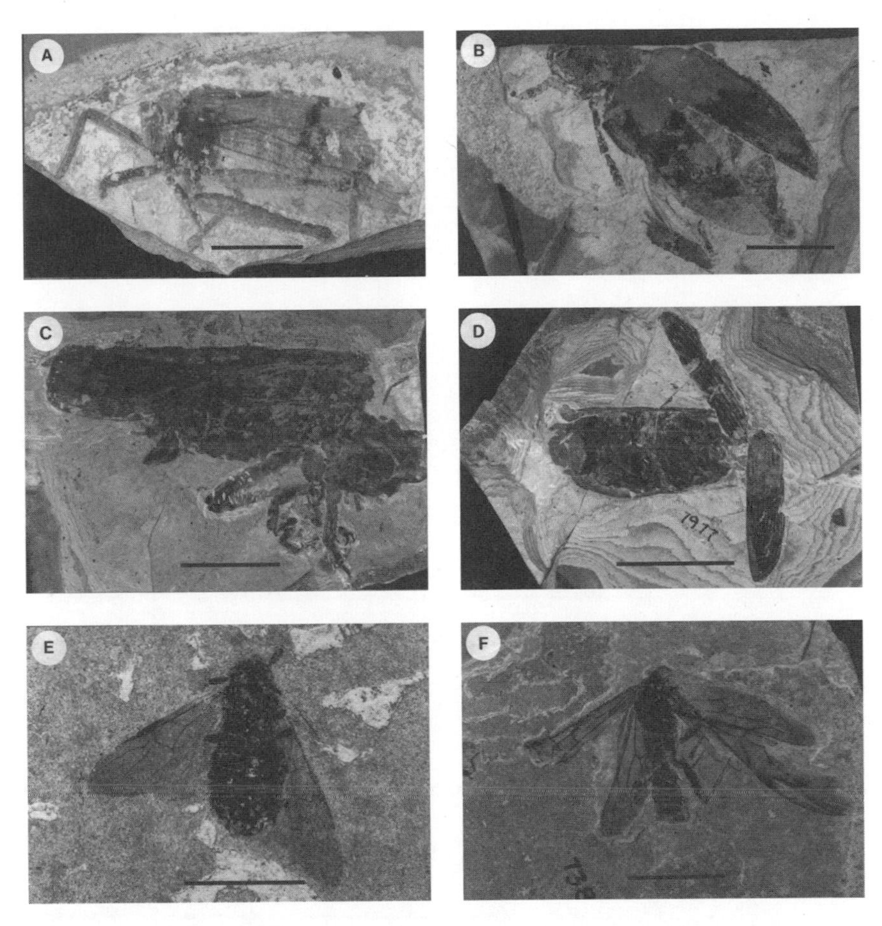

Figure 18.1 **Eocene insects of British Columbia.** (A) and (C) Grasshoppers (Orthoptera) from the Horsefly and Princeton areas. (B) and (D) Beetles (Coleoptera) from the Quilchena and Princeton areas. (E) Sawfly (Tenthredinoidea) from the Horsefly area. (F) Parasitic wasp (Ichneumonidae) from the Horsefly area. Scale bar lengths: (A) 20 mm; (B), (E), and (F), 10 mm; (C) and (D), 15 mm
Source: Fossil Insect Collections, University of Alberta

represented by these deposits, for beetles are an ancient group and are extremely common at fossil sites elsewhere. Where they do occur in the Eocene of British Columbia, the examples reported are of ground beetles (Carabidae), click beetles (Elateridae), stag beetles (Lucanidae), scarab beetles (Scarabaeidae), leaf beetles (Chrysomelidae), weevils (Curculionidae), and round-headed boring beetles (Cerambycidae). Some of the fossil beetles are surprisingly large (Figure 18.1B).

Overall, fossil deposits of Eocene age from British Columbia seem to contain greater numbers of insect species in certain groups and fewer in

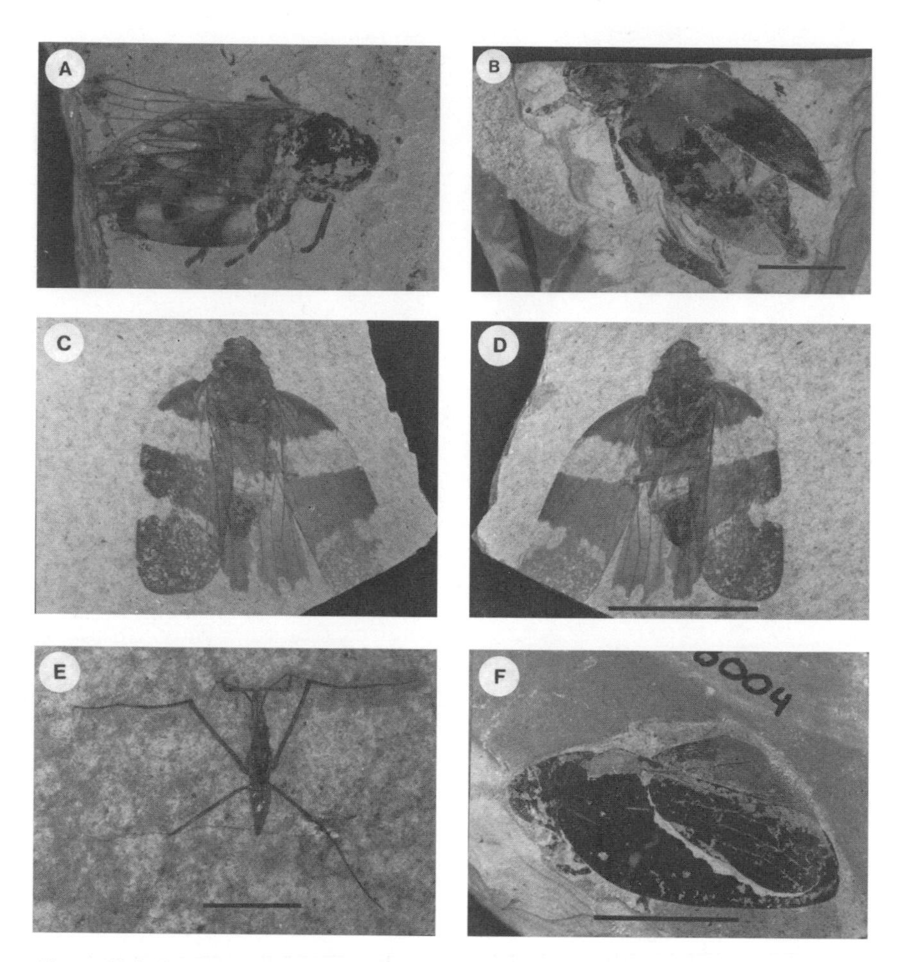

Figure 18.2 **Eocene insects of the order Hemiptera from British Columbia.** (A) to (D) and (F) Three examples of froghoppers (Cercopoidea), in part and counterpart, from the Horsefly and Princeton areas. (E) water strider (Gerridae) from the Smithers area. Scale bar lengths: (A-F), 10 mm
Source: Fossil Insect Collections, University of Alberta

others than do comparable modern environments. This disparity might be explained by a difference in the factors affecting the preservation of fossils or by a consistent difference in the environments being sampled. When we examine apparently similar fossil deposits of Paleocene, Eocene, Oligocene, and younger ages, however, we find what appears to be a consistent trend. Basically, the younger deposits and modern faunas contain much larger numbers of advanced groups of flies, wasps, and butterflies than do the older fossil deposits. These differences were recognized a century ago by Scudder and elaborated on half a century ago by Charles Brues, who tallied fossils of various groups from fossil localities

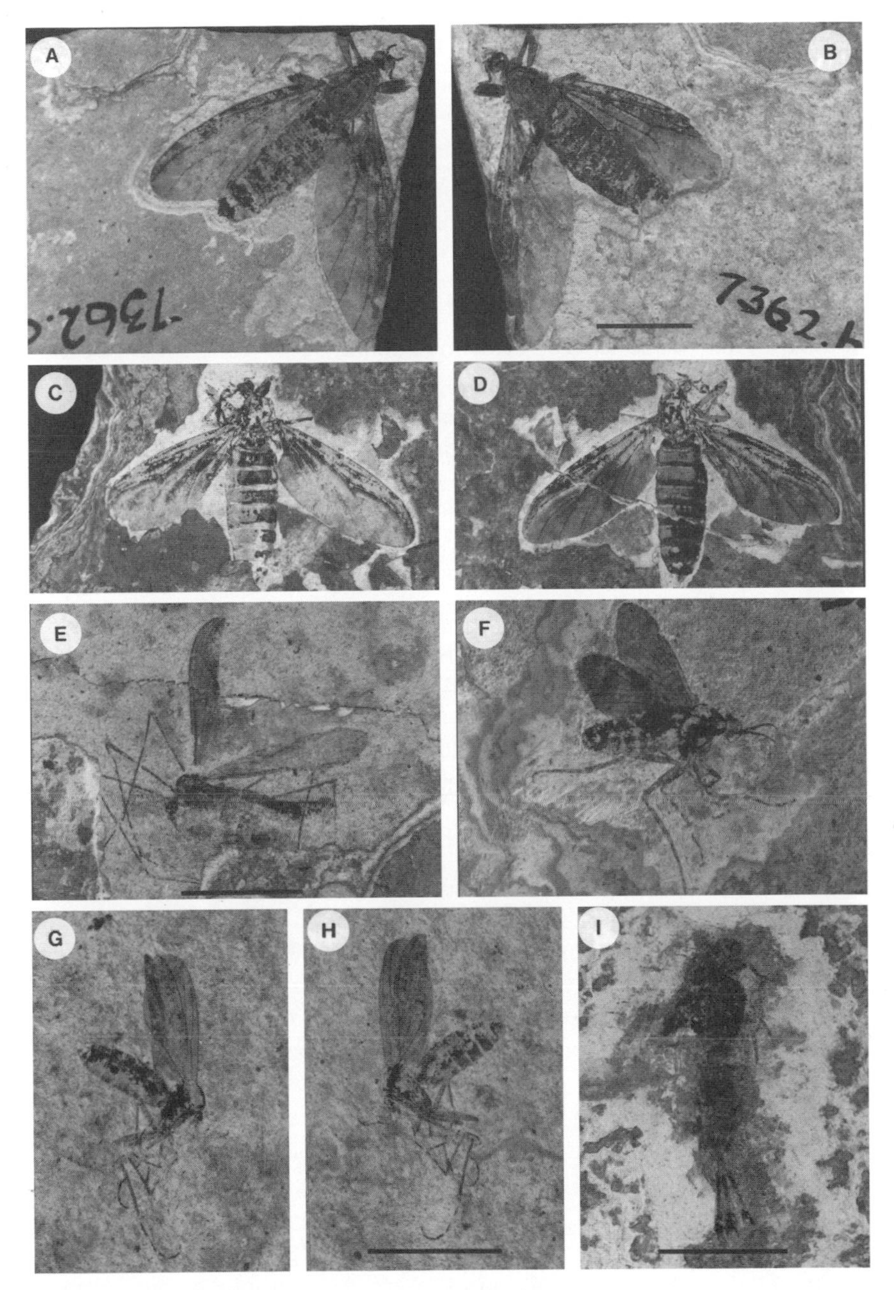

Figure 18.3 **Eocene insects, mostly flies (order Diptera), from the Horsefly area.** (A) to (D) Two March flies (Bibionidae), in part and counterpart. (E) Crane fly (Tipulidae). (F) Wood gnat (Anisopodidae). (G) and (H) Fungus gnat (Mycetophilidae), part and counterpart. (I) Mayfly aquatic nymph (order Ephemeroptera). Scale bar lengths: (A), (B), and (E-H), 5 mm; (C), (D), and (I), 10 mm

Source: Fossil Insect Collections, University of Alberta

and compared them to the modern fauna. I, too, have noted differences when comparing Paleocene fossil insects from Alberta with Eocene insects from British Columbia and with those from younger ages (Wilson 1982).

There is no obvious explanation for these observed differences in the overall composition of Eocene and modern insect faunas. If we assume that the differences are real and not just the result of biases in preservation, one possible cause is that the insect groups really did change over the past fifty million years. Perhaps some of the advanced kinds of flies, wasps, and butterflies had not yet evolved by Eocene times. A second possibility is that the advanced groups had already evolved by then but that individuals were much less numerous than now. If so, they would be less likely to be preserved in fossil deposits. A third possibility is that the advanced groups had evolved but had not yet spread to the BC area, or even to North America. The real explanation might well include a combination of all these factors.

Environmental Indicators

Insect fossils can reveal secrets of climate, local conditions, and even the changing of the seasons. Climatic indicators include the large body sizes of grasshoppers, bugs, and beetles mentioned earlier. Comparable body sizes are rarely seen in northern climates today, supporting the idea that climate at the latitude of British Columbia was much warmer during the Eocene than it is now. The abundance of March flies suggests much the same thing.

Local lake conditions are revealed by both the kinds of insects found and their state of preservation. Abundant aquatic larval and nymphal forms (for example, the Mayfly nymph in Figure 18.3I) or water-dwelling adults point to water conditions appropriate for insect growth and reproduction, as they do today. However, we learn from other groups of organisms that the environment favoured by modern species is not an infallible guide to environments preferred by extinct species, for such preferences can change.

A better environmental indicator might be the state of preservation of the fossils, part of the field of study called 'taphonomy,' the science of death and burial (Wilson 1988b). Many of the factors affecting preservation are unlikely to change with time either because they are physical and chemical or because they are the overall effect of numerous biotic agents. Many fossil insects are found intact, for example, whereas others have lost wings and legs through the actions of bacteria, water turbulence, and scavenging animals. Also, insects are more common in some fossil localities than in others. Both of these observations probably result from the fact that insects may be abundant near the shore but are unlikely to remain intact here because shallow water is warm, oxygen-rich, and turbulent. In contrast, insects may be less abundant in central areas of the lake but they are more likely to remain intact because deep water is cool, oxygen-poor, and quiet (Wilson 1988a).

Insects are also considered sensitive indicators of annual change, because modern insects are highly seasonal animals with pronounced yearly rhythms of reproduction, growth, and death. In the case of the annually layered fossil beds near Horsefly, the white, diatom-rich rock layers that contain most of the March flies (Figure 18.3C) and many other insects are likely to represent the spring and early summer seasons, whereas the thin, dark layers probably represent winter deposits.

Insect-Plant Interactions

One of the most interesting areas of study in evolutionary biology is the extent to which traits in animal species have evolved in response to changes in plant species, and vice versa. Among the relationships of particular interest are those of parasite and host, herbivore and food, and pollinator and flower. One line of research questions whether the animal and plant features, or the species that display them, are equally ancient; that is, did they originate together? Another asks whether both animal and plant have changed over time; that is, have they co-evolved since then?

Several fossil examples from the Interior of British Columbia and from Washington provide striking examples of insect-plant relationships that seem virtually unchanged since Eocene times (Lewis 1992). Then, as now, termites established colonies in decaying wood; caddisfly larvae built their cases out of tiny pieces of leaves; leaf beetles laid their eggs in characteristic rows on alder leaves; weevils had a characteristic snout; aphids infested flower spikes; leaf-cutter bees removed large chunks from leaves; and moths, flower flies, and probably many other insects pollinated flowers.

Many of these examples are so similar to their modern counterparts that there seems to have been little or no change since the Eocene. A closer look, however, reveals that, although the patterns of behaviour and the structural features of the insects and plants are similar, the species involved are different, however slightly, from living species. And, for the most part, the detailed studies needed to establish whether the insect, the plant, or both have changed in ways directly related to the interaction have yet to be conducted.

Rates of Evolution of Insects

Eocene insects seem so similar to living insects that one might well ask whether there has been any significant change since that time. At the level of species and genera there appears to have been little change. Most

Eocene species and genera are now extinct – or at least have no living North American representatives – as is the case for the fossil termites whose relatives survive today only in Australia. Even so, only minor changes distinguish the living from the fossil species.

How minor were some of these changes? A recent study of fossil and living water striders of British Columbia (Anderson et al. 1993) found that the Eocene species are nearly indistinguishable from their living relatives, and that this group was much older than biologists had previously estimated through study of genetic differences. Nevertheless, the fossils are exceptionally well preserved, and it is often possible to distinguish sexes and even to measure the lengths of individual segments of the antennae. A fossil of a male water strider from the Smithers area, for example, appears identical at first glance to a male of a species living today in British Columbia; but, unexpectedly, it has a terminal antenna segment longer than that of any of its living relatives. No other feature appears to differ, even slightly. Whether this small change in the antenna had any significance for the survival of these water striders is not known.

It has long been realized that insects are exceptionally diverse and abundant. Some entomologists have argued that, to achieve such diversity, insects must have evolved and produced new species very rapidly. Yet many insect groups possess great antiquity (Wilson 1983) and have experienced little morphological and genetic change (Anderson et al. 1993). These findings suggest that insect species are long-lived, lasting millions of years, and that even genetic molecules change slowly in some groups of insects. It seems a paradox that such a very diverse group can evolve so very slowly. This important lesson is one to be learned from the fossil insects of British Columbia.

Insects are phenomenally varied, numerous, and sensitive indicators of past environments and of the interdependence of animals and plants. Paradoxically, they appear to have evolved slowly, yet they dominate the terrestrial environment. Although they are among the most easily collected and intellectually rewarding of fossils, they are among the least studied. Imagine the interesting tales that BC fossil insects could tell if they received their fair share of attention.

References

Anderson, N.M., J.R. Spence, and M.V.H. Wilson. 1993. Fifty Million Years of Structural Stasis in Water Striders (Hemiptera, Gerridae). *American Entomologist* 39:174-6

Lewis, S.L. 1992. Insects of the Klondike Mountain Formation, Republic, Washington. *Washington Geology* 20:15-19

Wehr, W.C. In press. Middle Eocene Insects and Plants of the Okanagan Highlands. In *Contributions to the Paleontology and Geology of the West Coast in Honor of V. Standish*

Mallory, ed. J.E. Martin. Thomas Burke Memorial Washington State Museum Research Report 6.

Wilson, M.V.H. 1982. Early Cenozoic Insects: Paleoenvironmental Biases and Evolution of the North American Insect Fauna. Third North American Paleontological Convention. *Proceedings* 2:585-8

–. 1983. Is There a Characteristic Rate of Radiation for the Insects? *Paleobiology* 9:79-85

–. 1988a. Reconstruction of Ancient Lake Environments Using Both Autochthonous and Allochthonous Fossils. In *Paleolimnology: Aspects of Freshwater Paleoecology and Biogeography,* ed. J. Gray. Amsterdam: Elsevier. Reprinted from *Palaeogeography, Palaeoclimatology, Palaeoecology* 62:609-23

–. 1988b. Taphonomic Processes: Information Loss and Information Gain. Paleoscene #9. *Geoscience Canada* 15:131-48

19
Flowering Plants in and around Eocene Lakes of the Interior

Ruth A. Stockey and Wesley C. Wehr

BC fossils of Eocene angiosperms, or flowering plants, are among the best-preserved plant fossils in the world. Because many of the sites in the Interior were at one time upland freshwater lakes, they are also well known for their fossil fishes and insects. Most fossil plant localities in British Columbia preserve compression fossils; a few include petrified plant remains.

The area around Princeton, British Columbia, is particularly rich in fossil localities. Among these are Asp Creek (also known as China Creek), Whipsaw Creek, Coalmont Road (the Princeton-Tulameen Road), Thomas Ranch, One Mile Creek, Blakeburn Mine, and the Princeton Chert locality on the Similkameen River. Nearby fossil localities include Quilchena, Merritt, Chu Chua, and McAbee. Localities farther north in the province are Horsefly, Smithers, and Driftwood Creek. All of these sites are of Middle Eocene age, about forty-five to fifty million years old. Only a few have been studied in detail.

The Princeton Chert Locality

Of the numerous localities in British Columbia that contain Eocene angiosperm fossils, the Princeton Chert locality on the Similkameen River, part of the Allenby Formation, probably has the best-preserved material found in the province. In fact, few plant fossils elsewhere in the world exhibit such excellence in both preservation and diversity. Plants found here are permineralized; that is, they are preserved in three dimensions in a solid matrix of silica that formed from infiltration of the intercellular spaces and cells with silicic acid – a process that resulted from periodic volcanic eruptions. The Princeton Chert locality is about eight kilometres south of Princeton. It consists of forty-nine layers of chert that alternate with coals, and is exposed along the Similkameen River in a stratigraphic section about ten metres thick. Although little is visible on a fresh rock surface, preparation of the chert with acid in the laboratory reveals extremely well-preserved plants, many of which were growing in place at the time of their burial and preservation. This unusual situation allows studies that are not possible with isolated plant organs. Specimens are studied from cellulose acetate peels and by etching with hydrofluoric acid to remove the siliceous matrix. Recent research, particularly that at the University of Alberta, involves reconstructing these plants and studying their habitat, growth, and reproduction.

Table 19.1 lists the flowering plants currently known from the Princeton Chert. The dicots are those with two cotyledons (primary leaves of the embryo) and reticulate vein patterns in leaves. The monocots have one cotyledon and usually have parallel leaf veins. In addition to the angiosperms listed in Table 19.1, Princeton Chert flora also includes several types of conifers and ferns, about twenty types of undescribed dicot woods, at least nine unidentified monocots, and numerous fruits and seeds.

The Paleoenvironment

The Princeton Chert locality is thought to represent a series of shallow-lake deposits (Cevallos-Ferriz et al. 1991). Three lines of evidence indicate such an environment. First, the plants are associated with fossil remains of lake-dwelling animals. A partially articulated skeleton and several isolated bones of the freshwater fish *Amia* and disarticulated remains of *Amyzon* and *Libotonius* have been reported from the shales overlying the chert beds. This shale also contained the bones of a soft-shelled turtle.

The second line of evidence is the affinity of the fossil plants to living aquatic plant groups. Families such as the Nymphaeaceae, Lythraceae, Alismataceae, and Araceae include many aquatic plants.

The Nymphaeaceae (water lilies) are represented in the chert by numerous seeds, which bear the name *Allenbya* after the nearby abandoned mining town of Allenby. *Allenbya* seeds (Figure 19.1A) are ovoid and about seven millimetres long. They have an apical operculum, or cap, which was probably shed when the seed germinated. Isolated caps are often found in the chert, as are seeds without caps. Although young seedlings of these plants have not yet been identified, seeds have been shown to contain perisperm (tissue that may serve as a source of nutrition for the developing embryo).

The Lythraceae (loosestrife family), which is primarily terrestrial, contains some species that now occupy margins of aquatic systems. *Decodon verticillatus,* the swamp willow, grows today along riverbanks and lake shores with its stem bending toward the water. This plant forms a floating rhizome-like system that produces new roots and vertical shoots. Fruits and seeds of the fossil species *Decodon allenbyensis* have been described from the Princeton Chert. The fruits (Figure 19.1B) are capsules with numerous seeds that are very similar to those of the swamp willow.

Table 19.1

Permineralized angiosperms from the Princeton Chert flora
Dicots
Magnoliidae
Magnoliaceae (magnolia family)
Liriodendroxylon princetonensis
Lauraceae (laurel family)
Nymphaeaceae (water lily family)
Allenbya collinsonae
Incertae sedis (family unknown)
Princetonia allenbyensis, Eorhiza arnoldii
Rosidae
Rosaceae (rose family)
Paleorosa similkameenensis, Prunus allenbyensis, Prunus sp.
Lythraceae (loosestrife family)
Decodon allenbyensis
Myrtaceae (myrtle family)
Paleomyrtinaea princetonensis
Vitaceae (grape family)
Ampelocissus similkameenensis
Sapindaceae (soapberry family)
Wehrwolfea striata
Monocots
Alismatidae
Alismataceae (water plantain family)
Heleophyton helobiaeoides
Arecidae
Arecaceae (palm family)
Uhlia allenbyensis
Araceae (arum and calla lily family)
Keratosperma allenbyensis
Commelinidae
Juncaceae/Cyperaceae (rush/sedge families)
Ethela sargantiana
Liliidae
Liliaceae (lily family)
Soleredera rhizomorpha

Large numbers of these pyramidal seeds are found in the chert. The plant body has not yet been identified, but delicate fossil embryos (Figure 19.1C) are preserved in some of these seeds, emphasizing the unusual preservation found in the Princeton Chert plants.

The Alismataceae (water plantain family) is a monocot family that includes aquatic and semi-aquatic herbs. This family is represented in the chert by at least one described petiole (the stalk that attaches leaf to stem) of

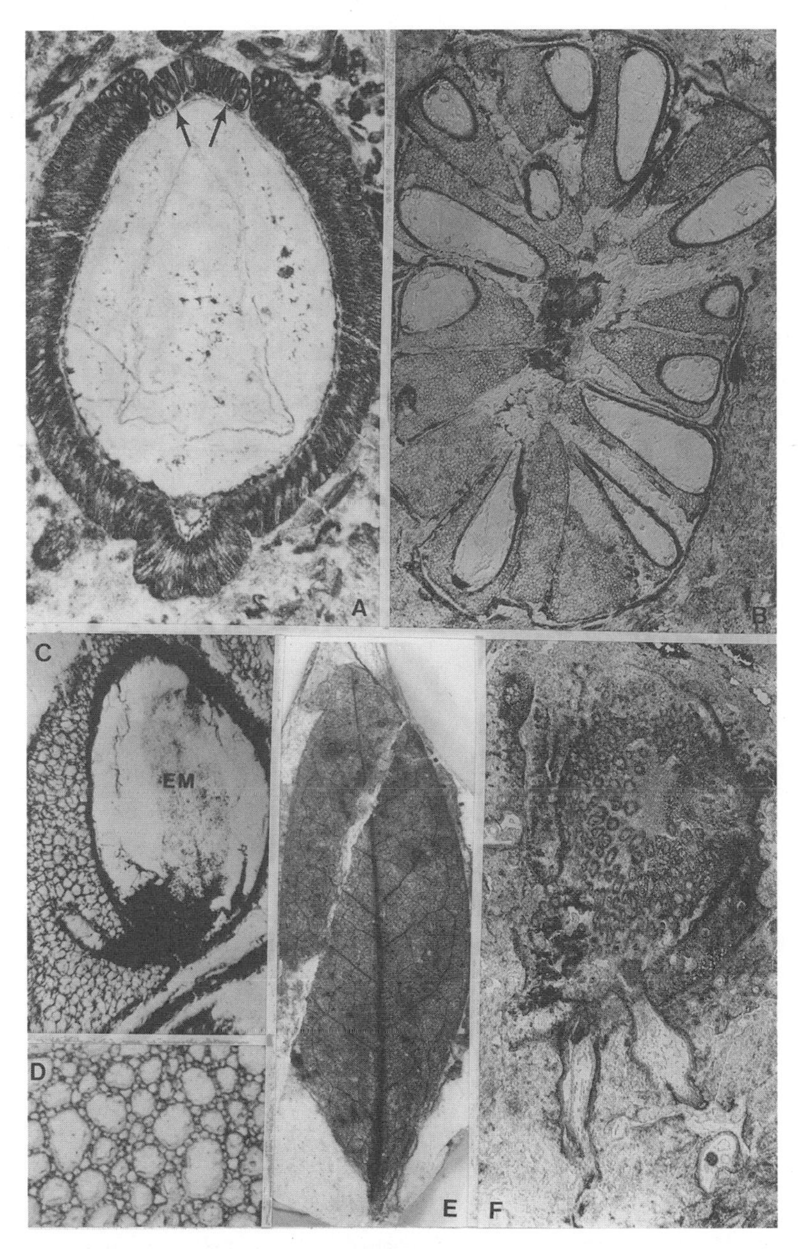

Figure 19.1 **Eocene aquatic angiosperms from British Columbia.** (A) *Allenbya collinsonae* seed. Arrows indicate operculum. (X 20) (B) *Decodon allenbyensis* fruit with pyramidal seeds. (X 25) (C) *Decodon allenbyensis* seed with embryo (EM). (X 130) (D) Aerenchyma tissue. (X 75) (E) *Decodon* sp. (X 1.5) (F) *Ethela sargantiana* stem with attached roots and leaves. (X 11)
Sources: (A-D, F) Paleobotanical Collections, University of Alberta; (E) Collections of Princeton Museum, Princeton

Heleophyton. The arrangement of the vascular bundles – the bundles of sap-carrying plant tissue) in *Heleophyton* is very similar to that of some living Alismataceae and is a good indicator of a shallow freshwater environment.

Stems of *Ethela sargantiana* with attached roots and leaves also occur in the chert (Figure 19.1F). These small plants were buried where they grew and rooted in peat that is now petrified. They are probably related to sedges (Cyperaceae) or rushes (Juncaceae). The leaf anatomy is similar to that of sedges, but no silica bodies have been demonstrated in the leaves and therefore the closely related rushes cannot be ruled out.

The third line of evidence indicating an aquatic habitat for many of the Princeton plants is that they include anatomical modifications that characterize modern aquatics. These modifications include the development of aerenchyma tissues (thin-walled cells with numerous air spaces between the cells) and reduced vascular systems. Aquatic plants frequently develop aerenchyma (Figure 19.1D) for aeration of the tissues and for buoyancy. Aerenchyma is found in many of the Princeton Chert plants including *Dennstaedtiopsis* (fern), *Eorhiza* (dicot) (Figure 19.2A), *Heleophyton,* and *Uhlia* (monocots), and in the tissues of several different types of seeds, fruits, and leaves.

There is abundant evidence that many of the Princeton Chert angiosperms are preserved in growth position. Rooted stems with small, delicate roots and often with lateral roots occur in *Eorhiza, Uhlia, Soleredera* (another small monocot) (Figure 19.2F), *Ethela* (Figure 19.1F), and one conifer, *Metasequoia.* Transport of plant material was minimal, as is evident in the preservation of organs with delicate tissues. Flowers and fruits of plants such as *Princetonia* (Figures 19.2B to 19.2E), and flowers of *Paleorosa* (Rosaceae) and *Wehrwolfea* (Sapindaceae) (Figure 19.3A), for example, are preserved completely with petals and sepals, and with stamens that contain pollen (Figures 19.2D and 19.4F).

In addition to the plants that lived in small ponds or lakes, many others are represented by small twigs, leaves, or seeds, but these remains are found in smaller numbers than those of aquatic plants. Woody twigs of *Liriodendroxylon* (magnolia family) and *Prunus* (rose family) (Figure 19.4E), as well as those of conifers, represent plants that probably grew nearby and dropped their parts into the basin. Also present are small numbers of seeds of the grape family, Vitaceae, for example, *Ampelocissus* (Figure 19.3I). Because their external surfaces were abraded prior to fossilization, it is believed that they may have passed through the gut of a bird before being dropped into the peat.

The Monocots

One of the most exciting areas of research involving Princeton Chert plants is the study of the remains of monocots. The monocots include

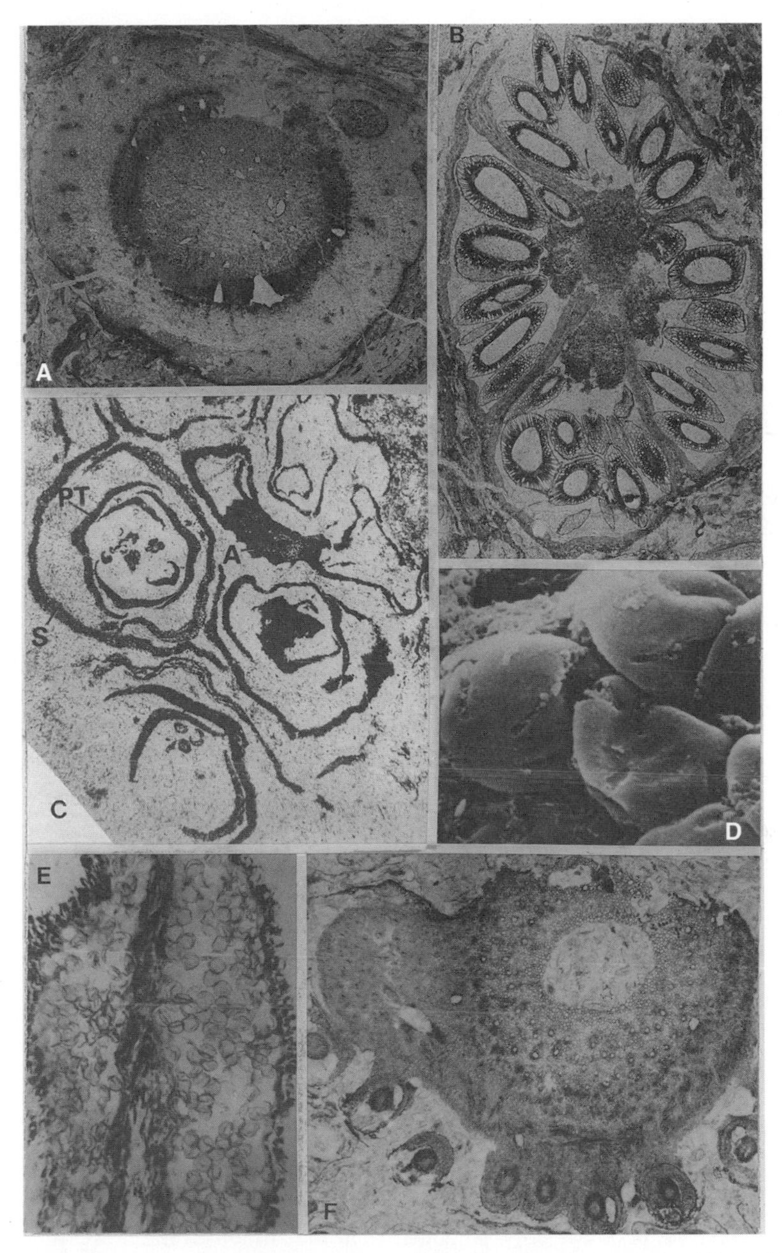

Figure 19.2 **Eocene aquatic angiosperms from the Princeton chert.** (A) *Eorhiza arnoldii* rhizome with branch trace. (X 2.5) (B) *Princetonia allenbyensis* fruit with seeds. (X 9) (C) *Princetonia allenbyensis* inflorescence axis with young attached flowers, A = axis, PT = petal, S = sepal. (X 25) (D) *Princetonia allenbyensis* pollen. (X 15,000) (E) *Princetonia allenbyensis* anthers with pollen. (X 140) (F) *Soleredera rhizomorpha* rhizome and branch with attached roots. (X 9)

Source: Paleobotanical Collections, University of Alberta

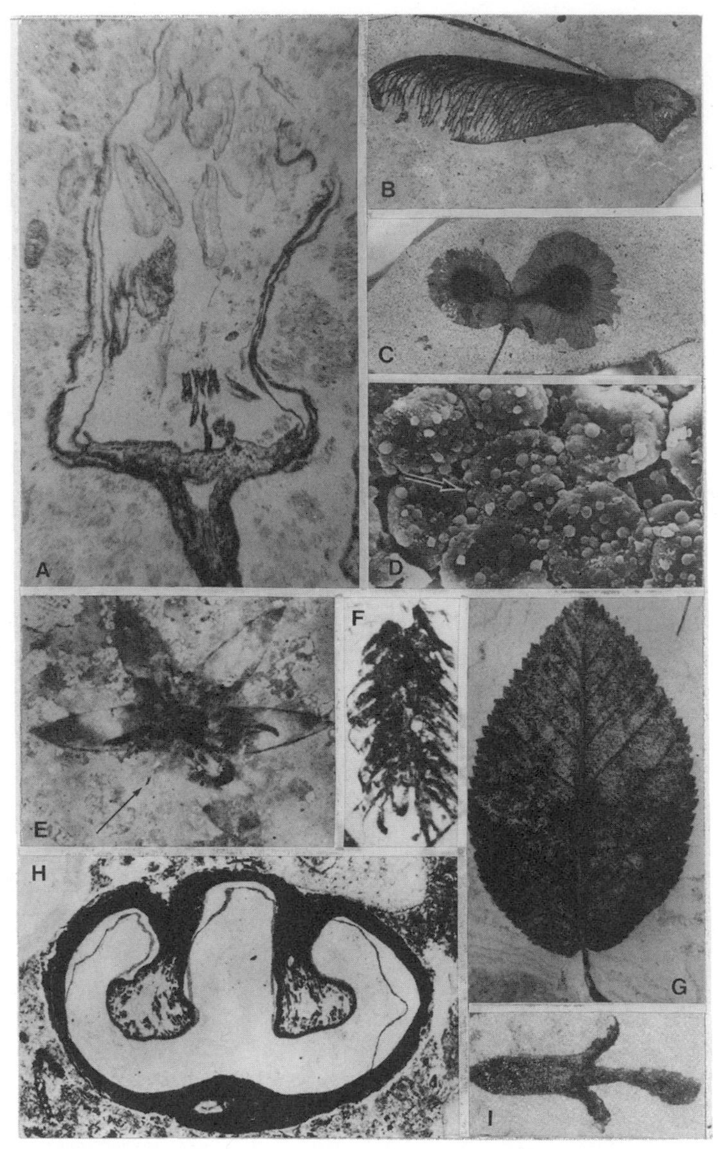

Figure 19.3 **Eocene angiosperms from British Columbia.** (A) *Wehrwolfea striata* flower. (X 33) (B) *Acer stewartii* fruit. (X 1.6) (C) *Bohlenia americana* fruits. (X 1.5) (D) *Pistillipollenites mcgregorii*-type pollen from the anthers of the flower illustrated in (E). (X 850) (E) *Pistillipollianthus wilsonii* flower. (X 2.7) (F) Infructescence associated with *Betula leopoldae*. (X 2) (G) *Betula leopoldae* leaf. (Actual size) (H) *Ampelocissus similkameenensis* seed. (X 17) (I) Bract associated with *Betula leopoldae*. (X 5)

Sources: (A, D, E, H) Paleobotanical Collections, University of Alberta; (B, C) Paleobotanical Collections, Burke Museum, University of Washington; (F, G, I) Field Museum, Chicago

Figure 19.4 **Eocene rosaceous angiosperms from British Columbia.** (A) *Ribes* sp. leaf. (X 1.5) (B) *Neviusia* sp. leaf. (X 1.3) (C) *Stonebergia columbiana* leaf. (Actual size) (D) *Amelanchier* sp. leaf. (X 1.5) (E) *Prunus allenbyensis* stem. (X 10) (F) Pollen from *Paleorosa similkameensis*. (X 1,020) (G) *Prunus* fruit with seed. (X 14) (H) *Prunus* sp. leaf. (X 1.2)

Sources: (A-D) Paleobotanical Collections, Burke Museum, University of Washington; (E-H) Paleobotanical Collections, University of Alberta

such living plants as grasses, palms, orchids, and lilies. In most compression floras, few monocots are preserved and these specimens are usually only leaf fragments (Table 19.2). The Princeton Chert provides a unique opportunity to study whole plants. The palm *Uhlia* has recently been reconstructed, based on remains that include stems with attached petioles and roots. Its leaves have also been identified on the basis of their anatomy. This small rhizomatous palm was probably a good indicator of a relatively warm climate during the Eocene in British Columbia.

The presence of smaller monocot plant bodies is even more unusual. In addition to *Ethela,* which has been mentioned already, *Soleredera* (lily family) has been described from specimens of rhizomes with attached leaves and roots. These rhizomes, only 1.3 centimetres in diameter with a hollow central region (Figure 19.2F), are the first anatomically preserved remains of the Liliaceae to be described in the fossil record of North America. Although the flowers and fruits of these monocots have not yet been described, large numbers of undescribed monocot plant organs have been noted in the chert, and work in this area continues.

Fungal Parasites

In addition to the flowering plants themselves, the Princeton Chert provides a rare opportunity to study the fungi that interacted with these plants in life. A new area of investigation in the chert involves the study of fungal parasites and of fungi that exist in a symbiotic relationship with their hosts. A preliminary study of fungi in the chert has revealed a wealth of fossil material for future work. Already these deposits have yielded the oldest known fossil record of the smut fungi (Currah and Stockey 1991). Well-preserved masses of fungal spores in the intact anthers (pollen sacs) of a small angiosperm flower indicate that interactions between fungi and angiosperms were already highly evolved by the Paleogene.

Compression Floras

Compared to the extensive and detailed studies of the permineralized plants in the Princeton Chert, relatively little has been published on Eocene compression floras of British Columbia. Until recently, the studies of Dawson, Penhallow, and Berry from the late 1800s and early 1900s represented most of what was known. In recent years, Rolf Mathewes has studied the Quilchena flora, and Jack Wolfe and Wes Wehr have published several studies of floras in nearby Republic, Washington, and in Princeton (Wolfe and Wehr 1987, 1988). All of these floras are approximately the same age. Table 19.2 is a summary of the angiosperm compression fossils of the Eocene in British Columbia. Although the remains listed are mostly leaves, a few plants are also known from flowers and fruits. Some, like *Decodon* (Figure 19.1E) and *Prunus* (Figure 19.4H), are

Table 19.2

Eocene compression floras from localities around Princeton, Quilchena, Chu Chua, and McAbee, and at Horsefly

Magnoliidae
  Lauraceae (laurel family)
    Sassafras hesperia, Sassafras sp., *Lindera* sp.
Hamamelidae
  Tetracentraceae
    Tetracentron sp.
  Trochodendraceae
    Unnamed genus
  Cercidiphyllaceae (katsura family)
    Cercidiphyllum sp.
  Platanaceae (sycamore family)
    Macginitiea gracilis, Platanus sp.
  Hamamelidaceae (witch hazel family)
    Langeria magnifica, Corylopsis sp.
  Ulmaceae (elm family)
    Ulmus sp.
  Juglandaceae (walnut family)
    Pterocarya sp.
  Myricaceae (wax myrtle/sweet fern family)
    Comptonia columbiana
  Fagaceae (oak family)
    Castaneaphyllum sp., *Fagopsis undulata, Fagus* sp., *Quercus* sp.
  Betulaceae (birch family)
    Alnus cremastogynoides, Alnus parvifolia, Alnus sp., *Betula leopoldae, Betula* aff. *papyrifera, Corylus* sp., *Palaeocarpinus* sp.
Dilleniidae
  Sterculiaceae (cocoa tree family)
    Florissantia quilchenensis
Rosidae
  Grossulariaceae (currant family)
    Ribes sp.
  Rosaceae (rose family)
    Physocarpus sp., *Spiraea* sp., *Amelanchier* sp., *Stonebergia columbiana, Photinia pageae, Rubus* sp., *Neviusia* sp., *Prunus* sp.
  Lythraceae (loosestrife family)
    Decodon sp.
  Cornaceae (dogwood family)
    Cornus sp.
  Davidiaceae (Chinese dove tree family)
    Tsukada davidiifolia
  Olacaceae (olax family)
    Schoepfia sp.

[continued on next page]

Table 19.2 [continued]

Eocene compression floras from localities around Princeton, Quilchena, Chu Chua, and McAbee, and at Horsefly
Rosidae *[continued]*
Vitaceae (grape family)
Vitis sp.
Sapindaceae (soapberry family)
Bohlenia americana, Bohlenia sp.
Hippocastanaceae (horse chestnut family)
Aesculus sp.
Aceraceae (maple family)
Acer princetonense, Acer rousei, Acer stewartii, Acer stonebergae, Acer stockeyae, Acer torodense, Acer washingtonense, Acer wehrii, Acer verschoorae, Acer sp.
Anacardiaceae (cashew family)
Rhus malloryi, Rhus sp.
Asteridae
Oleaceae (olive family)
Fraxinus sp.
Caprifoliaceae (honeysuckle family)
Sambucus sp.

known from leaves preserved as compressions but are also recognized by permineralized fruits, seeds, and wood.

Only a few detailed studies of individual fossils or localities have been completed to date. The Horsefly Beds, in particular, have great potential for the future. In an elegant multidisciplinary study, Mark Wilson (1980) determined the water depth and distance from shore of fish, insect, and plant assemblages and reconstructed the lake environments in general. The lake sediments at Horsefly are characterized by leaves and fish coprolites (fossilized excrement) in summer sediments and fish scales in winter sediments. A similar pattern is found at the Blakeburn Mine locality in southern British Columbia. The Horsefly Beds are also known for well-preserved flowers that contain pollen. Flowers of *Pistillipollianthus* (Figure 19.3E) are a good example – they contain the well-known pollen type *Pistillipollenites mcgregorii* (Figure 19.3D). Although this pollen was previously known only in a dispersed state, it was often used as a Middle Eocene index fossil. The Horsefly flowers provide the first evidence of the specific type of flower that produced these pollen grains.

Relationship to Modern Floras

Not long ago, paleobotanists thought that many of the fossil plants in the Paleogene were barely distinguishable from their modern counterparts. In many cases comparisons were based on superficial resemblances between the fossil and modern forms and on poorly preserved material. Recently, however, the superb preservation of compression plants at Princeton, McAbee, Quilchena, and Horsefly have shown that these plants often differ from modern genera of flowering plants. In fact, they almost always represent extinct species, often extinct genera, and occasionally extinct families. This is true in both the compression and permineralized floras. One such plant from Coalmont Road near Princeton, dubbed the 'Princeton taxon' by Jack Wolfe, probably represents an extinct order. These plants had leaves with a combination of characteristics unknown in any group of modern plants. The leaves show some resemblance to two angiosperm groups (the Eupteleales and the higher Hamamelidales) and may provide important evidence for evolutionary relationships among several major groups of flowering plants.

Another example is *Eorhiza* from the Princeton Chert (Figure 19.2A). These rhizomes with attached leaves have no living counterpart. In fact, they display the internal anatomical structure of dicot stems but have leaves like monocots, thus combining features of both major groups of angiosperms. The flowers and fruits of *Princetonia* (Figures 19.2B to 19.2E), which are thought by paleobotanists to belong to *Eorhiza,* are themselves unique (Stockey and Pigg 1991). The combination of characteristics exhibited by these remains is not known in living plants.

Leaves of the Mexican elm, *Chaetoptelea,* and the Chinese elm, *Zelkova,* are often so similar that it is difficult to identify isolated leaves with certainty. Fossil leaves of this general type are common at One Mile Creek near Princeton. Because the two modern genera are readily distinguishable by their different fruits, the recent recovery of fossil branches with attached leaves and fruits is important to the identification of the Princeton elm. The branches show winged fruits of the type found today only in true elms and are similar to the Mexican elm. This discovery proved that many, if not all, of the elm leaves found near Princeton belong to *Chaetoptelea.*

The Eocene was a time of rapid evolution and diversification in angiosperm families such as the Rosaceae (roses), Aceraceae (maples), and Sapindaceae (soapberries). Paleobotanists are interested in the origins of these families, their geographical distributions, and their habitats in the present and in the past.

The rose family (Rosaceae) is particularly well represented by many types of fossil remains, including wood (for example, of *Prunus,* Figure 19.4E), fruits (for example, of *Prunus,* Figure 19.4G), and flowers (for example, of *Paleorosa*) from the Princeton Chert. Numerous genera are represented by compressed leaves (Table 19.2). In fact, the oldest known fossils of cherry *(Prunus)* (Figure 19.4H), raspberry *(Rubus),* wild currant and gooseberry *(Ribes)* (Figure 19.4A), and Saskatoon berry *(Amelanchier)*

(Figure 19.4D) are preserved in the Princeton area, only a few dozen kilometres from today's fruit-producing areas of Penticton, Osoyoos, and Keremeos.

Alabama snow-wreath *(Neviusia alabamensis)* is a modern rosaceous shrub found natively in the southeastern United States. It is closely related to *Kerria japonica,* a rosaceous shrub native to Japan and China, and to *Rhodotypos,* another Japanese rosaceous shrub. Recently, a second species of *Neviusia* has been discovered, this one growing near Mount Shasta in northern California. When plants grow natively only in widely separated areas, as happens with these two species of *Neviusia,* plant geographers use the term 'disjunct' to describe their distribution. Recently, fossil leaves of *Neviusia* were found in the deposits at Princeton (Figure 19.4B) and at nearby sites in Republic, Washington. It is tempting to conclude that the oldest known fossil record of *Neviusia* in the Okanagan Highlands indicates its place of origin. Yet this is not at all certain. With two modern species of *Neviusia* in the southeastern United States and California and the closest living relatives in China and Japan, this is obviously not a simple story. The occurrence of this genus in the Okanagan Highlands fifty million years ago is both important and intriguing.

Other disjunct plants once grew in the Okanagan Highlands during the Eocene but are unknown there today – presumably because of climatic changes – and now grow only far from this region. *Photinia,* another rosaceous shrub, is found at Princeton and Republic as a fossil, and today grows only in Japan, Sumatra, and the Himalayas. *Stonebergia,* a genus now extinct, also occurs at Princeton and may have given rise to the two living western American rosaceous shrubs, *Chamaebatia* and *Chamaebatiaria.* Interestingly, *Stockeya,* a closely related genus to *Stonebergia,* occurs in Idaho and Montana, ten million years after the appearance of its ancestral stock at Princeton. This rosaceous lineage appears to have been moving south, like some of the fossil insects found in the Okanagan Highlands.

Although fourteen genera and over forty species of the rose family are known as fossils in the Okanagan, rosaceous pollen has not been recorded from compression floras. The rare occurrence of pollen in the flowers of *Paleorosa similkameenensis* from the Princeton Chert is especially noteworthy as it represents the oldest known record of rosaceous pollen and the first report of pollen preserved within a rosaceous flower (Figure 19.4F).

Like the Rosaceae, the Aceraceae and the closely related Sapindaceae underwent rapid evolution during the Eocene. Their remains are well represented in BC floras (Table 19.2). Of particular interest are the fruits and flowers from the Princeton region (Figures 19.3A to 19.3C). The extinct genus *Bohlenia,* originally described from nearby Republic, is rare at One

Mile Creek near Princeton (Figure 19.3C). This genus combines characteristics of maples (Aceraceae) and of some soapberries (Sapindaceae), suggesting that these two families may share a common ancestor.

BC deposits of Eocene fossils also contain the oldest known records of other plant groups. These include: Chinese dove tree *(Davidia/Tsukada),* ninebark *(Physocarpus), Spiraea, Amelanchier, Ribes, Rubus,* and *Neviusia.*

The locality at One Mile Creek is best known for its remains of birches and alders, the family Betulaceae. These remains are common at several localities in the province (Table 19.2). The best-studied plant is *Betula leopoldae,* which is also found at Republic. The material from Princeton, however, has provided the basis of a reconstruction of this species that includes leaves, fruits, flowers, and pollen. *B. leopoldae* is now one of the best-known species of Paleogene Betulaceae and the earliest record of the genus represented by both vegetative and reproductive structures (Crane and Stockey 1987).

Fossil sites in Eocene beds of British Columbia have enormous potential for elucidating the past. The angiosperms in particular are well represented with many unique and rare occurrences. Because BC fossil angiosperms, particularly those from the Princeton Chert, are among the best preserved in the world, future paleobotanical work looks very promising. With the help of paleobiologists and the cooperation of amateur paleontologists, these avenues of research will move forward in the future.

References

Cevallos-Ferriz, S.R.S., R.A. Stockey, and K.B. Pigg. 1991. The Princeton Chert: Evidence for *in situ* Aquatic Plants. *Review of Palaeobotany and Palynology* 70:173-85

Crane, P.R., and R.A. Stockey. 1987. *Betula* Leaves and Reproductive Structures from the Middle Eocene of British Columbia, Canada. *Canadian Journal of Botany* 65:2,490-500

Currah, R.S., and R.A. Stockey. 1991. A Fossil Smut Fungus from the Anthers of an Eocene Angiosperm. *Nature* 350:698-9

Stockey, R.A., and K.B. Pigg. 1991. Flowers and Fruits of *Princetonia allenbyensis* (Magnoliopsida; family indet.) from the Middle Eocene Princeton Chert of British Columbia. *Review of Palaeobotany and Palynology* 70:163-72

Wilson, M.V.H. 1980. Eocene Lake Environments: Depth and Distance-from-shore Variation in Fish, Insect, and Plant Assemblages. *Palaeogeography, Palaeoclimatology, and Palaeoecology* 32:21-44

Wolfe, J.A., and W. Wehr. 1987. Middle Eocene Dicotyledonous Plants from Republic, Northeastern Washington. US Geological Survey, Bulletin 1597, pp. 1-25

–. 1988. Rosacaeous *Chamaebatiaria*-like Foliage from the Paleocene of Western North America. *Aliso* 12:177-200

20 Eocene Conifers of the Interior

James F. Basinger, Elisabeth McIver, and Wesley C. Wehr

The evolution of modern Canadian vegetation began in the early Cenozoic. Many of the principal groups of plants seen today in British Columbia made their first appearance after the mass extinctions at the end of the Cretaceous. Some of their earliest fossil representatives can be found in the rich Eocene coal fields of central British Columbia.

The many Eocene sedimentary basins of the Interior of the province have been the source of spectacular fossil collections. As might be expected, they are familiar to many field naturalists. The best-known probably are the Princeton Basin, and the Kamloops, Okanagan, Quesnel, and Driftwood Creek areas (Figure 20.1).

These basins are commonly associated with extensive coal fields. Although none of these coal seams is currently mined in a major way, they were an important source of coal in the early part of this century. The open-cut coal mines that dot these areas are treasure troves of local history. Many once-prosperous mining towns are now only reminders of a lively episode in British Columbia's history. The village of Coalmont, near Princeton, is a classic example, with its sign warning all who enter of the peculiarities of its high proportion of bachelors.

All of these basins were formed during the Eocene, and all are about forty-five to fifty million years old. In each case, a small portion of the earth's crust sank slowly over millions of years; and into these depressions, sediment was deposited. Volcanism in the western Cordillera delivered extensive ash into these basins, so that many plants were encased in ash-rich sediments, a situation that results in well-preserved fossils.

Commonly, these low areas were covered by great coal swamps and lakes. It seems likely that the floors of some of these depressions were at moderate altitudes, giving us a rare opportunity to glimpse the seldom-preserved upland floras. Deposition in these small basins ceased about forty million years ago, after which rocks throughout the Interior were folded and deformed.

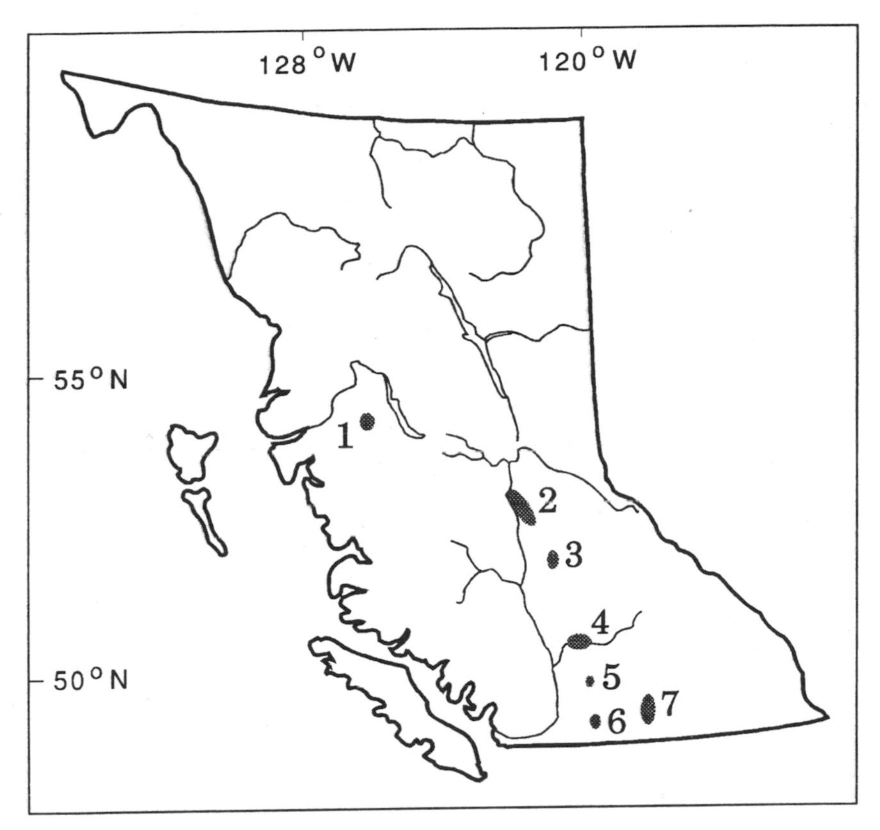

Figure 20.1 **Principal Eocene plant localities.** (1) Driftwood Creek, Smithers. (2) Quesnel. (3) Horsefly. (4) Cache Creek and Kamloops area, including the McAbee beds. (5) Nicola and Quilchena area. (6) Princeton and Coalmont area, including the Similkameen and Tulameen rivers. (7) Okanagan

In places the rocks are rich in fossils, many of which are familiar to collectors. Some fossil conifers represent genera now growing in the nearby mountains; however, others represent genera now extinct in North America and restricted to regions such as Japan and southeastern China.

During the early Cenozoic, the earth experienced global warmth, apparently lacking large ice sheets altogether. Even the most northerly lands of Spitsbergen and Ellesmere Island were home to rich forests of deciduous hardwoods and conifers. Vegetation resembling that of Georgia and the Carolinas covered all land north of the Arctic Circle and extended south into the Interior of British Columbia and Washington. Interestingly, the tropics apparently were not significantly warmer than today.

After the peak warmth of about fifty-five million years ago, global climate began to cool. Plants formerly restricted to cooler, mountainous habitats appeared in lowland environments, and became widespread to form the early equivalent of our modern mixed vegetation of evergreen conifers and deciduous hardwoods. The fossil record of this vegetation in the Interior of British Columbia includes some of the earliest records of spruce *(Picea)*, fir *(Abies)*, and hemlock *(Tsuga)*, as well as an abundance of pines *(Pinus)*. These evergreen conifers joined the primarily deciduous conifers of an earlier epoch, for example, dawn redwood *(Metasequoia)*, swamp cypress *(Glyptostrobus)*, and golden larch *(Pseudolarix)*, as well as the maidenhair tree *(Ginkgo)*, all of which later became extinct in North America. Other conifers, such as the cedars *(Thuja* and *Chamaecyparis)* and yews *(Taxus)*, do not fall into these patterns but have a long fossil record and persist in our modern forests.

This chapter discusses a few of the more common of these fossil conifers. The Eocene was an important time in the evolution of conifers; neither before nor since has the coniferous vegetation of British Columbia been so rich in diversity. As a result, there is still much to learn from these fossil remains.

Study of Eocene Conifers of the Interior

Sir J. William Dawson was a great Nova Scotian naturalist and paleobotanist and the father of George M. Dawson, who has been discussed in earlier chapters. During his career he became principal of McGill University and played a major role in the transition of McGill from a regional college to an international institution. He was a major figure in nineteenth-century paleobotany, and is best known as the founding father of the study of early land plants of the Devonian. Dawson was the first to publish reports of Cenozoic plants of British Columbia, basing his descriptions largely on collections made during surveys of western Canada by the members of the Geological Survey of Canada. These expeditions established the boundaries of Canada and opened the Interior to exploitation of gold and other minerals.

After Dawson's retirement, David P. Penhallow, also of McGill University, continued this paleobotanical research. Penhallow's report (1908) is the most comprehensive, albeit out-of-date, description of Cenozoic plants of British Columbia. Later studies of BC Cenozoic conifers include the work of Chester Arnold (1955). A recent report by Wehr and Schorn (1992) describes similar conifers found in beds of the same age at Republic, Washington.

Few references about these plants are readily available to the naturalist. Most publications are old and have been out of print for years. Others are highly technical articles not sufficiently broad in their treatment to be particularly useful as field guides. Because Eocene fossil conifers generally resemble their living relatives, however, a handbook of contemporary trees of western North America is generally adequate for generic identification. Some guides even include commonly planted exotics such as dawn redwood and ginkgo.

Dawn Redwood

The dawn redwood *(Metasequoia)* is one of the most easily recognized and beautiful of BC fossils (Figure 20.2). Its appearance commonly leads collectors to think they have found a fern, but this plant is a member of the redwood and cypress family (Taxodiaceae) and a close relative of the giant sequoias of California. *Metasequoia* is a true living fossil, a genus recognized first from the fossil record, and then discovered still living in the remote interior of China in the 1940s.

During the Eocene, the dawn redwood was abundant in lowlands throughout northern North America, Greenland, Spitsbergen, and Siberia,

Figure 20.2 **Leafy twigs of the dawn redwood *Metasequoia occidentalis.*** Allenby Formation, Princeton Basin. (X 1.3)
Source: Collections of the Department of Geological Sciences, University of Saskatchewan

where it was a major contributor to coal deposits. Because it grew in low, wet areas, where terrestrial fossil-bearing sediments are formed, dawn redwood is one of the most abundant of fossils. In river and roadside exposures of shale in the Princeton Valley, for example, which are commonly fossil bearing, the dawn redwood is found almost everywhere.

During the Neogene, which followed the Paleogene, *Metasequoia* was gradually excluded from North America, then from Japan and northeastern Asia, finally to be driven nearly to extinction in central China. Ironically, horticulturalists today favour this tree for its attractive shape and foliage, so that now, once again, it is distributed throughout the northern hemisphere and can be found growing in lawns and parks in southern British Columbia. Fossil and living dawn redwoods are so similar in appearance that scientists debate whether they are the same species.

The dawn redwood is a deciduous conifer. Like its relative, the bald cypress of the southeastern United States, it drops its foliage each year. Individual leaves remain attached to the branches, and entire branch units are shed. Branches bear two rows of closely spaced needle-leaves, and the genus is readily identified by the opposite, paired arrangement of the leaves. Seed cones are about the size of a thumbnail, perhaps a bit larger, and occur on a naked (almost leafless) stalk that is typically much longer than the cone. The cone is made up of several opposite pairs of scales that are expanded outward to form closely fitting, rhomboidal 'shields.'

Other Redwoods

Although foliage of the dawn redwood may be distinguished by its opposite pairs of leaves, foliage of other members of the redwood family (Taxodiaceae) is not so distinctive. One of the problems is that species such as swamp cypress *(Glyptostrobus),* bald cypress *(Taxodium),* redwood *(Sequoia),* and certain other living and extinct forms bear highly variable foliage on the same tree. This characteristic makes it almost impossible to know for sure which species, or indeed how many species, are present at a fossil locality.

Foliage of this type may have been deciduous, like the dawn redwood, and shed as whole leafy branches. The leaves are somewhat irregularly spaced, and they are generally shorter and typically more slender and tapered than leaves of the dawn redwood. We know from some cone-bearing fossil branches that an early species of *Sequoia* once grew in British Columbia (Figure 20.3). In the absence of cones, however, it is often impossible to identify unequivocally whether their foliage belongs to any one particular genus. As a result, paleobotanists refer to it as *Elatocladus,* which is a general repository for redwood-like foliage of uncertain affinity.

Plants that bore this type of foliage were abundant in North America

Figure 20.3 **Leafy twig with a small, globular cone characteristic of the redwood *Sequoia*.** McAbee beds near Cache Creek. Actual size.
Source: Collections of the Department of Geological Sciences, University of Saskatchewan

during the Eocene. Only the redwood and bald cypress survive today in North America – the others became extinct or restricted to eastern Asia. Apparently, climatic cooling and the drying of the interior of North America eliminated these conifers from most of their previous range.

Cedars

Foliage of cedars is readily distinguished by its small, scale-like leaves arranged in opposite pairs. Branchlets are typically arranged in a single plane to form flat sprays, and the entire spray is shed as a unit. When fossilized, these branches are laid out on the bedding plane. Foliage of this type looks much like the common garden pyramidal cedars *(Thuja occidentalis).* Indeed, fossilized foliages of many of the cedars, including western red cedar *(Thuja plicata),* yellow cedar *(Chamaecyparis lawsoniana),* and Alaskan cedar *(Chamaecyparis nootkatensis),* can scarcely be distinguished from each other. Cones of the various cedars are distinctive, and thus permit identification. Unfortunately, cones are rarely found on fossil foliage, so that the best identification may simply be 'cedar' (Figure 20.4).

Cedar seed cones are typically smaller than the nail of a little finger. If cones are round with only a few opposite pairs of shield-shaped scales, they are probably *Chamaecyparis.* If they are elongated with several pairs of somewhat overlapping scales, they are probably *Thuja.* Anyone finding cones on a fossil cedar should contact a paleobotanist or a museum, as the specimens may be scientifically significant.

Figure 20.4 **Cedar foliage and cones.** (A) Cedar foliage, unidentifiable without attached cones. (B) Globular cones, which, when attached to cedar foliage, identify *Chamaecyparis,* the yellow cedar. (C) Elongated cones with thin, overlapping scales, belonging to *Thuja*. McAbee beds near Cache Creek. (X 0.8)
Source: Collections of the Department of Geological Sciences, University of Saskatchewan

Pines

The pine family (Pinaceae) includes most of the familiar forest trees of modern British Columbia. In addition to pines *(Pinus),* the family includes spruces *(Picea),* Douglas fir *(Pseudotsuga),* hemlock *(Tsuga),* fir *(Abies),* and larch *(Larix)*. The various species of pines are the most widely distributed of modern conifers, and are adapted to a great variety of habitats. Also, they have the most extensive fossil record of any of the Pinaceae. Like other members of the family, they increased in prominence and diversity during the Eocene. Evidence of this diversification is well displayed in the fossil record of British Columbia.

Pines are an ancient group within the pine family, and they have characteristics that set them off from other genera. Accordingly, they probably should not be considered ancestral to any other living genus. Perhaps the most conspicuous feature, and one that is useful in identifying fossils, is the occurrence of leaves in small bundles, or fascicles, that are actually highly reduced leafy shoots. Leaves do not occur individually, as in the spruces and firs, for example. Typically they occur in fascicles of two, three, or five needles, although a minor proportion of the fascicles on a tree will include one more or one less needle than is typical. Only one species has fascicles with a single needle.

So many names have been applied to the remains of pines and pine-like fragments that a quick look through reference works only results in confusion. The most common types of fossil pine found in British Columbia are a three-needle pine called *Pinus trunculus* and a five-needle pine called *Pinus latahensis* (Figure 20.5). Because poorly preserved pine cones may resemble cones of other genera, they cannot be identified reliably. Well-preserved cones showing the internal structure necessary for proper identification are extremely rare, and are potentially important to science. Only a few specialists have the expertise to prepare these fossils properly.

True Firs

The first true firs of the genus *Abies* that are recorded reliably date from the Paleogene. Some of the earliest known remains are to be found in Eocene fossil deposits in the Interior of British Columbia.

Firs are typically found in mountainous regions, where they commonly occur at the tree-line. Some species also form part of the boreal forests of North America and Eurasia. All are evergreen, needle-leaved plants. The leaves are somewhat flattened and generally flexible, quite unlike the usually stiff, squarish leaves of the spruces. When the needles fall from the branch, they leave a circular scar that is flush with the bark, so that the

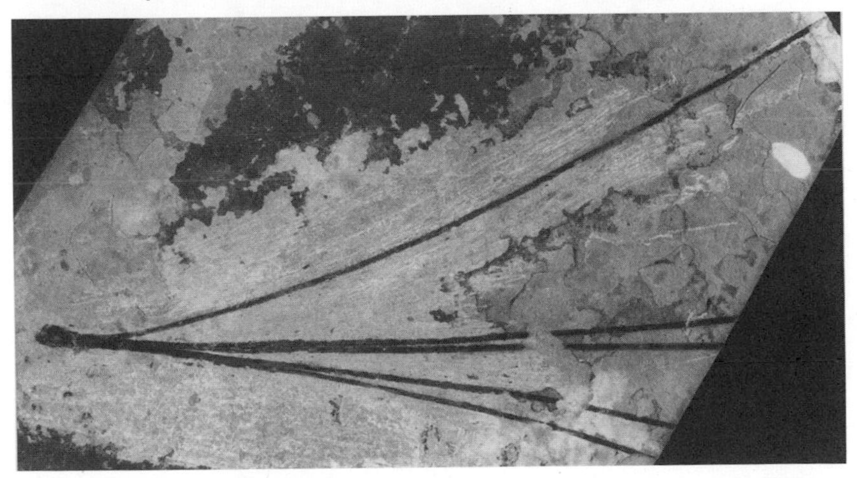

Figure 20.5 **The five-needle pine *Pinus latahensis*, a relative of living white pines.** McAbee beds near Cache Creek. Actual size.
Source: Collections of the Department of Geological Sciences, University of Saskatchewan

twigs appear smooth. Fossilized needle-bearing branches of firs may be distinguished from the more common redwood-like branches by the much bushier appearance of the fir foliage.

More recognizable in the fossil record are the cone scales of firs. Fir cones shed not only their seeds but also their scales at maturity. Essentially the cones fall apart, leaving behind the spike-like central axis. Fir cone scales are typically short and wide, with a broad, rounded outer margin (Figure 20.6).

The evergreen needle-leaved conifers such as the firs, spruces, and hemlocks that dominate much of British Columbia today are relative newcomers, first appearing in the Paleogene along with a number of other plants typical of our Canadian temperate forests. Their sudden appearance and rapid diversification were probably in response to the deterioration of global climate that began during the Eocene. This climatic decline ultimately led to the ice age, but nearly fifty million years ago some critical threshold of seasonality and temperature was reached that permitted these conifers to expand across the continent.

Figure 20.6 **A scale of the true fir *Abies*, shed from the seed-cone axis.** McAbee beds near Cache Creek. (X 4)
Source: Collections of the Department of Geological Sciences, University of Saskatchewan

The ancestors of these plants may have inhabited high mountains, where they would not be fossilized, a suggestion supported to some degree by the record of fossil pollen grains. Nevertheless, from what we know it is unlikely that they have a very long pre-Paleogene history. Instead, the diversification and spread of evergreen conifers probably parallels that of the temperate hardwoods, thus challenging our preconception of conifers as ancient relics.

Maidenhair Tree of the East

Few trees are more attractive than ginkgo, *(Ginkgo biloba),* the maidenhair tree of the East, with its soft-green, fan-shaped leaves. Its native habitat is western China, where the ancient peoples cultivated it and treated it as sacred. This was fortuitous – for us and for the ginkgo – because, as believed by many, the species probably became extinct in the wild centuries ago and thus only survived through its cultivation. Ginkgo, like the dawn redwood, is now grown in cities throughout North America, Europe, and Asia, where it shows a high tolerance for air pollution. Once again ginkgo is approaching the extraordinary breadth of distribution that it enjoyed during the Cretaceous and Paleogene.

Ginkgo is a common member of the mid-to-high-latitude fossil floras of the Cretaceous, although its ancestry can be traced back through earlier forms for nearly 300 million years. In fact, it is so distantly removed from any other living seed plant that it fits poorly into any system of plant classification. Paleobotanists include it with the conifers simply because, with them, it is least out of place.

By the end of the Cretaceous, ginkgos were already declining, and by the Paleogene only a few different 'types' remained. There has been much discussion about how many species should be recognized in the fossil record, because virtually all fossil leaf forms may be found in living ginkgo, often on the same tree. Differentiation of species typically is done at the microscopic level, which is impractical for field naturalists.

Ginkgo became extinct in North America during the Neogene, apparently the victim of the same climatic changes that eliminated dawn redwoods, swamp cypresses, and several other plants once common here.

Most fossil leaves of the Eocene basins of British Columbia have the bi-lobed, fan-shaped outline familiar to anyone who has seen a living ginkgo. These remains likely belong to the species *Ginkgo adiantoides*. Leaves may be highly dissected or merely notched – a search generally reveals every possible variation. An exception is a rare species found on occasion in British Columbia, whose leaves were *always* deeply dissected. Whatever the species found, these treasures are the most beautiful of fossil plants and the most unmistakable.

Most biologists are accustomed to thinking of the conifers as an ancient group of relics that were eclipsed in the evolutionary process by the enormously diverse and successful flowering plants. Residents of British Columbia may be excused for thinking otherwise, as the province is blanketed by coniferous vegetation. In many books on plant distribution, this situation is described as anomalous. However, by examining the Cenozoic history of plant life in British Columbia, we appreciate that, while some conifers have been lost, others, notably members of the pine and cedar families, have proliferated and come to play a leading role in the vegetation of mid-to-high latitudes. Conifers, indeed, are a thriving group that has continued to evolve, and nowhere is their success more evident than in British Columbia.

References

Arnold, C. 1955. Tertiary Conifers of the Princeton Coal Field of British Columbia. *Contribution of the Museum of Paleontology, University of Michigan* 12:245-58

Penhallow, D.P. 1908. *Tertiary Plants of British Columbia.* Geological Survey of Canada, Publ. 1013

Wehr, W.C., and H.E. Schorn. 1992. Current Research on Eocene Conifers at Republic, Washington. *Washington Geology* 20:20-3

21
Quaternary Animals: Vertebrates of the Ice Age

C. Richard Harington

Many interesting vertebrates lived in British Columbia during the ice age or Pleistocene – from 10,000 to two million years ago – for example, mammoths, mastodons, wild horses, muskoxen, and bison. Even walruses once lived on Vancouver Island and its coastal waters. Bear-sized ground sloths, small horses, and woolly mammoths roamed what is now the Cariboo District. Columbian mammoths that stood more than three metres at the shoulder and giant bison occupied central British Columbia about 30,000 years ago. And, until recently, caribou survived on the Queen Charlotte Islands. This chapter outlines vertebrate life of the ice age, as known from evidence found in various regions of British Columbia (Figure 21.1).

Vancouver Island and Vicinity: Mammoth, Mastodon, Muskox, and Walrus

The earliest recorded discoveries of ice-age mammals on Vancouver Island were pieces of the molars of the gigantic imperial mammoth *(Mammuthus imperator)* collected from James Island in 1895 and from Cordova Bay near Victoria in 1898. Most of the nearly two dozen specimens of ice-age mammals from Vancouver Island have come from gravel pits in the Saanich Peninsula. Mastodon teeth come from the Shawnigan Lake and Courtenay areas, a mammoth tooth from near Courtenay, and a walrus skeleton from Qualicum Beach.

A dozen or more fossils, including remains of imperial and Columbian mammoths *(Mammuthus columbi),* bison, and brown bear *(Ursus arctos)* have been found on the islands between the mainland and Vancouver Island. These fossils are also known from James Island, as well as from Whidbey, Sucia, Orca, Smith, and Protection islands on the American side of the border.

It is difficult to say how old these fossils are. Evidence gathered from studying the sequence of sediments at gravel pits on the Saanich Peninsula indicates that the animals lived near the peak of the last (Wisconsinan)

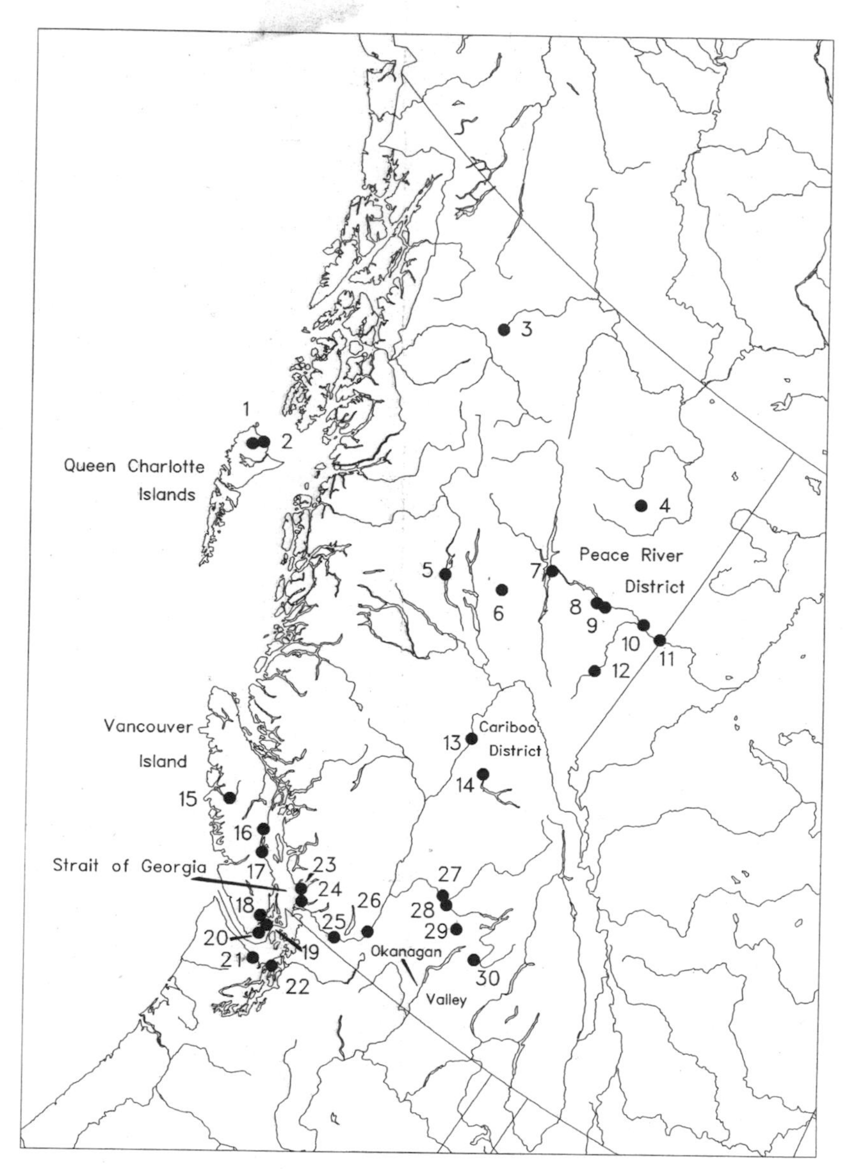

Figure 21.1 **Quaternary vertebrate sites of British Columbia.** (1) Graham Island, west of Naden Harbour. (2) Virago Sound. (3) Dease Lake. (4) Trutch. (5) Babine Lake. (6) Chuchi Lake. (7) Finlay Forks. (8) Portage Pass. (9) Hudson Hope. (10) Charlie Lake Cave, Fort St. John, and Taylor. (11) Clayhurst Crossing. (12) Tumbler Ridge. (13) Quesnel. (14) Quesnel Forks. (15) Gold River. (16) Courtenay. (17) Qualicum Beach. (18) Shawnigan Lake. (19) James Island, Saanich Peninsula, and Saanich Inlet. (20) Victoria and Beaver Lake. (21) Sequim (Washington State). (22) Whidbey Island (Washington State). (23) Bowen Island. (24) Vancouver. (25) Chilliwack. (26) Yale. (27) Tranquille and Hanging Valley creeks. (28) Kamloops. (29) Westwold. (30) Lumby.

glaciation about 20,000 years ago. This conjecture is supported by radiocarbon dates obtained from mammoth tusks and a humerus (upper foreleg bone), which indicate the materials are about 17,000 years old. Mammoths were also living around this time in the lower Fraser River valley near Chilliwack where fragmentary tusks from a gravel pit have been dated to about 22,000 years ago.

The largest species of mammoth known from North America is the imperial mammoth, which stood about four metres (Figures 21.2 and 21.3). It may have had some hair but probably much less than other mammoths, such as the woolly mammoths, which roamed the tundra or tundra-like terrain of northern Eurasia and North America, or the Columbian mammoths, which had adapted to life on the cool grasslands of North America.

The American mastodon *(Mammut americanum)* was more primitive than the mammoths, and more squat in appearance (Figure 21.4). Its thick-enamelled teeth (Figure 21.5) with paired cusps were adapted to browsing (that is, feeding mainly on trees and shrubs), in contrast to the series of compressed enamel plates in the cheek teeth of mammoths, which were adapted to grazing (that is, eating grasses). Studies of well-

Figure 21.2 **Largest kind of mammoth known from North America, the imperial mammoth**
Source: Ink sketch by Charles Douglas

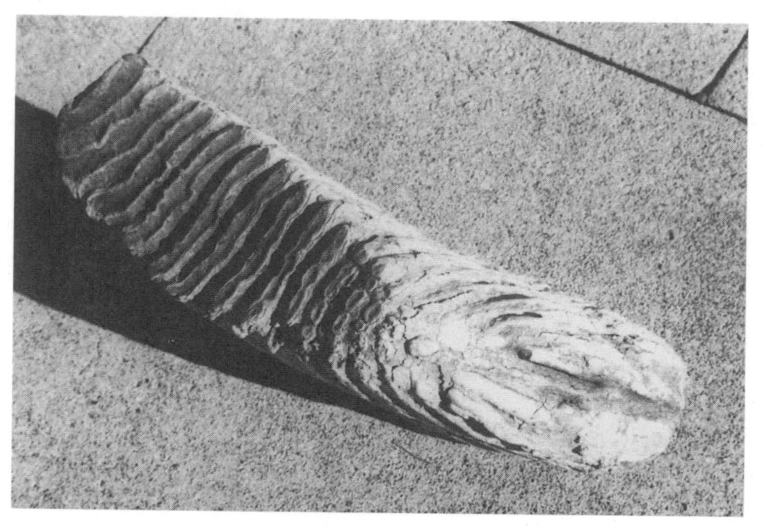

Figure 21.3 **Lower molar of an imperial mammoth found at Cordova Bay in 1971.** The enamel plates showing on the well-worn surface of the tooth are an adaptation to feeding on grasses.

Figure 21.4 **American mastodon.** The American mastodon was smaller than the imperial mammoth.
Source: Ink sketch by Charles Douglas

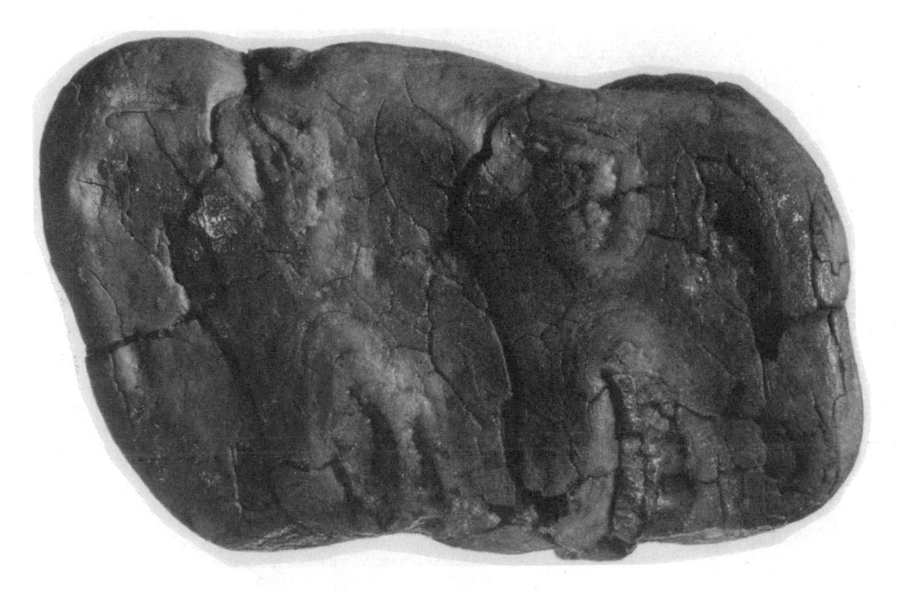

Figure 21.5 **Worn molar of an American mastodon.** The molar is 10 cm long and comes from near Courtenay (Tsolum River). One can see a series of paired cusps adapted for browsing on tender spruce twigs.

preserved American mastodons indicate that they lived mainly in marshy, open spruce woodlands and spruce forests, where they fed on tender twigs, and that they had coats of reddish-brown hair. In addition to the teeth found on Vancouver Island, part of a molar was collected near Trutch on the Alaska Highway. American mastodon fossils are rare elsewhere in the province and in other parts of western Canada.

Early people living in shrub-tundra surroundings hunted mastodons near Vancouver Island toward the close of the last glaciation, about 12,000 years ago. In 1977, mastodon bones were excavated from a peat bog near Sequim in Washington, about fifty kilometres southwest of Victoria. A bone spearpoint was lodged in one of the mastodon's ribs. Evidently a bison was also killed and butchered at this site.

One of the most exciting fossil finds on Vancouver Island is part of an extinct helmeted muskox skull (Figure 21.6). It was collected in a gravel pit on the Saanich Peninsula in 1969. The only other fossil of this species from British Columbia is a skull fragment from the vicinity of Dease Lake in the northern part of the province. The helmeted muskox *(Symbos cavifrons)* was taller and more slender than the tundra muskox that now lives in northern Canada and Greenland. Instead of the deep central crease between the hornbases of male tundra muskoxen, the bases were fused to form a massive helmet with flaring horns. As well, this ancient muskox was adapted to warmer conditions, probably living in parkland

Figure 21.6 **Partial skull of a helmeted muskox.** The 25-cm-high skull is seen from behind. It was collected in a gravel pit on the Saanich Peninsula.

or steppe-like grasslands (Figure 21.7). According to the number of specimens recovered, it was the most common North American ice-age muskoxen.

Remains of two extinct western bison *(Bison bison occidentalis)* were found in 1987 and 1988 near Victoria. The first, a partial cranium with horncores (the bony core of the horn), was uncovered during excavations to enlarge an irrigation pond in North Saanich. A radiocarbon date indicated that the animal died about 11,800 years ago. Other bison remains, probably of similar age, were found nearly four metres below the surface of a bog near Beaver Lake. Analysis of pollen remains associated with one of the specimens showed that pine and probably trembling aspen were common at that time. According to Richard Hebda of the Royal British Columbia Museum, the ancient landscape was similar to that of the Chilcotin country in the Interior.

Mammoths, mastodons, muskoxen, bison, and other ice-age mammals probably reached Vancouver Island by large flood plains that filled the Strait of Georgia from about 20,000 to 30,000 years ago. Earlier crossings may also have been possible. As mammoths occupied the Fraser lowland about that time, they represent a likely source for the 'island' mammoths.

Figure 21.7 **Extinct helmeted muskox bull attacking a wolf.** This species occupied southern Vancouver Island and the Dease Lake area during the ice age.
Source: Ink sketch by Bonnie Dalzell

The Vancouver Island marmot *(Marmota vancouverensis)* may be a living representative of another earlier crossing. It is the only mammalian species restricted to the Island, and differs from the North American hoary marmot *(Marmota calligata)* in its dark brown colour and small size. Presumably it evolved from paler hoary marmots of the mainland that reached what is now Vancouver Island during a relatively cold glacial phase of the ice age more than 130,000 years ago. Its arrival during a cold phase is suggested by its adaptation to alpine and high-subalpine conditions.

Except for the partial skull of a brown bear *(Ursus arctos)* from Whidbey Island, carnivore remains are scarce in the region. Even so, we know that black bears *(Ursus americanus)* occupied the mountainous part of central Vancouver Island near Gold River about 9,800 years ago. Remains of several of these bears and of deer were recovered by speleologists (scientists who study caves) in the Windy Link Cave complex (Figure 21.8).

Remains of several vertebrates have been collected from marine or coastal deposits on southern Vancouver Island and near Vancouver. They include walrus, Steller sea lion, baleen whale, and killer whale *(Orcinus orca)*. Even fishes are known from these deposits. Part of the skull of a Pacific sardine *(Sardinops sagax)* was recovered from a sea bottom core taken at Saanich Inlet. Radiocarbon dates indicate that it is about 10,000 years old.

Of particular interest is the skeleton of an adult female walrus *(Odobenus rosmarus)* discovered near Qualicum Beach on Vancouver Island. Found in glaciomarine clay from the Early Wisconsinan (about 70,000 years ago), this specimen is the best-preserved ice-age walrus from the west coast of North America (Figure 21.9). Paleoenvironmental, stratigraphic, and geochronological data on the Qualicum walrus agree with known habits

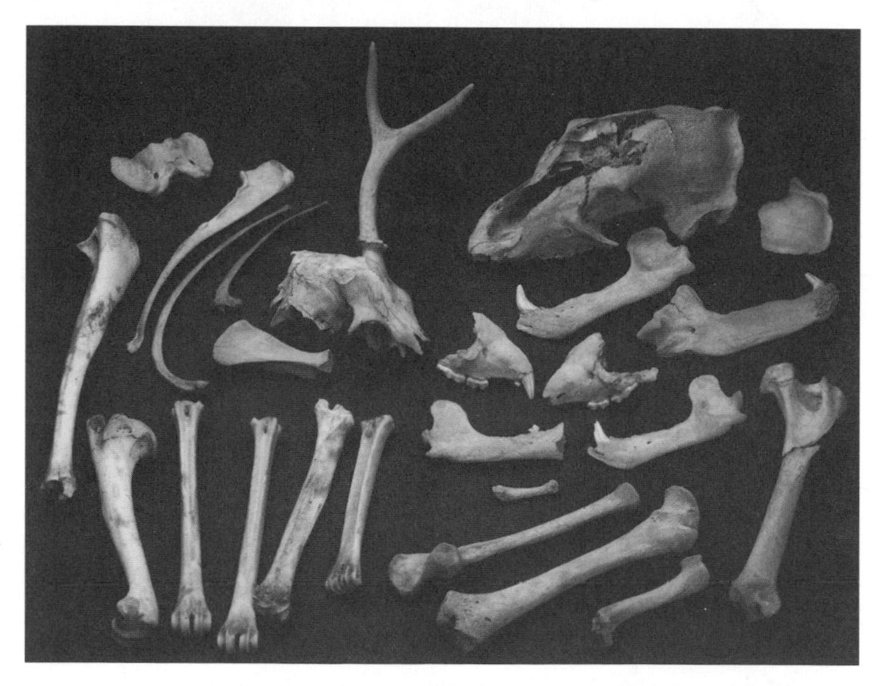

Figure 21.8 **Remains of black bear (right) and deer (left) from Windy Link Cave complex near Gold River, Vancouver Island.** Bones date from 9,800 years ago.
Source: Photograph courtesy of Grant Keddie, Royal British Columbia Museum

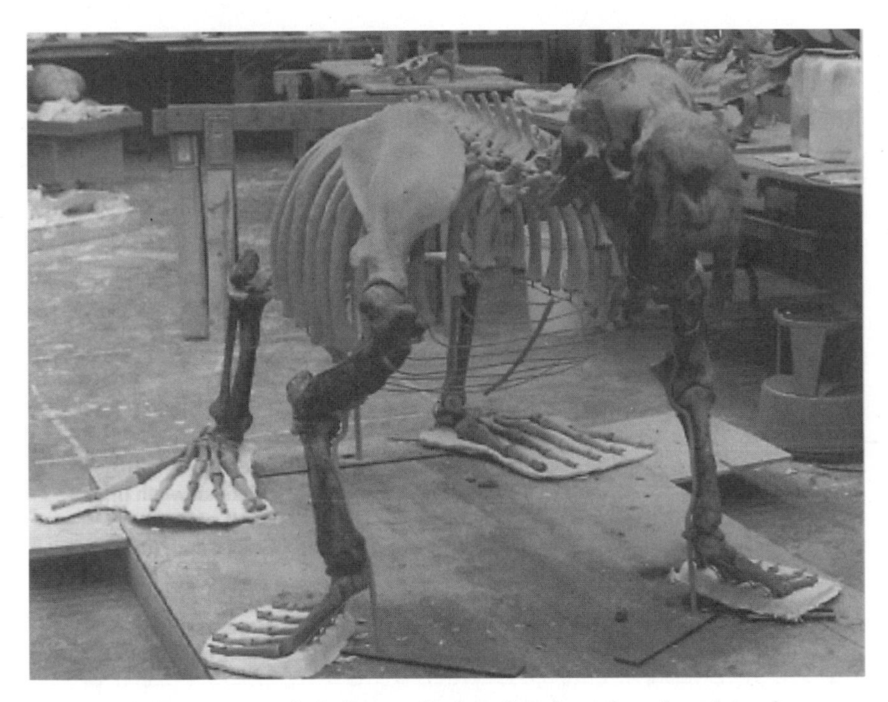

Figure 21.9 **Reconstructed skeleton of adult female walrus found in clay near Qualicum Beach, Vancouver Island.** Skeleton estimated to be 70,000 years old.

and habitat of modern walruses, as well as with our knowledge of the past distribution of the species. These data suggest that walruses made southerly advances during Pleistocene glaciations and northward withdrawals during the warmer interglacial periods (Harington and Beard 1992).

Steller sea lions *(Eumetopias jubatus)* lived in the region about 12,500 years ago, according to radiocarbon-dated organic material associated with a humerus (upper foreleg bone) of an adult female from Bowen Island, just across the Strait of Georgia from Qualicum Beach. The species still occupies this area during winter.

Ancient Fishes and Fishing

The earliest people to fish in the Cordillera were Paleo-Indians presumably adapted to life along rivers, particularly to salmon fishing. They occupied a site near Yale on the Fraser River approximately 9,000 years ago. Deposits along the Thompson River contain fossils of Pacific salmon *(Onchorhyncus)* that may date from the last glaciation, which spanned the period going from about 90,000 to 10,000 years ago. Several excellently preserved specimens of salmon have been found in calcareous concretions near Kamloops Lake. One specimen, comprising the head, pectoral girdle, and fins, shows the strongly hooked snout, or 'kype,' of breeding males. Geologists originally considered that these fossils represented ice-age salmon. However, analysis of fossil pollen in a rock sample containing the male salmon skull suggests that this fish lived much earlier: between two and twenty-two million years ago. So, salmon may have been spawning in coastal streams of British Columbia for millions of years.

Cariboo District: Bighorn, Ground Sloth, and the Earliest Mountain Goat

Very few fossils of larger alpine mammals have withstood the crushing weight of ice during the intense glaciation accompanying cold climatic phases. These bones were also rolled and battered in rushing mountain streams. A good example is a stream-eroded skull fragment of a mountain goat *(Oreamnos americanus)* from the Cariboo District found in 1932 by Erle Anness. At that time the manager of Morehead Mines Limited, Anness collected the back part of a mountain goat skull from a layer of deeply buried, gold-bearing gravel near Quesnel Forks. The fragment was beneath more than eighty-five metres of alternating beds of glacial and nonglacial gravel. This discovery suggests that mountain goats lived in the province more than 90,000 years ago during an interglacial period when the landscape appeared much the way it does today. The Quesnel Forks skull fragment is the oldest known mountain goat fossil (Harington 1971).

In 1969, a well-preserved cranium of a bighorn ram was found in a gravel pit near Finlay Forks in northern British Columbia. Radiocarbon analysis of bone from this specimen indicated that the species lived in this locality about 9,300 years ago (Rutter et al. 1972). Several partial skulls of bighorn sheep *(Ovis canadensis)* of the Early Holocene (about 5,000 to 10,000 years old) have been recovered from the beds of Tranquille Creek and adjacent Hanging Valley Creek farther south near Kamloops. In 1986, R.J. Fulton collected the partial cranium of an adult female mountain sheep from a Kamloops gravel pit. This specimen was found in an ancient channel of a large river and is presumably about 34,000 years old. Analysis of fossil pollen from the channel deposits suggests that the sheep lived among Engelmann spruce and subalpine fir.

In addition to the mountain goat fossil from Quesnel Forks previously mentioned, remains of other species have been reported from gold-bearing gravels of the Cariboo District. They include Columbian mammoth, woolly mammoth, an extinct small horse, mule deer, moose, caribou, bison, and ground sloth.

The long-haired, bear-sized Jefferson's ground sloth *(Megalonyx jeffersonii)* is represented by a claw that was found in river gravel near Quesnel. It is difficult to imagine this lumbering animal as a former member of the BC fauna. In more southerly regions of ice-age North America it is associated with forest-dwelling animals, and its broad blunt teeth suggest an adaptation to stripping leaves from trees. Of South American ancestry, *Megalonyx* seems to have reached northwestern North America during the last interglacial period. This animal became extinct about 9,000 years ago.

Peace River District: Charlie Lake Cave and Western Bison

Mammoths once lived in the Peace River District, as well as on southern Vancouver Island and in the Chilliwack area. The earliest known mammoth in the Peace River area probably dates from the last interglacial period, as wood from the enclosing gravel has been radiocarbon dated at more than 40,000 years old. This specimen, from near Hudson's Hope, consisted of a lower jaw, two molars, parts of all leg bones, a scapula, and a vertebra. Apparently one leg bone had a small patch of dried flesh and dark brown hair adhering to it.

Nearby at Taylor, a molar tooth of a woolly mammoth was found in a gravel pit. Radiocarbon analysis of this fossil indicates that these tundra-adapted mammoths lived in the region about 27,000 years ago, before the last glaciation. A radiocarbon-dated tusk from gravels at Portage Pass suggests that mammoths probably lived there as well but about 11,600 years ago.

Early postglacial bison, like the extinct western bison, also lived in this region. A skeleton of a bison *(Bison bison* cf. *occidentalis)* from gravels near Clayhurst Crossing represents the most complete Late Pleistocene bison reported from Canada. Radiocarbon dating of bone from this skeleton suggests an age of about 10,500 years ago. This fossil is probably closely related to bison specimens of a similar age from Taylor, Charlie Lake Cave, and Tumbler Ridge.

A wealth of information about vertebrates of northeastern British Columbia spanning the last 10,800 years has come from the Charlie Lake Cave site, which is about seven kilometres northwest of Fort St. John. Following the retreat of Glacial Lake Peace from this site shortly before 10,500 years ago, the gully in front of the cave began filling with sediments, ultimately reaching a depth of more than four metres. During this process, the site was visited occasionally by people who left stone artifacts and bones as early as 10,500 years ago. It was also visited by predators carrying their prey and by various animals who lived in the area. The result is a well-preserved and well-dated faunal assemblage.

The bottom of the sequence, which represents sediments covering the period from about 10,500 to 9,000 years ago, has produced many fossils. Included are remains of suckers and other unidentified fishes. Since few fishes of the Late Pleistocene and Early Holocene have been reported from western Canada, the presence of suckers is interesting. The frogs found also suggest nearby lakes or streams.

Of the birds found, evidently horned grebe, grouse and ptarmigan, coot, short-eared owls, and cliff swallows were most common. The horned grebe and coot suggest nearby wetlands or perhaps shallow lakes; the owl bones may have been collected by people. Cliff swallows probably nested at the site and died there naturally. Other birds found were western grebe, medium-sized grebes, surface-feeding ducks including green-winged teal and mallard, ruddy duck, small rail, small wader, and perching birds.

Among the mammals, snowshoe hare, ground squirrel, small rodents, and bison (probably a medium-horned bison like the western bison) were most commonly represented. Other mammals were large (possibly arctic) hare, woodchuck or marmot, deer mouse, Gapper's red-backed vole, meadow or long-tailed vole, chestnut-cheeked vole, other unidentified rodents, wolf or dog, least weasel, another small weasel, and a deer.

These fossils indicate that, during the period from about 10,000 to 10,500 years ago, the landscape was open, with some water, marshes, and patches of forest, changing to forest about 10,000 years ago. By 9,000 years ago, the fauna was modern.

Central Interior: The Babine Lake Mammoth and Giant Bison

An important discovery was made in 1971 when workers excavating an open-pit copper mine at Babine Lake found fossil bones of an elephant. Body and limb bones of an individual were found but no teeth, making identification difficult. Limb lengths and ratios do suggest, however, that the elephant was a mammoth about three to four metres high at the shoulder, like the Columbian mammoth (Harington et al. 1974).

An interesting feature of the discovery is that the specimen was found near the bottom of an ancient peat bog that was sealed in a bedrock depression by about one metre of fine gravel and twenty-four metres of boulder clay laid down during the last glaciation. This bog deposit of organic-rich silt held important information about the landscape of the region before the last glaciation. Radiocarbon dates on spruce and fir sticks from the bog indicated that they were about 43,000 years old and confirmed an earlier estimate of the mammoth's age based on the types and sequence of sediment layers. However, a younger date of about 34,000 years from bone of the mammoth indicated that the large animal may have lived more recently and sunk while struggling in the bog.

Analysis of fossil pollen from the clay in cavities in the mammoth bone suggests that vegetation near Babine Lake during the relatively warm interval before the peak of the last glaciation was similar to present-day shrub tundra just beyond the tree-line in northern Canada. Probably, birch and willow shrubs bordered the pond and were scattered over the landscape. Open ground supported abundant grasses, sagebrush, and other composites. Also present were various members of the rose family, pinks, willow-herb, and several other herbs. Although the exact contemporaneity of the plant and mammoth remains is questionable, both types of fossils suggest the presence of grassland near Babine Lake about 34,000 to 43,000 years ago.

The first evidence that giant bison *(Bison latifrons)* once lived in British Columbia came to light in 1993, although the bones had been collected about twenty-five years earlier by Peter Koropatnisky at Chuchi Lake. This specimen is similar in age (about 31,000 years old) to the Babine Lake mammoth, and was found only about 115 kilometres farther to the east. This spectacular animal had a horncore span more than two metres. Remains consist of partial horncores, neck vertebra, right foreleg bones, left hind-leg bones, and three toe bones. Giant bison have also been reported from Alberta and Saskatchewan, and farther south from Mexico and Florida to California and Idaho. The species lived from at least the second last (Illinoian) glaciation to late Wisconsinan time; the Chuchi specimen may be one of the latest known survivors.

Southern Interior: Traces of a Last Interglacial Fauna

Only traces have been found of the ice-age fauna that occupied the Southern Interior during nonglacial phases. Remains of fishes, rodents, small horses, and bison were collected from deposits covered by two glacial tills near Westwold in the northern Okanagan. We know, therefore, that these animals lived in the region during the last (Sangamon) interglacial, which reached maximum warmth about 130,000 years ago. About 120 kilometres southeast of Westwold, near Lumby, small horse, bison, and mammoth bones were found in sediments that seem to be mainly of Middle Wisconsinan age (that is, extending from more than 43,800 years ago to about 19,000 years ago).

The Queen Charlotte Islands: Dawson Caribou and Sticklebacks

One of the greatest mysteries of British Columbia wildlife is the fate of an extinct, small species of caribou that once lived in the Queen Charlotte Islands. How the caribou got to the islands and when they died out are questions with no certain answers. An hypothesis of isolation in a refuge on the Queen Charlotte Islands provides only partial explanation.

The name Dawson caribou *(Rangifer dawsoni)* is for G.M. Dawson, who first reported these animals (Figure 21.10). An unusual feature of this animal is that the females lacked antlers. Like the Vancouver Island marmot, the Dawson caribou seems to have developed from Mainland ancestors – in this case from woodland caribou. The caribou are thought to have migrated to the Queen Charlotte Islands during the second-last

Figure 21.10 **Dawson caribou that survived the last glaciation on Graham Island.** This species, unique to the Queen Charlotte Islands, has not been seen since about 1920.
Source: Ink sketch by Frank Beebe

glaciation, when the worldwide sea level is estimated to have dropped 140 to 160 metres. This drop probably exposed a broad shelf between the mainland and the islands, which the caribou could have crossed. Evidently they survived the last glaciation and, in historic times, were confined to northern Graham Island. In 1908, two Natives hunting in a large bog near Virago Sound shot three of a herd of four caribou. The skins and skulls were sent to the British Columbia Provincial Museum (now the Royal British Columbia Museum). As nothing definite has been seen of the species since about 1920, it is presumed to be extinct.

A similar situation surrounds the three-spined stickleback *(Gasterosteus aculeatus)* on the Queen Charlotte Islands. This fish is different from all other sticklebacks, exhibiting gigantism, complete loss of lateral plates, and loss of dorsal and pelvic spines. Analysis of DNA from sticklebacks suggests that this Graham Island fish diverged from ancestral marine forms more than 100,000 years ago – well before the last glacial advance. The evidence also suggests that the Queen Charlotte Islands subspecies was isolated in a freshwater refuge, although some genetic exchange may have occurred when sea levels rose about 8,000 years ago, submerging river valleys and giving marine sticklebacks access to the lake populations. Additional support for such a Queen Charlotte Islands refuge comes from plants, crustaceans, insects, and birds, as well as evidence that plants survived on Graham Island as early as 16,000 years ago.

British Columbia does not have Early Pleistocene vertebrates such as those found in the Old Crow Basin of the Yukon. Nor does it possess a thick layer-cake of sediment with vertebrate remains extending back to at least Middle Pleistocene time such as that in southern Alberta. It does, however, have rare, important specimens of alpine mammals; the most complete ice-age walrus skeleton from western North America; and an extremely rich and varied vertebrate fauna spanning the last 10,500 years from the Charlie Lake Cave site. Furthermore, there is growing evidence for important ice-age refuges on both Vancouver Island and the Queen Charlotte Islands. Who knows what paleobiological riches may be discovered tomorrow from British Columbia's ice-age deposits.

References

Cowan, I.M. 1941. Fossil and Subfossil Mammals from the Quaternary of British Columbia. *Transactions of the Royal Society of Canada,* 3rd series. 35 (Sect. 4):39-50

Driver, J.C. 1988. Late Pleistocene and Holocene Vertebrates and Paleoenvironments from Charlie Lake Cave, Northeast British Columbia. *Canadian Journal of Earth Sciences* 25:1,545-53

Harington, C.R. 1971. A Pleistocene Mountain Goat from British Columbia and Comments on the Dispersal History of *Oreamnos. Canadian Journal of Earth Sciences* 8:1,081-93

Harington, C.R., and G. Beard. 1992. The Qualicum Walrus: A Late Pleistocene Walrus *(Odobenus rosmarus)* Skeleton from Vancouver Island, British Columbia, Canada. *Annales Zoologici Fennici* 28:311-19

Harington, C.R., H.W. Tipper, and R.J. Mott. 1974. Mammoth from Babine Lake, British Columbia. *Canadian Journal of Earth Sciences* 11:285-303

Rutter, N.W., V. Geist, and D.M. Shackleton. 1972. A Bighorn Sheep Skull 9,280 Years Old from British Columbia. *Journal of Mammalogy* 53:641-4

22
Late Pleistocene Salmon of Kamloops Lake

Catherine C. Carlson and Kenneth Klein

Concretions eroding from glacial lake sediments along the south shore of Kamloops Lake contain bony skeletons of the only known salmonids of Late Pleistocene age in North America. The family Salmonidae includes salmon, trout, char, grayling, and whitefish. The late geologist and paleontologist Richard Hughes identified these fossils as *Oncorhynchus nerka* (sockeye salmon). Hughes originally collected at this site in the 1970s, acquiring specimens for the Cariboo College geology collection. His suggestion of a Late Pleistocene age for the salmon was undoubtedly based on the age of the host strata – a fine-grained, silty, grey clay extending along the Kamloops Lake shore from above the high-water line to below the low-water level.

This site has been known to local rockhounds for years, but the exact location had fallen into obscurity. Curiosity about the origin of Hughes's specimens led us, with the help of local resident Bill Huxley, to relocate the site in the fall of 1991. Between 1992 and 1994 we made several trips for collection and mapping, which resulted in the recovery of over 175 specimens, none representing a complete skeleton.

The age of these fish fossils is significant for reconstructing glacial environments at the end of the ice age because it suggests that the ice cover was much more limited than was commonly presumed, thereby allowing salmon to migrate into the watershed. Also, these fossils are substantially older than the oldest dated salmon from archeological sites, and hence are important for understanding the zoogeography of salmon and their potential for use by early humans.

The Fossil Salmon

Two kinds of concretions are found at the fossil site. The predominant type is roughly spherical in shape – cannonball-like with an equatorial ridge located at the plane of deposition (Figure 22.1A). These concretions are extremely hard and upon cracking show no evidence of organic remains. The centres appear somewhat coarser than the outer layers, and some show the star-shape mineralization common to concretions from other locales and ages. The other type is distinctly different. These concretions are relatively thin and take the shape of the salmon remains that they contain (Figure 22.1B). Both types of concretions weather out of the fine grey clay, which when wet resembles potter's clay in consistency and when dry becomes very hard. X-ray fluorescence analysis carried out on this host clay showed no abnormal elements that could affect the radiocarbon dating process. Except for the salmon, no other organic remains have been identified.

The best-preserved parts of the fish are those with the greatest bone mass – the skull area (Figures 22.2 and 22.3), backbone segments, and basal portions of the fins and tail (Figure 22.4). Many of the backbone segments include well-preserved rib sections. Although some concretions encase a single well-preserved fin, most contain body sections conspicuously lacking the distal portions of the fins. Only one specimen recovered to date appears to be over 90 per cent complete, with only the peripheral fin portions and tail missing. This specimen measures forty-two centimetres from the snout to what appears to be the base of the tail. Direct evidence of scavenging, such as tooth impressions or masticated bone piles, is not evident. Because so few specimens are nearly whole, however, it is possible that scavenging did occur, much as it does today in spawning environments.

Figure 22.1 **Concretions in Pleistocene clay on south side of Kamloops Lake.** (A) Hard cannonball concretions about 10 cm across. These concretions lack organic remains. (B) Irregularly shaped concretions. Such concretions may contain fossil salmon.

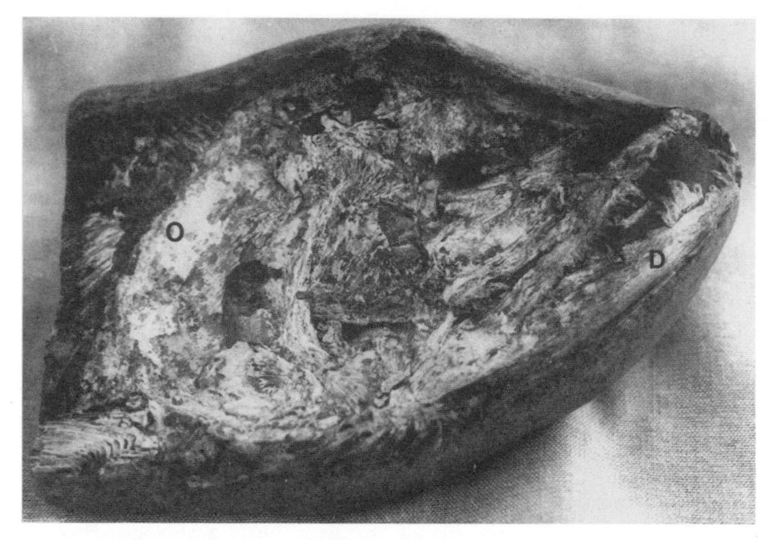

Figure 22.2 **Largest male salmon skull collected, 11.5 cm long.** Note the protruding hooked jaw, or dentary (D), with teeth and the opercule (O), or gill cover. *Source:* Collections of the Geology Department of the University College of the Cariboo, Kamloops

Figure 22.3 **Salmon skull 7.5 cm long showing the well-preserved nature of the cranial area.** Cranial area includes the eye socket, opercule, and dentary with teeth. *Source:* Collections of the Geology Department of the University College of the Cariboo, Kamloops

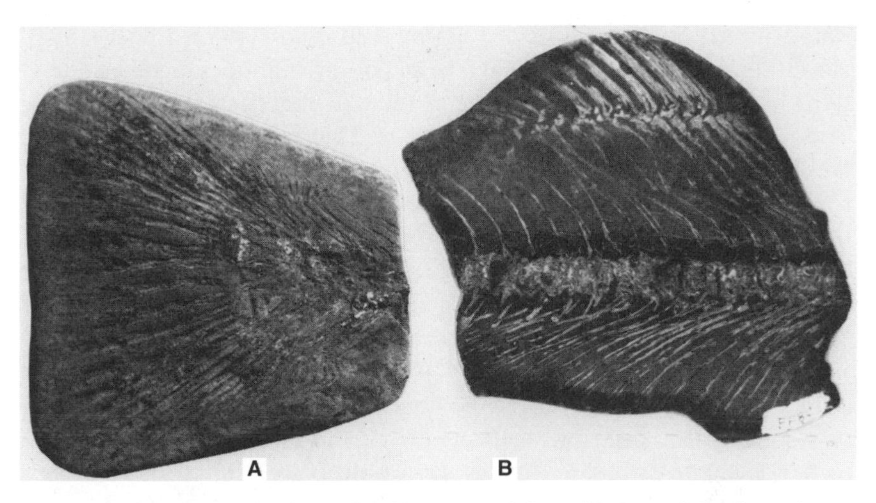

Figure 22.4 **Sections of salmon backbones and fins.** (A) One of the few salmon tail sections collected. Specimen is 7.5 cm long. (B) A vertebral section of a salmon with part of the dorsal fin preserved. Specimen is 9 cm long.
Source: Collections of the Geology Department of the University College of the Cariboo, Kamloops

Three species of salmon presently migrate to the Thompson River system: Coho, Spring, and Sockeye (including Kokanee, the landlocked form). The presence of a prominent hooked jaw on many of the specimens suggests that these were spawning fish that died along the lake shore, thus accounting for the concentration of fish in one locality, consistent with modern salmon-spawning behaviour. We agree with Richard Hughes that the fish in these nodules is *Oncorhynchus,* but we are uncertain of the species. The five living species of Pacific Salmon cannot be differentiated solely on the basis of external bone morphology. We suspect, however, from two lines of evidence, that these specimens are Kokanee, the landlocked and stunted form of Sockeye. First, the heads are small; none that we have collected is longer than 11.5 centimetres. Second, some evidence indicates that these fish never fed in the ocean. Chisholm and others (1982) have shown that the carbon-13 isotope ratio in bone can be used to distinguish marine and terrestrial sources of protein in an organism's diet. Carbon-13 measurements derived from bone collagen in the Kamloops Lake salmon samples show that these salmon did not eat protein from marine sources; therefore, they must have been a landlocked freshwater form.

Dating and Geology

Hughes considered the fossil salmon to be Late Pleistocene in age. Geological reconstructions of Pleistocene lake environments of the

Kamloops area have been carried out by Fulton (1969) and Tipper (1971). Nevertheless, the stages of glacial lake evolution and subsequent ice retreat remain undated, although it is estimated that deglaciation in the Thompson River region began 11,000 to 12,000 years ago in the Late Pleistocene (Fulton 1969, Clague 1989). Current thinking, therefore, is that environmental conditions in the vicinity of Kamloops were harsh in the Late Pleistocene. How could salmon have existed in an area believed to be covered with glacial ice? Did Hughes overestimate the age of the specimens? Are they really of Holocene age?

To answer these questions, we submitted samples for radiocarbon analysis of bone collagen. The fossil fish skeletons are not completely mineralized, and the presence of organic bone collagen permitted radiocarbon dating. Our first specimens, collected from the surface, were determined to be 15,480 years old. Since this age would place salmon in Kamloops Lake when the valley was presumably still full of ice, we suspected that the figure was too high, perhaps due to surface contamination of the sample. However, samples from buried specimens recovered in subsequent field work yielded an even greater age – 18,110 years old. These carbon isotope tests therefore confirm Hughes's initial age estimate, placing the samples firmly in the Late Pleistocene at a presumed glacial peak. The research of Carlson (1992) and of Hocutt and Wiley (1986) indicates that these remains are the only known salmonid fossils of Pleistocene age in North America. As such, they constitute a significant find.

Implications

The Late Pleistocene age of the Kamloops Lake specimens is important for understanding the paleozoogeography of salmon in British Columbia. It also has implications for dating Quaternary glacial events and for determining archeological reconstructions of past fish resources. The accepted hypothesis, as proposed by Fladmark (1975), is that salmon was not available in large quantities for human consumption in the Interior until the mid-Holocene (approximately 5,000 years ago). Prior to the discovery of the Kamloops Lake fossils, salmon remains older than 3,500 years had not been found in the Interior of British Columbia in either paleontological or archeological contexts. Cultural reconstructions of prehistoric Native peoples in the Interior plateau have correlated the presumed mid-Holocene expansion of salmon, an important resource for these people, with increased population sizes, the construction of large permanent villages, and the development of social complexity (Hayden and Spafford 1993). Our documentation of the presence of salmon in the Interior at least 10,000 years before the mid-Holocene may require the modification of this cultural model.

These salmon specimens have Late Pleistocene radiocarbon ages.

According to Fulton (1969), as the ice began to recede in a northwest direction from the South Thompson area, the main drainage channels were initially to the south through the Nicola area and then to the east to a developing Lake Thompson, which ultimately expanded to encompass the areas occupied by present-day Kamloops Lake and the southern portion of the North Thompson River. Isostacy – the uplifting of land that occurs over long periods after the immense weight of glacial ice has been removed – split Lake Thompson into Lake Deadman to the west and the ancestral Shuswap Lake to the east, leaving the two connected by the South Thompson River. Drainage was still to the east and it was through this system that the salmon must have entered Lake Deadman, the immediate predecessor of Kamloops Lake. Continued isostacy in the South Thompson ended the eastern drainage of Lake Deadman. Until isostacy later caused westward drainage to develop, the fish were isolated, a situation resulting in the appearance of landlocked, stunted salmon known as Kokanee.

With an 18,000-year-old radiocarbon age, these salmon fossils are the only salmonids of Late Pleistocene age known in North America. This date is significant for understanding glacial environments in the central Cordillera at the end of the last ice age because the presence of fishes at that time demonstrates that the Thompson Valley was not covered by glacial ice, as previously thought. Waterways for salmon migration at this time appear to have linked the Thompson River system with the unglaciated Columbia River system through the Okanagan Valley. These fossils and their age demonstrate that salmon have been available for human consumption for a much longer time than presumed by archeologists and, further, that humans would not have found the Late Pleistocene climate in southern British Columbia particularly harsh.

Acknowledgments

We are indebted to the University College of the Cariboo for funding the initial stages of this study through a Scholarly Activity Grant. We also thank Catherine Klein, Douglas Baleshta, and Donald Coolidge for their valuable assistance with the project.

References

Carlson, C.C. 1992. The Atlantic Salmon in New England Prehistory and History: Social and Environmental Implications. PhD dissertation, Department of Anthropology, University of Massachusetts, Amherst

Chisholm, B.C., D.E. Nelson, and H.P. Schwarca. 1982. Stable Carbon Isotope Ratios as a Measure of Marine versus Terrestrial Protein in Ancient Diets. *Science* 216:1,131-2

Clague, J.J. 1989. Quaternary Stratigraphy and History: Area of the Cordilleran Ice Sheet. In *Quaternary Geology of Canada and Greenland,* ed. R.J. Fulton, 48-58. Vol. 1 of *Geology of Canada*

Fladmark, K. 1975. *A Palaeoecological Model for Northwest Coast Prehistory.* Mercury Series, National Museum of Man, Ottawa, No. 43

Fulton, R.J. 1969. *Glacial Lake History, Southern Interior Plateau, British Columbia.* Geological Survey of Canada, Paper No. 69-37

Hayden, B., and J. Spafford. 1993. The Keatly Creek Site and Corporate Group Archaeology. *BC Studies* 99:106-39

Hocutt, C.H., and E.O. Wiley, eds. 1986. *The Zoogeography of North American Fresh Water Fishes.* New York: John Wiley and Sons

Tipper, H.W. 1971. *Glacial Geomorphology and Pleistocene History of Central British Columbia.* Geological Survey of Canada, Bulletin 196

23
Quaternary Plants: Glimpses of Past Climates and Landscapes

Richard J. Hebda

Many people think of fossils as remains or traces of life that have to be hammered or chipped out of rocks. Indeed, most of the spectacular fossils like dinosaurs and petrified tree trunks are uncovered this way. Yet fossils begin their 'lives' once organisms die and become entombed in some type of soft sediment, such as lake or marine muds, river gravels, or bog peat. From the moment they become buried, these remains are technically 'fossils' – remains of once-living organisms.

Billions of tonnes of sediment accumulate every year in British Columbia, burying countless fragments of plants. Remains from the past two million years, the period of time known as the Quaternary, record the momentous changes that altered the life, land, and climate of British Columbia, as sheets and tongues of ice covered and licked the landscape many times. Almost no sediments and fossils of the first 1.9 million years of this two-million-year period are known. As we approach the present time, more fossil deposits preserve the detail of plant life that once covered British Columbia. This chapter describes the types of Quaternary plant fossils found in this province and the sites that contain them. It also gives a selective account of changes to the vegetation throughout the interval, as revealed by plant fossils.

Plant Macrofossils

Quaternary paleobotanists generally divide plant fossils into two categories: those that can be seen with the unaided eye, called plant macrofossils; and those that must be examined with a microscope, called plant microfossils.

Macrofossils are those parts of plants with which people are most familiar and which are immediately recognizable in a sediment exposure. As might be expected, they consist of pieces of wood, seeds, fruits, cones, leaves, and needles. Most often macrofossils become buried and preserved in peat, clay, or silty sediments deposited close to the original plant communities. The best macrofossils are found in sediment deposited in floodplains, deltas, lakes, and other wetlands.

In wetland settings such as bogs and swamps, plant parts simply fall to the wet ground, where decomposition is slow or absent, and become buried by more plant parts and occasional mineral sediments. In lakes, plant remains fall into the basin, and are blown in by wind or washed in by streams. Parts of aquatic plants, in particular, fall to the lake bottom becoming entombed in the ooze. Floodplain and delta environments often preserve great quantities of plant remains, especially wood. A visit to a floodplain or delta during a spring or summer flood shows how much wood is afloat and washed into drainage systems to become sediment buried in its lowest reaches. These sediments are exposed when sea levels and lake levels drop, and watercourses and waves cut into the deposits.

Aside from this normal sediment burial of macrofossils, there is the peculiar preservation of plant parts by woodrats (commonly called packrats) in middens and nests. In British Columbia, bushy-tailed woodrats *(Neotoma cinerea)* construct nests and produce middens of plant debris, which are cemented together by solidified amberat, the concentrated urine of the rats. In dry climates, like those of the southern Interior, the plant material in the middens and nests is perfectly preserved. Needles, seeds, and other parts can be removed by soaking the middens in warm water. So far, BC middens have yielded plant remains only a thousand years old, but similar middens in the southwest United States reach back more than 20,000 years.

Plant Microfossils

Microfossils are tiny but recognizable bits of plant matter such as pollen grains and spores, the walls of algae and fungi, and the remains of microscopic animals, all invisible to the unaided eye but easily identified under a microscope. Of the two plant fossil categories, microfossils are more widespread and more informative, occurring in many kinds of sediment. Pollen provides a good example. Each year, at least one airborne pollen grain or spore lands on almost every spot of the land surface of British Columbia. As long as there are sediments with the right chemical makeup to bury that pollen grain quickly, it will be preserved.

Microfossils occur in thousands of different forms. The most common of these are:

- pollen grains produced by advanced vascular plants (mainly flowering plants and conifers)
- spores produced by many groups of primitive plants
- reproductive structures of fungi
- walls of microscopic algae.

Pollen grains are the male reproductive bodies of advanced plants. Their outside walls are remarkable in that they are composed of a nearly indestructible substance called 'sporopollenin,' which ensures the preservation of these bits under many kinds of circumstances. Even though the living insides often are eaten by other organisms, the sporopollenin coat survives.

Most important, different types of plants produce different types of pollen grains, which can be identified readily (Figure 23.1). Pollen grains have distinctive shapes, sizes, structural features, and especially, sculpturing. Many coniferous trees, for example, pine and spruce, produce pollen with floats or bladders. Several types of deciduous trees produce pollen grains with openings or pores.

Variation occurs not only in the basic structure of pollen grains but also in their surface features (Figure 23.2). Pollen of the hardhack *(Spiraea douglasii),* for example, a member of the rose family, often bears numerous ridges. Using images of 8,000-year-old hardhack pollen obtained from a scanning electron microscope, graduate student Greg Allen and I discovered that this swamp species grew in the bottom of what is today a lake on south Vancouver Island. Because swamp plants cannot grow underwater in the bottom of lakes, the climate must have been much drier than it is today.

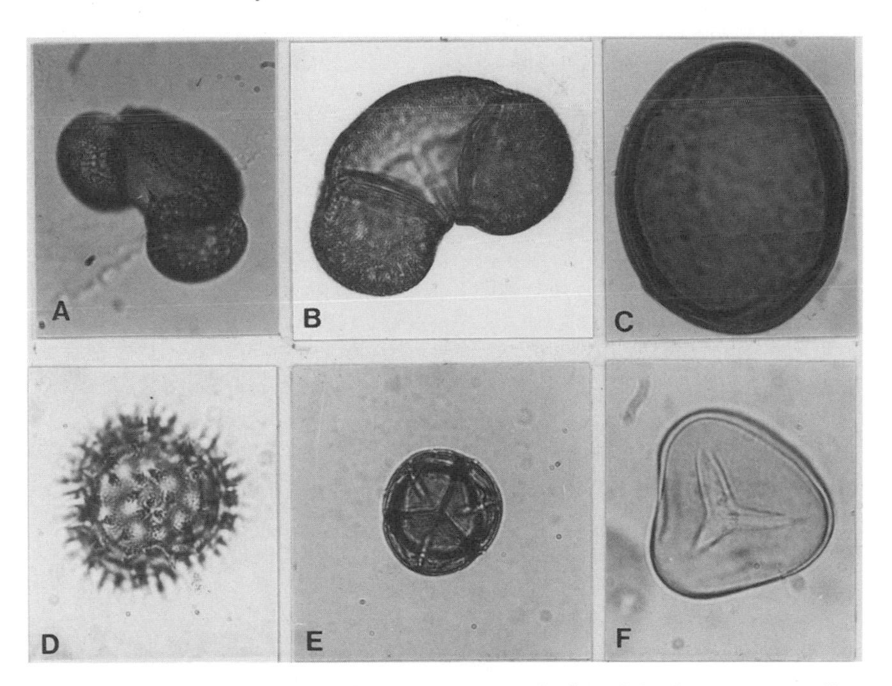

Figure 23.1 **Selected pollen and spores commonly found in Quaternary sediments of British Columbia.** (A) pine (B) spruce (C) Douglas fir (D) aster family (E) heather family (F) bracken fern spore

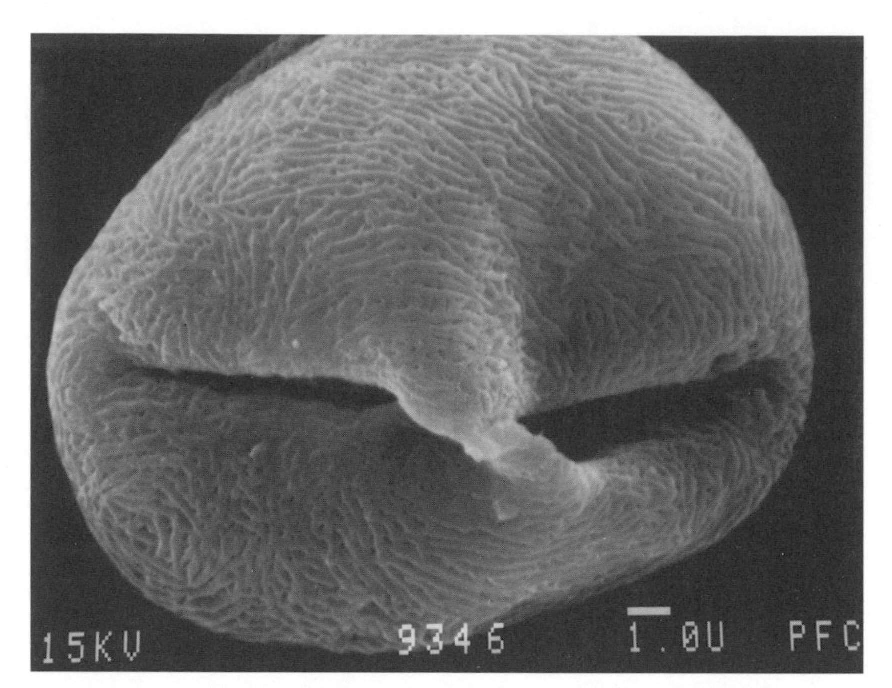

Figure 23.2 **SEM photograph of 8,000-year-old pollen of hardhack *(Spiraea douglasii)* from lake-bottom sediments on south Vancouver Island**
Source: Photograph by G. Allen

Spores are the single-cell asexual reproductive bodies of primitive plants and fungi. Like pollen grains, spores vary distinctly in basic form and in sculpturing. Some are kidney-shaped, whereas others bear prominent three-armed marks. Microfossils from algae also come in a variety of curious but distinctive forms. For information on plant microfossil analysis, see the publications by Moore et al. (1991) and by Faegri and Iversen (1975).

Like macrofossils, microfossils are best preserved in peats, clays, and silts. Microfossils also survive in organic soils, however, and will be preserved if those soils become buried. Unlike macrofossils, which are concentrated near the source of the material, microfossils are easily transported because of their tiny size and thus can be preserved in any part of a body of water, including the middle of large lakes and the bottom of the ocean.

The identification and study of microfossils requires a microscope that will magnify the sample to 400 times its actual size. Those learning the technique usually start by collecting living pollen from a tree and putting it in water on a microscope slide. A thin cover slip is added and the specimen is examined under the microscope. The outline of the pollen grains is usually visible, although not the inside, which may contain living tissue.

To study fossil pollen and spores, the researcher scoops up a sample of muck peat from a swamp or bog, or plunges a tube deep into the muck at the bottom of the lake. A bit of the muck from the retrieved tube is smeared on a microscope slide, a cover slip is added, and the slide is examined under appropriate magnification. As long as the material under the cover slip is spread thinly enough, a variety of microscopic plant debris, including pollen grains, will be visible among the finer particles. We normally use a series of strong chemicals to remove all of the fine debris, leaving behind mostly pollen and spores.

Finding Quaternary Plant Fossils

Microfossils usually occur anywhere there is sediment. Every lake, swamp, bog, fen, and year-round wet soil collects microfossils as part of a process that has been ongoing for more than 10,000 years. Even the limy deposits around hot springs and the debris in caves may contain preserved microfossils. In almost all of these situations, you will find the 'Mickey Mouse' hat-shaped pollen grains of the common conifers such as pine *(Pinus)* and spruce *(Picea)* (Figures 23.1A and 23.1B). The pollen of these trees travels great distances because of the bladders or floats attached to the main body. Conifer pollen of this kind occurs not only in Quaternary sediments but also in rock as ancient as the Permian.

Macrofossils occur in the same settings as microfossils, for example, lake and bog sediments. It is generally more difficult to obtain good samples of macrofossils than of microfossils, because exposures or special coring equipment are needed. The most accessible sites are traditional geological exposures such as excavation faces, river banks, and cliff faces, from which plant macrofossils can be extracted with a shovel or hammer.

In British Columbia the chance of finding Quaternary plant macrofossils is greatest along the sea coast. In this setting, waves erode cliff faces, exposing beds with macrofossils. Occasionally waves slowly expose macrofossil beds in the intertidal zone as they cut a platform. Several such situations occur on the south end of Vancouver Island where plant beds several thousand years old crop out in the intertidal zone. Sometimes the stumps of ancient forests can be seen emerging from the beach surface. By clearing away the thin layer of sand and gravel over the plant beds, we may even see remnants of the ancient forest floor.

What Quaternary Plant Fossils Tell Us

Quaternary plant fossils, especially pollen grains and spores, provide one of the most powerful tools for understanding the history of modern terrestrial ecosystems and the processes that shape them. Sooner or later any change in the landscape changes the plant species that grow there.

Climate is the primary factor that controls the vegetation. As a result, plant fossils provide a record of changes in climate. Quaternary plant fossils document a time when climatic change was particularly dramatic. During the present time of uncertainty about future global climate, many scientists are studying Quaternary microfossils to gain a better understanding of the processes and rates of climate change. By simply documenting the history of vegetation in a region, we gain insight into the origin and longevity of ecosystems. Although much of the coast of British Columbia today is covered by cool, temperate coniferous rain forests of a similar nature, for example, these rain forests arose in different ways from region to region.

Micro- and macrofossil studies also tell us when various plant species came to the province and by what route (Ritchie 1987). Although all the major tree species of south Vancouver Island have been there for at least 10,000 years, Garry oak *(Quercus garryana),* for example, did not arrive until 7,000 to 8,000 years ago.

Microfossils also provide insight into changes in landforms. Pollen analysis of peaty silts and clays along the coast reveals distinct golf-ball-like pollen grains of the goosefoot and amaranth families (Figure 23.3). These pollen grains originated from plants such as saltworts *(Salicornia),* which grew in saltmarshes. Thus they indicate sea levels occurring since the last glaciation that were higher than present-day sea levels.

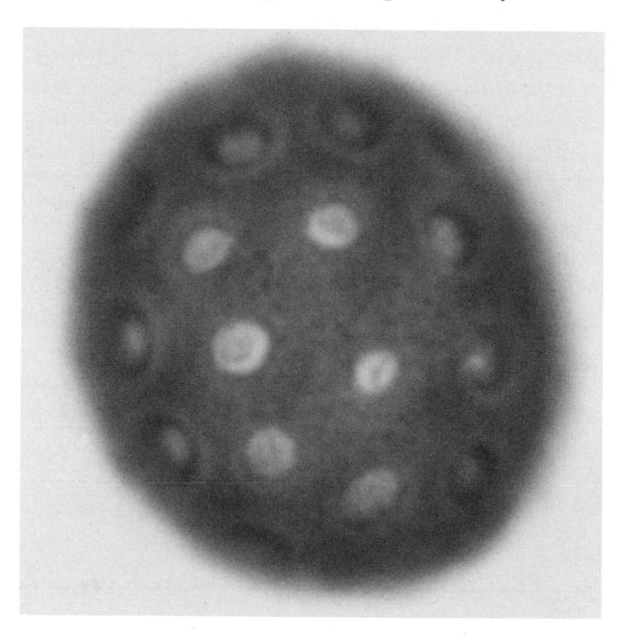

Figure 23.3 **Goosefoot family pollen, an indicator of saltmarshes in sediments of coastal British Columbia**

Quaternary Vegetation of British Columbia

Most of Quaternary time is poorly represented by plant fossils. Many sites and studies record only the last 15,000 years. Nevertheless we know that the plants from the Quaternary resemble those growing in British Columbia today, but their distribution has changed frequently and dramatically.

Our only insight into the first half of the Quaternary in British Columbia comes near Dog Creek, south of Williams Lake on the Fraser River. Sediments from this area were studied by Glenn Rouse and Bill Mathews, emeritus professors of the University of British Columbia. A wide range of pollen grains from conifers and flowering plants is preserved within these deposits. Pollen from pine, spruce, western hemlock, true fir, and larch occurs in combination with that of goosefoot especially, as well as of alder, birch, and sage. A variety of fern spores is also recorded. Despite the diversity of conifers, the vegetation is interpreted to have been tundra-like, as confirmed by the associated evidence of glacial activity.

A peek into slightly younger Quaternary environments comes from the work of Neville Alley and Stephen Hicock of the University of Western Ontario, who studied sediments exposed in sea cliffs near Sooke on Vancouver Island. These sediments are interpreted to represent an interglacial interval, perhaps the last one in North America, called the Sangamon Interglacial. The earliest layers record abundant Douglas fir pollen and indicate a grassy floodplain bounded by dry hills covered in nearly pure Douglas fir stands, with cedars and western hemlocks in moist settings. Cones of Douglas fir, western red cedar, and Sitka spruce from the sediment beds confirm the tree species present. Moisture-loving species such as alder, western red cedar, cattails, and yellow pond lily that appear later in the interglacial interval record expansion of the wetland environments.

Several exposures in the Fraser Lowland and the east coast of Vancouver Island reveal life during the major non-glacial interval, about 30,000 to 50,000 years ago, before the last series of ice advances. Stephen Hicock and I discovered clubmoss, grass, and diverse herb pollen and spores that indicate a treeless terrain similar to alpine and subalpine meadows for the area when the interval began. Soon after, lodgepole pine woodland covered the land, followed quickly by spruce and mountain hemlock forest. About 40,000 years ago the forests consisted of western hemlock, mountain hemlock, and lodgepole pine, which grew under a moderate but moist climate. This climate was not quite as warm as present-day conditions, however. Soon after, as the climate cooled, western hemlock declined, and spruce and mountain hemlock prevailed. At the end of the interval cold meadows returned, as glaciers began to build up again.

Several sites, again mainly from southwestern British Columbia, reveal that 20,000 to 30,000 years ago the landscape was covered mainly by open-terrain plant communities. Rolf Mathewes of Simon Fraser University found abundant sedge, grass, and herb pollen at the site of the University of British Columbia, indicating sedge-dominated wetlands on a floodplain, with rich and diverse herbaceous communities on higher land.

Sediments from Port Moody containing both macrofossil and microfossil remains provide a brief glimpse of plant communities and conditions 17,000 to 18,000 years ago, just before the last great ice sheet buried the land. The area was covered by forests of subalpine fir *(Abies lasiocarpa)* and spruce and by parkland, indicating that cold, humid continental climate predominated. The mean annual temperature must have been about eight degrees Celsius colder than today and the tree-line 1,200 to 1,500 metres lower. Abundant needles of Pacific yew *(Taxus brevifolia)* indicate that this relatively uncommon tree formed extensive thickets. This discovery of the distinctive Pacific yew needles was particularly fortunate because pollen of Pacific yew is rarely found.

Investigations of sea cliff exposures on Graham Island in the Queen Charlotte Islands provide insight into conditions at the time when ice covered much of British Columbia. According to Rolf Mathewes, who has spent many years studying sites on the Queen Charlotte Islands, patchy meadows and bare soil covered the land from 16,000 years ago to 13,000 years ago. The recovered seeds and pollen provide insight into the diversity of herb types, including a member of the pink family (Caryophyllaceae), as well as dock *(Rumex)* and pearlwort *(Sagina)*. Micro- and macrofossils such as net-leaved dwarf willow *(Salix reticulata)* in this interval and slightly later show that plants typical of alpine environments once grew near sea level and that the climate was likely much colder than today.

Studies of lake and bog sediments disclose the history of our landscape since the melting of the last ice sheets about 13,000 years ago (Ritchie 1987; Pielou 1991). Pollen and spores have been identified and counted from a series of samples representing levels in a core or exposure (Figures 23.4 and 23.5). Along the coast, lodgepole pine forests predominated during the first 2,000 or so years of this recent period. The east side of Vancouver Island seems to have been more open than other regions. Near Sidney, large now-extinct bison roamed a landscape that may have been covered by aspen groves. At the same time the southern Interior of British Columbia was open, possibly covered by sage, grasses, and scattered mixed shrubs on upland sites. Cattails and sedges grew in some of the lakes. The first tree to appear at several of the sites may have been aspen. Nevertheless, large glacial lakes and masses of ice remained in the heart of the Interior plateau. Conifers, predominantly pine, only began to appear between 10,000 and 11,000 years ago.

Figure 23.4 **Section of lake sediments exposed by draining and excavation of Heal Lake near Victoria.** The deposits consist almost entirely of plant micro- and macrofossils, and record vegetation and climate since the melting of Pleistocene ice more than 13,000 years ago. The white horizon in the middle of the exposure is the Mazama Ash, deposited as air fall when a volcano at the present-day site of Crater Lake in Oregon erupted 6,800 years ago.

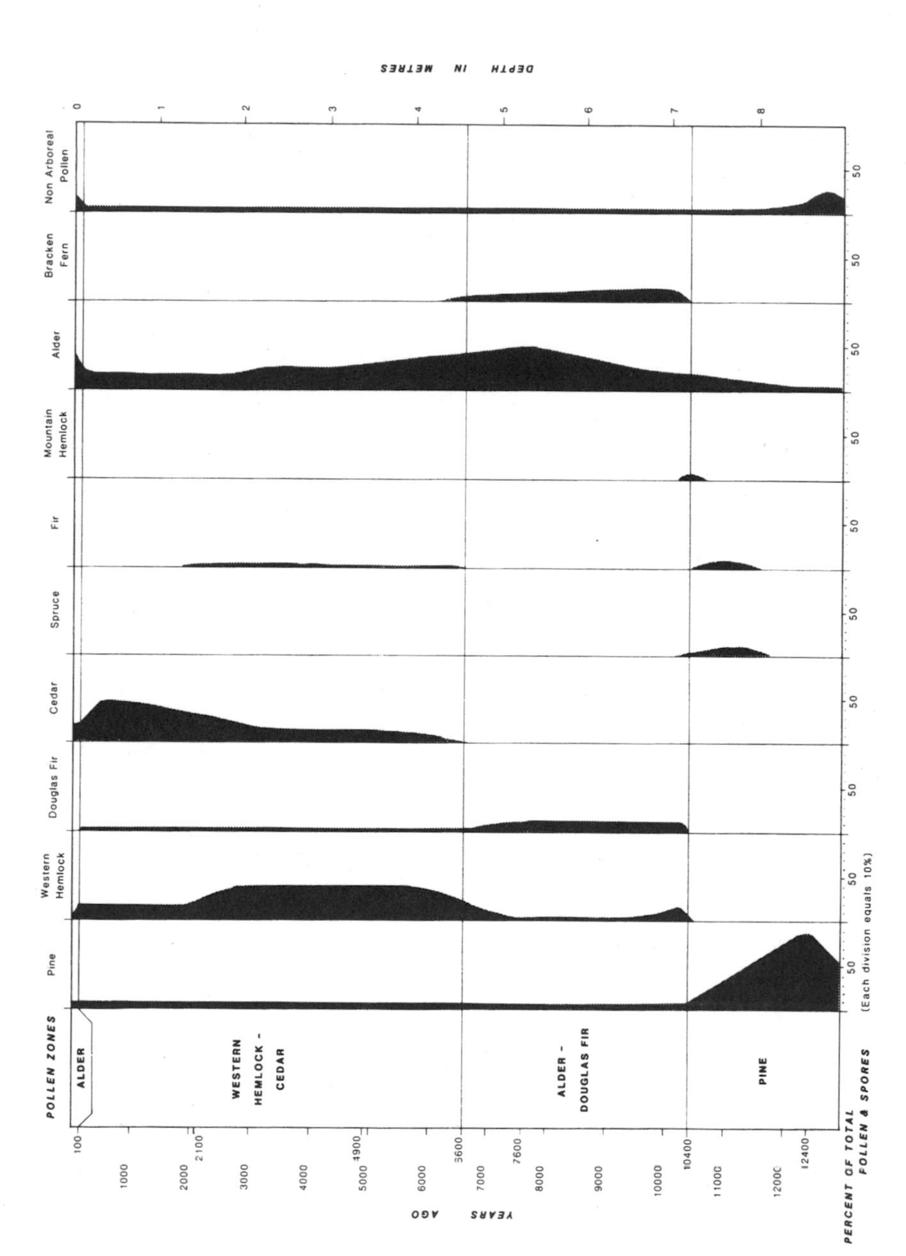

Figure 23.5 **Simplified pollen diagram from Fraser Valley based on a core from Marion Lake.** Diagram is the result of studies by Rolf Mathewes of Simon Fraser University. Only the major species have been included. Zone names are derived from dominant pollen types.

About 10,000 years ago major rapid warming brought about significant adjustments in the vegetation. On the southern coast Douglas fir, recognizable by its distinct sphere-like pollen (Figure 23.1C), spread along the coast, becoming a major forest element as far north as northern Vancouver Island. Familiar conifers like western hemlock and spruce occurred 7,000 to 10,000 years ago but played different roles from those they do today. Spruce, for example, assumed to be Sitka spruce, was a major forest species on northern Vancouver Island, whereas western hemlock was much less abundant than it is today. On the southeastern side of Vancouver Island there may have been little forest cover at all; instead, grasses and wildflowers likely predominated. Western red cedar, now a common coastal tree species, seems to have been only a minor constituent of the forests.

East of the Rocky Mountains, the Early Holocene was a time of hot and dry climate. Many of the valleys were covered by grassland or sagebrush. Open vegetation probably extended up the slopes into the present-day alpine zone. Many small lake basins held little or no water, being no more than seasonal ponds. The principal conifer species of today grew in the region but mainly at high elevations or on moist north-facing slopes.

Over the course of 3,000 years, from 7,000 to 4,000 years ago, forest composition changed as the area of open terrain shrank. On the coast, first western hemlock and later western red cedar flourished, at the cost of Douglas fir and spruce. Lake basins began to fill with water and wetlands expanded. In the rain-shadow of south Vancouver Island, Garry oaks formed a major part of the tree cover. Increased moisture and cooling encouraged the forests to expand down slopes and into mid-elevation valley bottoms in the Interior of British Columbia. Grassland and sagebrush plant communities shrank. Lakes expanded, and the landscape took on an appearance much as it has today.

Until recently plant macro- and microfossils helped us reconstruct Quaternary environments for intervals of hundreds of years or more. Now, however, new work is providing information on a much finer scale. Excavation of sediments at Heal Lake near Victoria has revealed hundreds of superbly preserved logs (Figure 23.6), some nearly 10,000 years old. Features of the tree rings of these logs provide significant information about climate and other events on an annual basis. It is likely that most lakes in the coastal and interior regions of British Columbia contain equally great troves of ancient wood, although not easily accessible. Through the study of logs from Heal Lake and perhaps others, we may come to understand climate and climatic processes on a detailed scale never before possible in northwestern North America.

Much remains to be discovered about Quaternary life in our province. Thousands of sites containing billions of plant fossils await study. Every

Figure 23.6 **Cross-section of 10,000-year-old fossil log of Douglas fir recovered from bottom of Heal Lake near Victoria**

lake tells a story. Once we understand the processes of our recent past and how they shaped British Columbia today, we may be able to prepare better for events and changes that lie ahead.

References

Faegri, K., and J. Iversen. 1975. *Textbook of Pollen Analysis.* New York: Hafner Press

Moore, P.D., J.A. Webb, and M.E. Collinson. 1991. *Pollen Analysis,* 2nd ed. Oxford: Blackwell Scientific Publications

Pielou, E.C. 1991. *After the Ice Age: The Return of Life to Glaciated North America.* Chicago: University of Chicago Press

Ritchie, J.C. 1987. *Postglacial Vegetation of Canada.* Cambridge: Cambridge University Press

24
Epilogue: The Cordillera through the Mists of Time

Christopher R. Barnes

During the last two summers I have had the opportunity to do paleontological field work in the remote northern Interior of British Columbia. The first summer involved collecting for conodont microfossils in the northern Rocky Mountains along the high-alpine ridge crests in Kwadacha Wilderness Park. Last summer the research was extended farther northwest, where collections were made on Gataga Mountain and then across the Rocky Mountain Trench to the Cassiar Platform – a segment of crust that has moved some 700 kilometres from the southeast along a linear fault system. The most exposed areas are always the best collecting sites, so the helicopter pilot dropped us off above the tree-line, high in the clouds at about 2,300 metres. We were at least 200 kilometres from the nearest road and surrounded by one mountain range after another as far as the eye could see. This pristine world, scarcely marked by human foot, is composed almost entirely of sedimentary strata with potential for yielding new and significant fossil discoveries. It was truly impressive to see such a vast, unexplored expanse of British Columbia, let alone to consider that its paleontological heritage had not yet been revealed.

This past summer, while in the main base camp, my colleagues working for the BC Geological Survey discovered a number of well-preserved graptolite fossils in rocks thought to be of Devonian age. Graptolites are complex, extinct colony-forming animals that have a deceptively simple pencil-streak appearance when preserved in shale. They were found by a particularly enthusiastic student field assistant from the University of Victoria who made a habit, if not a religion, of systematically scouring each and every rock surface for fossils whenever an opportunity arose. Accepting an invitation to examine the fossils, I decided, however, that they were much older – about eighty million years older – and likely of Early Ordovician age, an estimate later confirmed by a graptolite specialist in the Geological Survey of Canada. This student's discovery, therefore, suggested a very different stratigraphic and structural interpretation for the area.

A similar experience occurred three summers ago when I was sampling for Lower Paleozoic conodonts in the southern Cassiar Platform, west of Williston Lake in north-central British Columbia. One of the province's largest mining companies was searching for lead-zinc deposits, and I was given permission to sample many of the cores for conodonts. Initially I was advised that the rocks were of Silurian and Devonian age. During our sampling program, however, an eagle-eyed senior company geologist spotted an archeocyathid fossil in one core near the top of the section. This distinctive sponge-like fossil only occurs in Lower Cambrian rocks. Again, an unanticipated fossil discovery forced a drastically different geological interpretation for a region.

I use these recent experiences to illustrate that paleontology in British Columbia is still largely in a pioneering stage, with much to discover yet much to offer our sister earth-science disciplines. These examples also illustrate how enthusiasm, persistence, and knowledge can lead to important fossil discoveries.

Through the Mists of Geologic Time

Much of the landscape art of the province portrays coastal mountains shrouded in mist, with successively more distant ranges fading into different tones of blue-grey. To me this image is analogous to the province's paleontological record, which gives a reasonably complete picture of recent geologic life and progressively more obscure views of life in the distant past. These receding mists through geologic time are the theme for this epilogue.

Geologically, the province retains some ancient rocks in its eastern portion, which represents the edge of Laurentia (the North American craton) after the rifting of the ancient supercontinent of Rodinia some 750 million years ago. Eastern Australia and Antarctica were then located where much of British Columbia now lies, but they and other crustal blocks rotated away from Laurentia to form a new huge continent – Gondwana – later in the Paleozoic. The newly rifted margin of Laurentia was passive for many millions of years but then went through sequences of extensions and contractions, which created a series of mountains. These mountains, however, were but a pale shadow of those that developed later during the Mesozoic and Cenozoic eras, when successions of microcontinents and island arcs moving across the Proto-Pacific collided with the margin of Laurentia. From knowledge of the rate of sea-floor spreading and the amount of time necessary for subduction along the west coast, it has been estimated that over 13,000 kilometres of ancient sea-floor has been subducted under the region since the beginning of the Cretaceous. The western edge of North America became the repository for all the sweepings of microcontinents and scraped-off sediment, which were plastered successively to form a new outer margin to the continent. These collisions were rarely head-on; instead they were oblique and followed by lateral fault systems that allowed large areas to slide northwest toward Alaska, like giant cruise ships.

Much of British Columbia now consists of a series of large and small exotic (or suspect) terranes, some of which merged to form large superterranes out in the Paleo-Pacific Ocean prior to their eventual collision with Laurentia. The land that now makes up British Columbia has, therefore, been derived from several directions and by various styles of deformation at different times in the geological past to create a wonderfully diverse and supernatural landscape, within which are buried treasure troves of fossils.

Together, the chapters in this book have charted the amazing diversity of these fossil treasures and their occurrences across the province. This diversity is a boon to fossil collectors, yet the very abundance and variation displayed by ancient life makes it difficult to discern the broader patterns of evolutionary change and the underlying paleoecological and paleobiogeographical patterns. In truth, British Columbia is a place of both feast and famine for paleontologists and fossil collectors. This volume describes numerous exceptional fossil discoveries, all of which were made despite the fact that British Columbia presents many obstacles to paleontologists – it is host to the largest body of granite in the world; it has extensive belts of altered metamorphic rocks resulting from terrane collisions; and it is home to huge tracts of dense forests that obscure outcrops. Much of the province, therefore, is inaccessible to collectors and paleontologists.

The Phanerozoic Pattern

In this concluding chapter, key fossil discoveries of this province are placed within the context of the evolution of life through time. Nearly two hundred years ago, early professional geologists and paleontologists recognized the major divisions in the history of life; and somewhat later, these intervals became the criteria for establishing the Paleozoic, Mesozoic, and Cenozoic eras of the geologic time scale. Since then, the record of Cryptozoic life has been outlined, and the successive changes in metazoan (multicelled) life have been named the Cambrian, Paleozoic, and Modern evolutionary faunas by Jack Sepkoski of the University of Chicago.

Good Cryptozoic fossils are difficult to find here. Rocks of this age are largely confined to the eastern third of British Columbia and comprise thick sequences of sandstones and conglomerates that filled the rift valleys as the supercontinent Rodinia broke apart and Laurentia drifted free. Unstable sediments and rapid subsidence conspired to conceal fossil occurrences. Better discoveries are now being made in more inland areas in the Mackenzie Mountains of the Northwest Territories and around Waterton Lake and Glacier national parks in Montana and southern Alberta. Here, algal stromatolites (fossilized mats of algae) and trace and body fossils are contributing significantly to the global picture of early evolution of algae and metazoans.

The Cambrian Evolutionary Fauna began to develop in the Vendian, diversified in the Cambrian, and faded rapidly in the Early Ordovician, when it was replaced by the more diversified Paleozoic Fauna. The Cambrian Fauna is dominated by a wide variety of arthropods, together with early experimental forms of several different phyla. Cambrian trilobites are known in the Cordillera, especially from the Cranbrook area. A window into the Cambrian world was first provided by Charles Walcott's discovery of soft-bodied faunas in the Burgess Shale near Field, now in Yoho National Park. This fauna is of Middle Cambrian age. Similar discoveries in the Lower Cambrian of China and Greenland over the last few years document the nature and evolution of early metazoan life and the development of mineralized exoskeletons. Today, British Columbia is a mecca for such studies. The painstaking studies first organized by the Geological Survey of Canada and by the Royal Ontario Museum have documented fossil sites at several different stratigraphic levels. These studies help to clarify the evolutionary relationships among many bizarre groups of animals. The book *Wonderful Life* by S.J. Gould, published in 1989, greatly publicized the Burgess fauna. The locality has been designated a World Heritage Site by UNESCO, where collecting is only possible by special permit. Now, a new research field station and museum are being planned for Field to aid in future work and in communicating the significance of these discoveries to the public.

The cause of the rapid replacement of the Cambrian Evolutionary Fauna by the Paleozoic Evolutionary Fauna remains a principal paleontological enigma. The number of families in the geological record almost quadrupled over the seventy-million-year interval of the Ordovician Period. Thereafter, diversity seemed to stabilize through the rest of the Paleozoic until the profound extinction at the Permian-Triassic boundary. In British Columbia, representatives of the Paleozoic Fauna are found in Ordovician and Silurian strata in the Rocky Mountains, for example, the spectacular stromatoporoid (sponge-like) colonies at Top-of-the-World Provincial Park near Cranbrook. The new fossil groups found in these strata are biostratigraphically important: conodonts abound in most carbonates; graptolites occur in many black shales (for example, the Glenogle Formation in the southern Rocky Mountains); and trilobites, brachiopods, and corals are particularly common in the Lower Silurian Nonda Formation south of Muncho Lake in the northern Rockies.

The younger phase of the Paleozoic Evolutionary Fauna is well represented in British Columbia, because the fossil record of many exotic terranes in the western half of the province starts in the Devonian. Many of the Carboniferous and Permian rocks originated well out in the Proto-Pacific Ocean, and some have strong Asiatic faunal affinities. Conodont, foraminiferan, and radiolarian microfossils have become extremely important in resolving the structural and stratigraphic relationships among the various terranes of the Cordillera. More endemic fossils, such as brachiopods, corals, crinoids, and mollusks, are of particular value in determining the original location of the microcontinents in the Paleo-Pacific Ocean. Middle and Late Paleozoic faunas occur in many parts of British Columbia. They include fishes from Devonian beds near Muncho Lake; fusulinid foraminiferans and radiolarians from Carboniferous and Permian rocks near Cache Creek, Liard River, and Wapiti Lake; and diverse shelly fossils and conodonts from Permian limestones in Strathcona Provincial Park on Vancouver Island.

The term Modern Evolutionary Fauna refers to fossils of largely Mesozoic and Cenozoic age, although ancestors of many groups can be traced back to the Lower Paleozoic. This fauna diversified rapidly through the Mesozoic and was affected by mass extinctions at the top of the Cretaceous. The Modern Fauna is characterized not only by a high species diversity among invertebrates but also by diversification among plants and vertebrates.

Throughout the world, the Early Triassic is an interval of low diversity and low abundance of nearly all types of fossils, as a result of the Permian-Triassic extinctions. The recovery phase later in the Triassic is well preserved in British Columbia. The Lower and Middle Triassic marine rocks near Wapiti Lake have yielded a remarkable suite of fossil fishes, sharks, coelacanths, ichthyosaurs, and other marine reptiles. Triassic strata in the Peace River area, superbly exposed in the high banks above the river and near Williston Lake, have produced the most complete sequence of ammonoid and conodont faunas for this time interval found anywhere in the world. Higher strata of Early Cretaceous age revealed extensive trackways of dinosaurs. Although these impressive discoveries suffered the fate of drowning when Williston Lake was created, energetic staff and willing helpers from the Royal Tyrrell Museum of Palaeontology managed to map many and to excavate a few for posterity. Important marine successions of Triassic and Jurassic fossils are present on the Queen Charlotte Islands, but these are hard won from the low coastal cliff exposures.

The Jurassic and Cretaceous interval was a period of major mountain building and granitic intrusion in the Cordillera. New sedimentary basins formed outward along the coast. Their tilted strata are particularly well exposed in the Gulf Islands, where rocks of Late Cretaceous age can be sampled through a spectrum of shallow to deep-water environments. A diverse fossil assemblage has emerged, including ammonoids, clams, gastropods, lobsters and crabs, fishes, elasmosaurs, and mosasaurs, as well as plants in the shallower environments toward the west.

Tectonic forces and terrane collisions in the late Mesozoic converted the western margin of North America into an emergent mountain belt. Alternating periods of tectonic extension and uplift in the Cenozoic created internal depositional basins with non-marine deposits and eventually produced the high mountain ranges that today reach elevations exceeding 2,500 metres. These rising mountain belts separated the open Pacific Ocean from a great interior seaway in the Prairie region that extended from the Gulf of Mexico to the Arctic Ocean. Into this seaway fell huge volumes of sediments stripped off the rising mountains. Marine and non-marine deposits here yield the amazing ammonoid and dinosaur remains so characteristic of the Alberta Badlands. Within the mountain basins in British Columbia, remarkable suites of insects, plants, and freshwater fishes have been recovered. Eocene deposits are common, although other series of the Cenozoic are also represented. Together, these deposits provide key evidence for the different altitudes and climates that prevailed in the Cordillera during this time of dynamic uplift. These changes correlate with the climate record interpreted from the rich vertebrate and plant fossil records in the Prairie Provinces, where there was increasing aridity through the Cenozoic, as the warm, wet westerlies were forced to drop their moisture along the western front of the rising Cordillera. Along this margin other deep- and shallow-marine sedimentary basins developed. Many are now submerged under the Pacific and coastal waters, but some have been thrust above sea level, notably on Vancouver Island. From Sooke, west along the Pacific Rim National Park, coastal cliffs and wave-cut terraces yield rich mollusk faunas, as well as a variety of marine mammals. These fossils tell us that earlier coastal waters were somewhat warmer than those that challenge hardy swimmers today.

The dramatic changes in climate that affected the world during the Quaternary resulted in most of the province being buried under ice caps. Massive glacial erosion gave a new scenery to the province and, in the process, scoured away many fossil deposits. In some areas, Pleistocene and Holocene (postglacial) deposits produced diverse biotas – both marine and non-marine. Walrus, bison, mammoth, and mastodon are among the fossil vertebrates present. Of particular importance is the recent discovery of what appears to be a virtually complete postglacial sequence of deposits at Heal Lake, north of Victoria. This small lake basin was excavated to provide a larger repository for Victoria's garbage and unexpectedly exposed a record of about 10,000 years, with a huge concentration of well-preserved logs throughout the sequence. Radiocarbon dating is presently under way, as well as a detailed study of the pollen in the sediment. By comparing the record of spores and pollen in a variety of old lakes on Vancouver Island, it will be possible to reconstruct the nature of climate change in the region over the last several thousand years.

Other potential settings for preserving the recent geologic record are the many deep fiords along the coast of British Columbia. Saanich Inlet is the most intensively studied to date. New shallow cores are revealing immaculately preserved varves that not only show winter and summer deposits but also finer details of the seasons. The oxygen-poor bottom conditions excluded the benthic (bottom-living) organisms that burrow and churn up the sediment. An exciting development occurred in the fall of 1996, when the huge scientific drilling ship *JOIDES Resolution* will drill four cores, each about 250 metres deep. This is the first time in the program of the International Ocean Drilling Program that near-coastal drilling has been undertaken, and the potential significance of the anticipated scientific discoveries is tremendous. The *JOIDES Resolution* will make a scheduled port call to Victoria, so British Columbians will be able to see this important scientific exploration program up close.

In this brief excursion through the mists of geologic time, I have endeavoured to place the fossil discoveries in British Columbia within the wider context of evolution. The record of life through time demonstrates the sensitivity of the earth's flora and fauna – and a certain resilience – to the periodic disruptive processes that result in major extinctions. The scale of these perturbations that so affected ecologic systems in the past should show us that we still do not understand the natural processes operating on our planet. The fossil record is perhaps the most obvious and interesting story from which the public can fully appreciate the precarious balance that exists between life and a changing physical environment.

In a series of focused, easily read chapters, this book has documented the rich natural heritage represented by the fossils of British Columbia. Why has this heritage been hidden? Where are the public and educational exhibits displaying this wealth of fossil diversity and the message it articulates? The displays in museums in British Columbia are but sad representations of this wealth and testify to the lack of commitment by successive provincial governments to scientific, cultural, and educational institutions. In other provinces, institutions such as the Royal Ontario Museum and the Royal Tyrrell Museum of Palaeontology attract hundreds of thousands of visitors each year. In British Columbia, tourism is the third largest industry – British Columbia truly is supernatural. But why promote only our scenery, when there is a remarkable story below the surface? Why not display our fossil treasures to the world? In particular, why is the outstanding Burgess Shale and its soft-bodied fossils a World

Heritage Site, yet tourists and British Columbians must travel to Toronto, Washington, DC, or London, England, to see a suitable display?

The protection, curation, study, and display of British Columbia's wonderful fossil heritage is a challenge to us all. The recent formation of several local paleontological societies and museums is a remarkable development, and one supported by the newly formed British Columbia Paleontological Alliance. These groups must take a lead in lobbying for change if we are to alter the scientific and cultural myopia that pervades many recent decisions by federal and provincial governments. Paleontological ventures should be part of a new strategy for tourism and regional development in British Columbia. In addition to many academic and industrial applications, fossils clearly retain a deep fascination for the public, particularly for children. With a modest, long-term investment, British Columbia could use its rich fossil heritage to reap substantial educational, cultural, and economic dividends for many years to come.

Contributors

Christopher R. Barnes is with the School of Earth and Ocean Sciences of the University of Victoria.

James F. Basinger is with the Department of Geological Sciences of the University of Saskatchewan.

Lisa L. Bohach is with the School of Earth and Ocean Sciences of the University of Victoria.

Catherine C. Carlson is with the Department of Social and Environmental Sciences of the University College of the Cariboo.

Elizabeth S. Carter is with the Department of Geology of Portland State University.

Desmond H. Collins is with the Department of Invertebrate Palaeontology of the Royal Ontario Museum, Toronto, Ontario.

Philip J. Currie is with the Royal Tyrrell Museum of Palaeontology, Drumheller, Alberta.

James W. Haggart is with the Geological Survey of Canada, Vancouver, British Columbia.

C. Richard Harington is with the Canadian Museum of Nature, Ottawa, Ontario.

Richard J. Hebda is in Botany and Earth History at the Royal British Columbia Museum, Victoria, British Columbia.

Giselle K. Jakobs is with the Geological Survey of Canada, Vancouver, British Columbia.

Kenneth Klein is with the Department of Physical Sciences and Engineering of the University College of the Cariboo.

Rolf Ludvigsen runs the Denman Institute for Research on Trilobites on Denman Island, British Columbia.

Lee McKenzie McAnally is with the School of Earth and Ocean Sciences of the University of Victoria.

Elisabeth McIver is with the Department of Geological Sciences of the University of Saskatchewan.

James W.H. Monger is with the Geological Survey of Canada, Vancouver, British Columbia.

Andrew G. Neuman is with the Royal Tyrrell Museum of Palaeontology, Drumheller, Alberta.

Michael J. Orchard is with the Geological Survey of Canada, Vancouver, British Columbia.

David M. Rudkin is with the Department of Invertebrate Palaeontology of the Royal Ontario Museum, Toronto, Ontario.

Scott D. Sampson is with the New York College of Osteopathic Medicine, Old Westbury, New York.

David A.E. Spalding is with Brandywine Enterprises/Kanata Heritage on Pender Island, British Columbia.

Ruth A. Stockey is with the Department of Botany of the University of Alberta.

E. Tim Tozer is with the Geological Survey of Canada, Vancouver, British Columbia.

Wesley C. Wehr is with the Burke Museum at the University of Washington.

Mark V.H. Wilson is with the Department of Zoology and Laboratory for Vertebrate Paleontology of the University of Alberta.

Index

Set in Stone Serif and Stone Sans by Brenda and Neil West, BN Typographics West
Printed and bound in Canada by Friesens
Copy-editor: Grace D'Alfonso
Proofreader: Gail Copeland